Lecture Notes in Networks and Systems

Volume 846

The series "Lecture Notes in Networks and Systems" publishes the latest developments in Networks and Systems—quickly, informally and with high quality. Original research reported in proceedings and post-proceedings represents the core of LNNS.

Volumes published in LNNS embrace all aspects and subfields of, as well as new challenges in, Networks and Systems.

The series contains proceedings and edited volumes in systems and networks, spanning the areas of Cyber-Physical Systems, Autonomous Systems, Sensor Networks, Control Systems, Energy Systems, Automotive Systems, Biological Systems, Vehicular Networking and Connected Vehicles, Aerospace Systems, Automation, Manufacturing, Smart Grids, Nonlinear Systems, Power Systems, Robotics, Social Systems, Economic Systems and other. Of particular value to both the contributors and the readership are the short publication timeframe and the world-wide distribution and exposure which enable both a wide and rapid dissemination of research output.

The series covers the theory, applications, and perspectives on the state of the art and future developments relevant to systems and networks, decision making, control, complex processes and related areas, as embedded in the fields of interdisciplinary and applied sciences, engineering, computer science, physics, economics, social, and life sciences, as well as the paradigms and methodologies behind them.

Indexed by SCOPUS, INSPEC, WTI Frankfurt eG, zbMATH, SCImago.

All books published in the series are submitted for consideration in Web of Science.

For proposals from Asia please contact Aninda Bose (aninda.bose@springer.com).

Mustika Sari · Anastasia Kulachinskaya
Editors

Digital Transformation: What are the Smart Cities Today?

 Springer

Editors
Mustika Sari
Center for Sustainable Infrastructure
Development
Universitas Indonesia
Depok, Indonesia

Anastasia Kulachinskaya
Graduate School of Industrial Economics
Peter the Great St.Petersburg Polytechnic
University
St.Petersburg, Russia

ISSN 2367-3370   ISSN 2367-3389 (electronic)
Lecture Notes in Networks and Systems
ISBN 978-3-031-49389-8   ISBN 978-3-031-49390-4 (eBook)
https://doi.org/10.1007/978-3-031-49390-4

This Springer imprint is published by the registered company Springer Nature Switzerland AG
The registered company address is: Gewerbestrasse 11, 6330 Cham, Switzerland

Paper in this product is recyclable.

Editorial Note

Smart City Development for Optimizing City Efficiency, Improving Quality of Life, and Fostering Digital Economy

Dear colleagues,

The smart city concept is widely recognized as a technology-based urban paradigm that utilizes innovative solutions to integrate people, information, and various city elements. The overarching objectives are to optimize the efficiency of city operations and elevate the overall quality of citizen's life, thus enhancing productivity and competitiveness (Lin et al., 2019). This concept has gained traction and is being adopted in numerous countries worldwide, both developed and developing economies. There were an estimated total of 143 smart city projects back in 2013, distributed across North and South America, Europe, Asia, the Middle East, and Africa, which include Ottawa's Smart Capital and Quebec City in Canada and various initiatives undertaken in European cities of Barcelona, Amsterdam, Berlin, Manchester, Edinburgh, and Bath. Meanwhile, China has included smart cities in its 51 cities and 6 provinces, with a notable utilization of information and communication technology (ICT) in the growth of its major cities, including Beijing, Shanghai, and Shenzhen (Albino et al., 2015).

A smart city embodies an urban environment in which the networks and services are enhanced regarding flexibility, efficiency, and sustainability by utilizing digital, information, and telecommunication technologies. Its objective to improve the operations of the city for its inhabitants' benefit is achieved by incorporating a range of smart components and exchanging intelligent information that flows between its various city aspects, which collectively contribute to the smartness of the city (Caragliu et al., 2011; Zygiaris, 2013). These smart cities require several attributes, including data instrumentation, aggregation, flexible data presentation, real-time knowledge access, analytics, decision support systems, automation, collaborative spaces, and digital-physical integration. ICT is pivotal in realizing them, as it serves as the backbone of urban intelligence, providing smart solutions and data analytics to support city development, improving urban services, and expanding stakeholder access to information. Therefore, the global emergence of smart cities is characterized by comprehensive investments in ICT aimed at fostering technological innovation, further facilitating the growth of new sectors, and stimulating economic growth while ensuring environmental sustainability and improving the overall well-being of the inhabitants (Law & Lynch, 2019).

A smart city is viewed as a strategy to address various urban issues that adversely impact society and the environment. The multifaceted urban challenges vary from rapid urbanization, which strains the existing urban infrastructure and services provided, leading to increased energy consumption and heightened levels of environmental pollution. Moreover, smart city development aims to tackle global climate change by implementing green and sustainable practices. On the other hand, scarce natural resources, depleted by growing urban populations, pose a critical concern that smart cities seek to address. It also aims to overcome pressing social issues, such as inequality, to foster more inclusive urban societies. It is evidenced that, with smart city development, inhabitants' quality of life can be improved, digital-driven new sectors can be boosted, and resource efficiency can be optimized, along with other better social, economic, and environmental benefits that also can be provided (Berawi et al., 2022).

Built upon the combination of smart solutions providing data-driven decision-making, a smart city has the potential to enhance the quality of services offered to its inhabitants, as substantial advancements are evident in several domains, including smart governance and smart mobility. Within the smart governance domain, integrating Information Technology (IT), Information and Communication Technology (ICT), and data analytics plays a crucial role in optimizing the efficiency and effectiveness of public sector services. Smart mobility, on the other hand, is a vital component of the paradigm shift in urban mobility that aims to alleviate the burdens associated with traditional transportation systems by mitigating traffic congestion, reducing trip durations, and reducing road accidents, all while providing passengers with greater travel flexibility and concurrently contributing to improved air quality (Bıyık et al., 2021). Key features of the smart mobility domain include real-time traffic monitoring, adaptive smart traffic lights, and intelligent transport systems that utilize data for exact route and timetable optimization.

The pursuit of environmental sustainability is an objective in contemporary urban development, and smart city initiatives are emerging as pivotal enablers in this endeavor. These initiatives, grounded in data-driven technology, hold the potential to usher in a new era of environmental governance and bolster public health outcomes. Smart cities wield transformative potential in reducing carbon emissions; facilitating economic growth; and creating vibrant, community-centric urban spaces (Wang et al., 2022). The strategic deployment of renewable energy sources is one example of smart city ingenuity driving environmental sustainability. These clean sources not only mitigate the adverse environmental impacts of fossil fuels but also lay the groundwork for establishing robust and eco-friendly energy ecosystems (Lund et al., 2015). Moreover, the implementation of automation for enhancing energy efficiency, consisting of a diverse array of smart services such as smart grids, smart metering systems, smart buildings, and smart environmental monitoring, plays a pivotal role in improving environmental sustainability within smart cities by significantly boosting energy efficiency, reducing pollution, promoting the integration of renewable energy sources, and delivering real-time high resolution (Bibri & Krogstie, 2020).

Implementing smart city technologies can also contribute to the growth of the digital economy by utilizing innovative digital transformation and advanced infrastructure. They provide robust digital infrastructure, promoting high-speed internet access and empowering economic activities. Smart cities encourage innovation and cross-sector collaboration, fostering dynamic ecosystems crucial to digital economy sustainability. These cities are pioneering digital transformation by optimizing operations, enhancing services, and accelerating business digitalization. In addition, they unlock latent digital economic potential, as shown by e-commerce, digital social platforms, and mobile applications that connect communities and enable circular economy practices. These innovations offer economic value while aligning with smart city sustainability goals (Lekan & Rogers, 2020). The smart economy domain of smart cities thrives on innovation, entrepreneurship, a robust economic image, labor market flexibility, global market integration, and adaptability (Monfaredzadeh & Berardi, 2015).

The smart city framework is a crucial guide for governments and society in navigating the complexities of digital transformation in urban development. It enables governments to make informed and efficient decisions in city management, ensuring that smart city initiatives are executed with clear structures and measurable outcomes. It plays a pivotal role in guiding the city's digital transformation journey as an indispensable tool that steers urban digitization and facilitates the harmonious convergence of various technological, societal, and economic elements. Furthermore, smart city frameworks aid in the effective allocation of resources by delineating clear priorities and objectives. This targeted resource allocation helps maximize the impact of digital transformation initiatives, leading to efficient and cost-effective urban development. In practice, digital transformation encompasses allocating resources toward information technology solutions and systems to update software and hardware within government agencies. This initiative facilitates online government services, enhancing communication and service delivery between public workers and citizens, thus empowering cities to become smarter, more efficient, and better equipped to address urban challenges.

One exemplary smart city framework is the European Cities' Smart City Model, encompassing six essential domains: Environment, Governance, Economy, People, Life, and Mobility (Giffinger & Gudrun, 2010). Similarly, the Smart Dubai Framework is structured around six domains: Smart Economy, Smart Living, Smart Governance, Smart Environment, Smart People, and Smart Mobility. Meanwhile, London follows the Smart London Plan, focusing on four key domains: Smart Infrastructure, Smart Environment, Smart Mobility, and Smart Society. In Indonesia, the smart city framework for the new capital city comprises six primary domains: Smart Governance, Smart Transportation and Mobility, Smart Living, Natural Resources and Energy, Smart Industry and Human Resources, and Built-Environment and Infrastructure. These frameworks play a crucial role in guiding the implementation of technologies and policies, ultimately contributing to the realization of benefits for both citizens and the environment. It revolutionizes the urban landscape and offers a promising path to enhancing society's quality of life, ensuring environmental sustainability, and fostering the digital economy.

This book has international researchers and scholars with strong interest and expertise in smart cities as the contributors. There are 23 chapters discussing the digital transformation of smart cities on such elements as: smart governance, smart buildings, smart transportation, smart tourism destinations, smart economy, and smart environment.

Mustika Sari
Anastasia Kulachinskaya

References

Albino, V., Berardi, U., & Dangelico, R. M. (2015). Smart cities: Definitions, dimensions, performance, and initiatives. *Journal of Urban Technology, 22*(1). https://doi.org/10.1080/10630732.2014.942092

Berawi, M. A., Susantono, B., & Sari, M. (2022). Fostering resilient and smart cities to enhance sustainable urban development. In *The Routledge Handbook of Sustainable Cities and Landscapes in the Pacific Rim* (pp. 704–719). Routledge. https://doi.org/10.4324/9781003033530-60

Bibri, S. E., & Krogstie, J. (2020). Environmentally data-driven smart sustainable cities: applied innovative solutions for energy efficiency, pollution reduction, and urban metabolism. *Energy Informatics, 3*(1), 29. https://doi.org/10.1186/s42162-020-00130-8

Bıyık, C., Abareshi, A., Paz, A., Ruiz, R. A., Battarra, R., Rogers, C. D. F., & Lizarraga, C. (2021). Smart Mobility Adoption: A Review of the Literature. *Journal of Open Innovation: Technology, Market, and Complexity, 7*(2), 146. https://doi.org/10.3390/joitmc7020146

Caragliu, A., del Bo, C., & Nijkamp, P. (2011). Smart cities in Europe. *Journal of Urban Technology, 18*(2). https://doi.org/10.1080/10630732.2011.601117

Giffinger, R., & Gudrun, H. (2010). Smart cities ranking: an effective instrument for the positioning of the cities? *ACE: Architecture, City and Environment, 4*(12), 7–26. https://doi.org/10.5821/ace.v4i12.2483

Law, K. H., & Lynch, J. P. (2019). Smart City: Technologies and Challenges. *IT Professional, 21*(6), 46–51. https://doi.org/10.1109/MITP.2019.2935405

Lekan, M., & Rogers, H. A. (2020). Digitally enabled diverse economies: exploring socially inclusive access to the circular economy in the city. *Urban Geography, 41*(6), 898–901. https://doi.org/10.1080/02723638.2020.1796097

Lin, C., Zhao, G., Yu, C., & Wu, Y. (2019). Smart City Development and Residents' Well-Being. *Sustainability, 11*(3), 676. https://doi.org/10.3390/su11030676

Lund, P. D., Mikkola, J., & Ypyä, J. (2015). Smart energy system design for large clean power schemes in urban areas. *Journal of Cleaner Production, 103*, 437–445. https://doi.org/10.1016/j.jclepro.2014.06.005

Monfaredzadeh, T., & Berardi, U. (2015). Beneath the smart city: dichotomy between sustainability and competitiveness. *International Journal of Sustainable Building Technology and Urban Development, 6*(3), 140–156. https://doi.org/10.1080/2093761X.2015.1057875

Wang, L., Xie, Q., Xue, F., & Li, Z. (2022). Does Smart City Construction Reduce Haze Pollution? *International Journal of Environmental Research and Public Health, 19*(24), 16421. https://doi.org/10.3390/ijerph192416421

Zygiaris, S. (2013). Smart City Reference Model: Assisting Planners to Conceptualize the Building of Smart City Innovation Ecosystems. *Journal of the Knowledge Economy*, *4*(2). https://doi.org/10.1007/s13132-012-0089-4

Contents

Smart Cities: Development of a Model for Managing the State of the Social Environment

Dmitriy Rodionov, **Irina Baranova**, **Evgenii Konnikov**, **Darya Kryzhko**, and **Liudmila Mishura**

Abstract In the context of the transition to a new type of economy, the digital economy, the question of systemic management of the entire environment is becoming increasingly relevant. The COVID-19 pandemic has caused all decision-making entities to switch to digital communication, resulting in the formation of a ubiquitous digital ecosystem. The social environment is the most significant in this case, but the models of social environment management currently in place are ineffective in the context of digital transformation. In this study, the authors set out to develop an effective model for managing the state of the social environment of the smart cities. To achieve this goal, they developed a model of interaction between indicative indicators of the basic components of the social environment of the city, which was transformed into a program-targeted graph model for managing the state of the social environment of the region. The findings of this study have significant implications for representatives of state and municipal management, as well as researchers in the field of social development. By using the program-targeted graph model for managing the state of the social environment of the cities, they can effectively manage the social environment in the context of digital transformation. Overall, this research provides valuable insights into the development of an effective model for managing the state of the social environment of the cities in the context of digital transformation. The findings of this study can inform the practical activities of decision-making entities and researchers in the field of social development.

D. Rodionov · E. Konnikov · D. Kryzhko (✉)
Peter the Great St.Petersburg Polytechnic University, Polytechnicheskaya, 29, 195251 St., Petersburg, Russia
e-mail: darya.kryz@yandex.ru

I. Baranova
State Institute of Economics, Finance, Law and Technologies, 188300, Roschinskaya St. 5, Gatchina, Russia

L. Mishura
Saint Petersburg National Research University of Information Technologies Mechanics and Optics, Kronverksky Pr. 49, Bldg. A, 197101 St. Petersburg, Russia

1

Keywords Smart Governance · Smart City · Digitalization · Social environment · Digital transformation · Systemic management

1 Introduction

The problem of developing the social environment, creating the social prerequisites for sustainable social and economic development of national and regional systems is currently receiving considerable attention, both in numerous scientific research and government and management bodies [1–3]. This is associated with the formation of a new vector of social development—the transition to an innovative economy based on knowledge and requiring the modernization of literally all spheres of life [4]. The development, implementation and introduction of innovations, the use of new technologies, including organizational ones, requires a high level of development of human resources, which become the main driving force of the new economy. This means the humanization of all aspects of social life, the active development of the national and regional social infrastructure, and the shifting of focus to the social environment [5–7].

The unique mediator of this process was the world pandemic COVID-19, the result of which was the world quarantine. This quarantine has caused a universal need to create new, sustainable digital communication channels at all levels of both professional and social activity. The result of this transition was the immediate immersion of all participants in the social communication into a single digital environment, and the formation of a single digital ecosystem [8–10]. This digital ecosystem has no natural restrictions, resulting in all participants having access to high-speed communication channels and a massive amount of information. This system is decentralized, which significantly reduces its manageability [11]. Under the conditions of universal and high-speed exchange of social information, management subjects have the opportunity to instantly observe the reaction to the managerial decisions taken by them, as well as timely analyze the state of the social environment and react to its fluctuations [12, 13]. This circumstances requires an examination of the essence of the region's social environment, its structure, approaches to management, and the development of an effective model of social environment management [14, 15]. Therefore, the chosen topic of the research is currently relevant. The object of this study is the social environment of the region, and the subject is the organizational and managerial relationships arising in the process of forming and using the social environment.

The aim of the research is to develop an effective model for managing the state of the region's social environment.

The goals of the research are:

1. To critically examine the theoretical basis, formed by world scholars, in the field of determining and assessing the state of the region's social environment;
2. To develop a methodology for forming a multidimensional model for assessing the state of the region's social environment;

3. Forming a theoretical model for managing the state of the region's social environment;
4. Mathematical formalization of the system of interactions of elements of the region's social environment;
5. Designing a graph-targeted program model for managing the state of the region's social environment.

2 Literature Review

The state of a region's social environment is a complex phenomenon, and scientists have yet to reach an agreement regarding the indicators of its state. Currently, various complex indicators of the state of the social environment have been created around the world. One of the most commonly used indicators is the Human Development Index [16, 17], which evaluates three components: (1) the index of expected life expectancy, measured by the average expected life expectancy at birth; (2) the index of education, measured by the average expected duration of schooling of school-age children and the average duration of education of the adult population; (3) the index of gross national income, measured by the size of gross national income per capita in U.S. dollars at purchasing power parity. Despite its relative universality, the main disadvantage of this index is its extreme generalization and orientation to the assessment of macro-systems. The results obtained by calculating this index are the result of many systemic processes in the region, which in turn does not allow to effectively establish the cause and determine the directions of changes in the region in order to improve this index. This index is also criticized for the subjectivity of its components, in particular the index of gross national income [18].

More comprehensive is the Better Life Index of OECD, combining 11 elements [19, 20]. This index is more oriented towards managing the state of the region's social environment, however, not all used indicators are universal. In particular, such indicators as the time devoted to work, the level of voter activity, and expenses on housing have a national specificity. Consequently, the application of this index to any region allows for obtaining non-objective results. Also, critics note that this index does not take into account economic inequality and the level of poverty, does not assess the state of the family institute [21], and uses only relative rather than absolute estimates [22]. It is worth noting the International Happiness Index [23, 24]. Three indicators are used to calculate the index: the subjective satisfaction of people with life, the expected life expectancy, and the "ecological footprint". All the shortcomings outlined for the Human Development Index are applicable to this index, and the "ecological footprint" is often criticized. The general subjectivity of using GDP [25, 26] as one of the indicators in the methodology of calculation should also be noted. It should be noted that GDP was not intended to indicate all aspects of the social environment, as it only reflects the economic result [27]. This fact determines the need for gradual departure from the paradigm of using GDP as a universal indicator of the state of the region's social environment [28].

According to researcher Kostina, the state of the social environment can be measured by three fundamental factors: material well-being, health, and security. This author distinguishes between 'quality indicators' and 'living standards' such as access to educational services, medical care, cultural and sports facilities, general living standards, legal protection, and availability of a safe environmental environment [29].

In the work of Nagaytsev and Pustovalova, the differentiation of the complex indicator of the state of the region's social environment is more detailed and implies the selection of economic, legal, socio-psychological, ecological, technological and other indicators. The main consequence of the deterioration of the social environment is social tension. The main precondition for the emergence of social tension is a stable and long-term situation of contradiction between the interests, social expectations of the entire mass or a significant part of the region's population and the measure of their actual satisfaction. The authors assume that the most important indicators characterizing the growth of people's discontent and social tension in different spheres of the regional socium are the main ones. The following indicators can be proposed as the main ones: dissatisfaction with the level of personal well-being, dissatisfaction with working conditions, the emergence of a negative attitude of the region's population towards representatives of other nationalities arriving in the region for work and permanent residence, group hooligan actions and violations of public order, high level of unemployment and high level of morbidity [30].

In opposition to this view, Vasova believes that among the most important approaches to determining social potential one can single out the economic approach, which studies the issues of the most optimal and efficient distribution of material goods in society [31]. Kirhmeier, who is in solidarity with this approach, argues that not only the state of the region's social environment affects such factors as the expansion of the possibilities of hired workers and the stimulation of labor productivity. He also mentions the ecological component. For the characterization of ecological disasters, it is still customary to present the damage in monetary equivalent. The damage in social expression is reflected in the increase of morbidity, the reduction of life expectancy, the emergence of pathologies, the increase of mortality and population migration. Under the environmental-social development of the region, the author understands the progressive qualitative and quantitative change in the social parameters of life of the region's population, accompanied by economic growth against the background of a decrease and reduction to zero of the load on the environment [32].

According to a number of scholars, the key indicator of the state of the region's social environment is social cohesion, first of all, within the framework of family and kinship relations, as well as governmental policies aimed at forming social infrastructure for older people in the sphere of stimulating joint activities, independence, and autonomy in everyday life [33]. Thus, we are talking about such spheres as the social sphere and the sphere of health care [34, 35]. In addition, among scholars, special attention is paid to social protection of the population as one of the most important links in the system of managing the state of the social environment. Sudareva gives justification for social protection as one of the most important functions of the state. According to the author, the creation of an effective system of social protection is an integral condition for the development of any society, especially in conditions of market relations, is an indispensable pay society and business for social peace, stability of the social system and the possibility of normal economic activity [36].

Kugan considers the realization of the social-pedagogical function of the education sphere as one of the key elements of the social environment. According to the author, the meaning of the social pedagogue is to create conditions for directed socialization of the individual, and social-pedagogical professionalism is an indicator of social stability in society [37]. Chan and Rozhdestvenskaya agree with the author and study the factors of increasing social stability of the region in the conditions of the economy based on production and the increase of knowledge. The authors considered the factors of development of the social environment, among which the cognitive capital was especially singled out as the main factor. The knowledge economy implies the basic components of development: effective state institutions, high quality of life; high-quality education; effective fundamental science; effective venture business of scientific and technical nature; high-quality human capital in its wide definition; knowledge and high technology production; information society or knowledge society; infrastructure for realization and transfer of ideas, inventions and discoveries from fundamental science to innovative production [38].

Babayan and Pashinin believe that the condition for social development of a region is following the path of modernization of the social and economic sphere. Social development, as well as modernization, includes two basic characteristics—sociocultural and economic prosperity, which can be achieved through active labor in the conditions of introducing innovative production technologies, redistributing goods and services, and subsequent consumption of benefits, only under the condition of transforming social relations as a result of value transformations. The authors underscore the role of young people in promoting the tasks of innovative development, modernization, development of high-tech projects, professional self-improvement, and readiness to cooperate for achieving a common goal [39].

Therefore, based on all of the above, it can be assumed that the state of the region's social environment depends on a set of factors belonging to the following areas of societal life: socio-demographic, economic, environmental, technological, health, and education. However, at this time, no comprehensive model of management of the state of the social environment, including indicators from all of these spheres, has been developed yet.

3 Research Methodology

Literary analysis allowed to identify a number of indicative parameters in each of the indicated spheres, in the amount of 27 titles. For example, for the socio-demographic sphere such indicators as the level of unemployment and the index of prices on the housing market were selected, for the economic—the turnover of organizations, the number of profitable enterprises and organizations, for the ecological—the use of fresh water, for the technological—the coefficient of inventive activity, for health—the morbidity per 1000 people, for education—the number of governmental educational organizations of primary professional education, etc. The full list of indicators is presented in Table 1. These indicators represent separate quantified elements reflecting the state of the region's social environment and, based on their dynamics, an weighted assessment model of this state as a complex phenomenon can be formed. Nevertheless, these indicators are not isolated in terms of their influence on each other. These indicators change with the change of each other and, in fact, form complex wave oscillations of the state of the region's social environment over time.

The consequences of negative oscillations of the indicators of the educational sphere can have a delayed effect on the indicators of the economic sphere, while the positive dynamics of the indicators of the health sphere will invariably affect the social and economic spheres, and changes in the indicators of the ecological sphere will affect the indicators of the health sphere. Thus, the architecture of the proposed model can be of a closed nature, without the selection of any exclusively endogenous or exclusively exogenous indicators. Within the framework of this model, exogenous-endogenous indicators can also be distinguished, which can be affected by indicators, but at the same time, mediate the changes of other indicators. Based on formal-logical analysis and heuristics, a conceptual model was formed, presented in the form of a system of theoretical functional connections, presented in Table 2. It should be noted that at the level of the conceptual model, many indicators are in superposition and mediate changes in some indicators both directly and indirectly (through the influence of exogenous-endogenous indicators). It is also necessary to note the potential cyclicity.

For the purposes of increasing efficiency, the identified factors were coded. Table 1 lists the codes and names of the indicators, as well as the units of measurement. It should be noted that the statistical array was formed on the basis of aggregate information on the Russian Federation [40]. This is due to the significant differentiation of the regions of the Russian Federation, both from a natural and historical, as well as from a social and economic point of view.

The links provided within the conceptual model can be mathematically formalized as potential dependency functions. These functions are presented in Table 2.

To construct the corresponding functions, the regression analysis toolkit was automated through MS Excel and IBM SPSS. The following criteria were applied for variable selection in the regression analysis:

1. The equation was excluded if its coefficient of determination (R^2) was lower than 0.7. Equations with coefficients of determination between 0.7 and 1 were considered for review. For paired regression, the ordinary R^2 was considered, for multiple regression—the normalized R^2.
2. Factors with a P-level higher than 0.05 (5%) were excluded from the model in the case of multiple regression. As factors were excluded from the models according to the P-level criterion, either those with this indicator in the specified range were left, or the number of factors was reduced to one and further paired regression analysis was performed.

Table 1 A set of indicators reflecting the state of the social environment of the region

Code	Decryption of code (indicator)	Units of measure
A	Unemployment rate in the region	Percent for the year
B	Index of housing prices in the region (secondary housing market, all types of apartments)	Ercent for the first quarter compared to the corresponding quarter of the previous year
C	Real average monthly salary of employees of enterprises in the region	Percent compared to the previous year
D	Number of children of the region's families that went to summer camps	People
E	Expenditures on labor payments in the region's enterprises	Thousands of rubles
F	Number of state educational organizations of primary professional education in the region	Units
G	Number of profitable enterprises in the region	Units
H	Index of consumer prices for goods and services in the region	Percent compared to the previous year
I	Average number of employees in the region's enterprises	People
J	Amount of goods exported from the region	Thousands of rubles
K	The volume of voluntary insurance payments	Thousands of rubles
L	Debt for payments to budgets of all levels	Thousands of rubles
M	Turnover of organizations registered in the region	Thousands of rubles
N	The amount of debt for credits and loans of large and medium enterprises and organizations of the region	Thousands of rubles
O	Expenditures for technological innovations of the organizations of the region	Thousands of rubles
P	Number of advanced production technologies developed in the region	Units
Q	Inventive activity coefficient in the region (number of domestic patent applications for inventions filed in Russia)	Units per 10 thousand people
R	Incidence rate per 1000 people in the region	People
S	Number of doctors of all specialties in the region	People
T	Power of outpatient and polyclinic institutions in the region	Visits per 10 thousand people
U	Number of outpatient and polyclinic institutions in the region	Units
V	Number of abortions	Units per 100 births
W	Expenditures on land reclamation	Thousands of rubles

(continued)

Table 1 (continued)

Code	Decryption of code (indicator)	Units of measure
X	Number of objects with stationary sources of air pollution in the region	Units
Y	Emissions of pollutants from stationary sources into the atmosphere	Thousand tons
Z	Use of fresh water	Million cubic meters
Z1	Discharge of polluted wastewater into surface water objects	Million cubic meters

Table 2 Potential functional links

No	Equation	No	Equation
1	$A = F(F,G)$	15	$O = F(N,L,Z)$
2	$B = F(G,C,Y)$	16	$P = F(Q,M)$
3	$C = F(H,E)$	17	$Q = F(F,T,V)$
4	$D = F(C,V)$	18	$R = F(T,A,D,F,W,Z,Я,Y,X)$
5	$E = F(G,I,L,K)$	19	$S = F(F,C)$
6	$F = F(V,C)$	20	$T = F(P,S,E)$
7	$G = F(P,O,N,L,W)$	21	$U = F(S,R)$
8	$H = F(M,A)$	22	$V = F(Я,H,G,B,T)$
9	$I = F(V,A)$	23	$W = F(O,N,L,Я)$
10	$J = F(G,P)$	24	$X = F(O,G)$
11	$K = F(R,C)$	25	$Y = F(P,X,O)$
12	$L = F(N,U)$	26	$Z = F(P,G,U)$
13	$M = F(J,O,Q,I)$	27	$Z1 = F(N,G,Q,P)$
14	$N = F(M,P)$		

3. The approximation error for each equation accepted for further research should be between 0 and 15%. Usually, the standard for the approximation error is accepted as 5–10%, however, in this work, the upper limit of this indicator was increased due to the frequent excess of the norm by this indicator when performing regression analysis.

The resulting model from this study will be able to identify the key mediators of social environment transformation, and the targeted impact on those mediators will allow for effective management of the social environment. As a result, the target control model will take the following form (see Fig. 1).

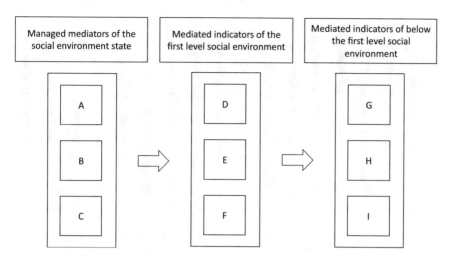

Fig. 1 Theoretical model of regional social environment control

4 Results

As a result of the conducted analysis, the following system of regression equations was obtained.

$$
\begin{cases}
A_n = 10^{0.52+(5.85E-06)\times G_{n-1}-0.077\times Z_n} \\
B_n = 1407.18 - 26.48 \times C_n + 0.134 \times C_n^2 \\
E_n = -(1.522E+11) - 0.356 \times G_n^2 + 3339625363 \times \ln L_n + 4733275290 \times \ln K_n \\
F_n = 10^{8.22-0.11\times C_n+0.0006\times C_n^2+2.65\times \log V_n} \\
G_n = 10^{9.06-0.49\times \log L_n} \\
J_n = 10^{9.13-0.38\times \log G_n+0.82\times \log P_n} \\
K_n = 10^{-5.78+6.3\times \log R_{n-1}-2.125\times \log C_{n-1}} \\
L_n = 996277668 + 0.027 \times N_n - 170030.49 \times U_n \\
M_n = (3.4122E+10) + 9,54 \times J_n - (1.446E+10) \times Q_n \\
N_n = 10^{-10.14+2.08\times \log M_n-0.718\times \log P_n} \\
P_n = 10^{2.69+(4.973E-12)\times M_n} \\
R_n = 2361.02 - 17.335 \times T_n + 0.0003 \times X_n + 0.107 \times A_n \\
\quad +(4.1014E-07) \times D_n + 0.0006 \times T_n^2 + 0.038 \times W_n^2 - (4.047E-09) \times A_n^2 \\
S_n = 40640 + 1.74 \times F_n + 0.14 \times C_n^2 - 0.0004 \times F_n^2 \\
T_n = 311.93 - 8.57 \times \log E_n \\
U_n = 5697167.12 - 258.55 \times S_n + 0.003 \times S_n^2 \\
V_n = 126.2 + 0.00017 \times G_n - 2.5 \times H_n + 0.026 \times T_n \\
\quad -(2.337E-05) \times B_n^2 - (1.459E-09) \times G_n^2 + 0.0155 \times H_n^2 \\
W_n = -5321608,6 + 0.017 \times L_n - 0.000391 \times N_n
\end{cases}
$$

As can be seen, not all of the described relationships are exclusively linear, and not every one of the described indicators acts as an endogenous or exogenous variable. The obtained equations can be classified according to their determination coefficients,

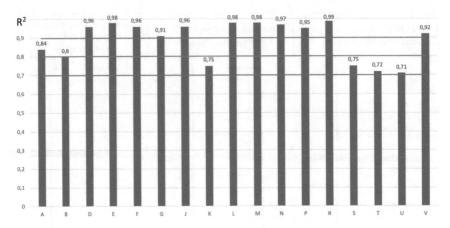

Fig. 2 Distribution of equations according to the value of the determination coefficient

namely: $0.7 < R^2 < 0.8$—acceptable strength of connection; $0.8 < R^2 < 0.9$ – high strength of connection; $0.9 < R^2 < 1.0$—extremely high strength of connection. The distribution of equations according to the determination coefficient is shown in Fig. 2.

As can be seen, the quality of the presented system of equations is high, and the average value of the determination coefficient is 0.89. Graphically, this system of equations can be represented in the form of an updated conceptual model (Fig. 3).

As can be seen, this model is highly complex. Many of the indicators have cyclic relationships, in particular, the indicator E_n is cyclically dependent on itself through the successive influence of the indicators T_n, R_n and K_n, and the indicator Jn is cyclically dependent on itself through just two indicators. Furthermore, many of the indicators are in superposition, significantly increasing the interferability of this model. However, for the purposes of this research, the reverse regression conclusions are much more significant, namely, conclusions regarding the influence of which of the indicators on the one being studied. It should be noted that, with each new iteration of influence, the share of explained variance is significantly reduced. It would be reasonable to limit the length of the graphs being studied. Thus, in order to study the strength of the influence of the indicators, it is necessary to consider individual graphs of each of the studied indicators (both endo- and exogenous, as well as endo-exogenous) of fixed length. As the length parameter of the graph, it would be reasonable to use 3, since on this span the share of explained variance is sufficient for each of the studied indicators, and no cyclic relationships are manifested. This analysis will help to identify the key mediators of social environment transformation, and form a model of regional social environment control.

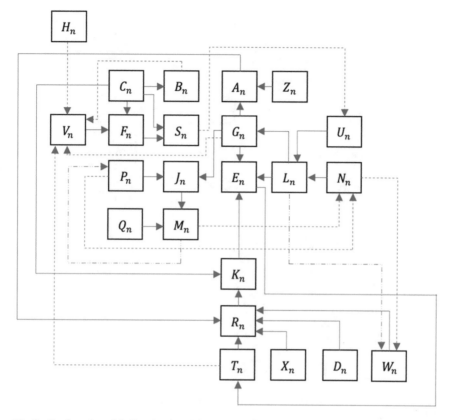

Fig. 3 Confirmed model of regional social potential indicators interaction

5 Discussion

For the purpose of identifying the key mediators of social environment transformation, it is necessary to return to the previously highlighted spheres. Each of these spheres is a unique aspect of the regional social potential, and should be considered separately. To simplify cluster perception, each of the designated spheres is assigned a color code for clustering:

1. Social-demographic sphere—blue;
2. Educational sphere—gray;
3. Economic sphere—orange;
4. Technological sphere—white;
5. Health sphere—yellow;
6. Environmental sphere—green.

The elasticity coefficient is used as a measure of influence to show how much the endogenous indicator will change when the exogenous variable is changed by one

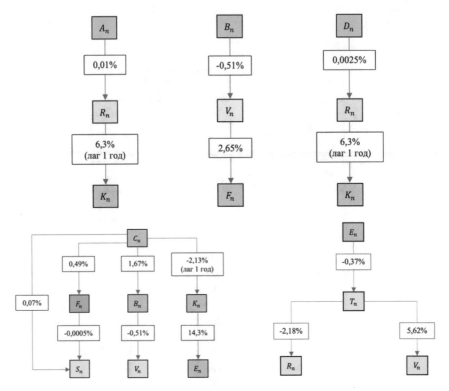

Fig. 4 Graphs of the social and demographic sphere impact on the regional social environment sphere

percent. The primary task is to identify patterns of interaction between the spheres. First of all, let's look at the social-demographic sphere (see Fig. 4).

As can be seen, the social and demographic sphere primarily mediates the sphere of health care, both at the first level of the graph and at the second. This may be due in large part to the consequences of the social and marginal situation in the region for the health of citizens. Practically all indicators of the health sphere are mediated, which allows one to conclude about the key importance of controlling the social and demographic sphere of the region for ensuring the stability of the health sphere. At the same time, three out of five indicators also mediate the amount of voluntary insurance payments related to the economic sphere.

However, this indicator is boundary and can also be referred to the sphere of healthcare in many aspects. Also, let us note the influence of the social-demographic sphere on the sphere of education, which can be determined by the same mechanisms as its influence on the sphere of healthcare. The most significant in this regard is the indicator C_n (real average monthly salary of the employees of the region business), which is exclusively exogenous. This indicator is in the superposition, and its influence on the sphere of healthcare is duplicated through the influence on the sphere of

Fig. 5 Graph of the
education sphere influence
on the regional social
environment

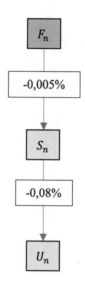

education. At the same time, it mediates other elements of the social-demographic sphere, which, in turn, mediate the sphere of healthcare. Thus, it can be concluded that the key indicator of the social sphere is the real average monthly salary of the employees of the region business and it is the influence on it that will allow to provide a significant share of the region's social potential.

Next, we consider the sphere of education, which, according to the results of modeling, is restricted solely to one indicator (Fig. 5).

As it can be seen, the education sphere only has an influence on the healthcare sphere. The mechanisms of this influence were described above. It should be noted that the strength of this influence is extremely insignificant, and it is duplicated by the mediation of the social-demographic sphere, namely the real average monthly wages of the region's enterprise employees. Thus, the direct influence of the education sphere on the development of the social environment of the region can be neglected and limited to the influence on the social-demographic sphere of the region.

Next, consider the most extensive economic sphere of the region (Fig. 6).

As can be seen, the influence of this sphere is extensive and heterogeneous. It is necessary to single out the indicators G_n (the number of profitable enterprises in the region), L_n (debt on payments to budgets of all levels), M_n (turnover of organizations registered in the region) and N_n (debt on credits and loans of large and medium enterprises and organizations of the region). The influence of the indicators M and N is largely duplicated, so they can be considered as a separate linked group. The indicator L is in a superposition, and partially mediates G, which in turn has the greatest influence on all spheres, including the economic one. This indicator also additionally mediates the socio-demographic sphere, which in turn causes the mediation of the health care and education spheres. Through the mediation of the indicator M, the influence on this indicator can also indirectly affect the ecological and technological

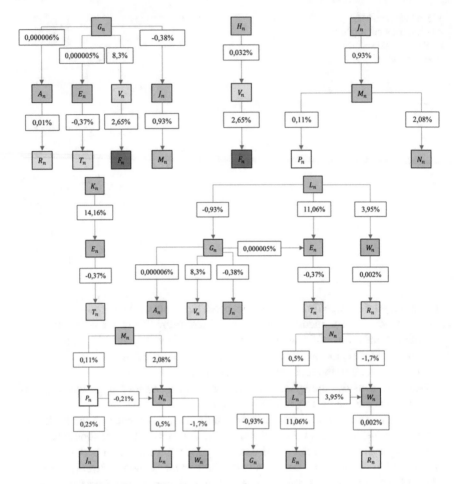

Fig. 6 Graph of the economic sphere influence on the regional social environment

spheres. Thus, by influencing the number of profitable enterprises in the region, it is possible to maximize its social potential.

Next, consider the technological sphere of the region (Fig. 7).

As can be seen, the technological sphere is limited to only two indicators, which primarily mediate the economic sphere. The nature of this influence is determined by the economic return from innovations. However, it should be noted that technological innovation is only one of the possible directions of mediation of the economic sphere of the region. Moreover, many of the identified links are duplicated by the indicators specified in other spheres. Thus, direct influence on the technological sphere for the development of the social environment can be ignored.

Next, consider the health care sphere (Fig. 8).

As can be seen, the health care sphere primarily mediates itself, thus duplicating the influence of the socio-demographic sphere. This mechanism of influence is largely

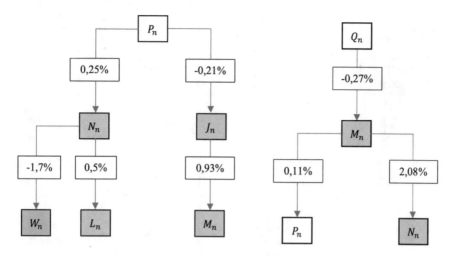

Fig. 7 Graph of the technological sphere influence on the regional social environment

determined by the service role of the health care sphere, as a result of which direct influence on the health care sphere for the development of the social environment can be ignored.

Finally, consider the ecological sphere (Fig. 9).

As can be seen, this sphere has a very limited effect on the social environment of the region, meditating exclusively in the sphere of health care, due to the extremely logically formalized mechanisms of ecology influence on human health. This influence is fully duplicated by the social environment, so direct influence on the ecological sphere for the purposes of developing the social environment can be disregarded.

Considering all 5 spheres of the region's social environment, it was established that the key mediating roles are played by the socio-demographic and economic spheres. These spheres do not bear a service nature. Within each of these spheres, key mediating indicators were identified, the management of which will allow effective control of the entire state of the social environment. For the socio-demographic environment, this indicator is the actual average monthly salary of the region's enterprise employees (C), and for the economic environment—the number of profitable businesses in the region (G). The graph of the complex influence of these indicators on the social environment of the region is presented in Fig. 10.

As can be seen, the influence of these indicators is diverse and partly overlaps, thus enhancing their superposition. Of course, the specifics of regions and the specialization of the model must be taken into account, however, the general mechanism of influence of key mediators remains unchanged. There are many ways to influence the identified mediators, but they are primarily based on the specificity of the region, its external environment and internal connections.

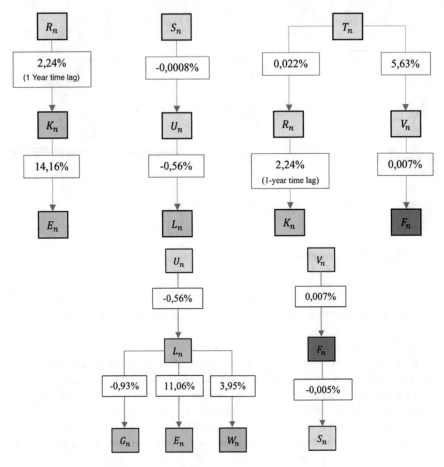

Fig. 8 Graph of the health care sphere influence on the regional social environment

Fig. 9 Graph of the ecological sphere influence on the regional social environment

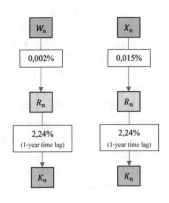

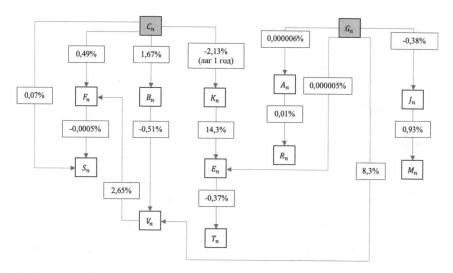

Fig. 10 Graph of the regional enterprises real average monthly employees salary and the profitable enterprises number in the region influence on the regional social environment

6 Conclusion

In this article, the authors developed a discussion-based but potentially effective model to manage the state of the region's social environment. A hierarchical system of indicative indicators reflecting the comparative characteristics of the region in the context of assessing the state of its social environment was formed in the process of development. This system reflects the state of the social environment as a result of the interaction of the indicative indicators of 6 basic components: the social sphere, the economic sphere, the ecological sphere, the technological sphere, the health care sphere and the sphere of education. The authors mathematically formalized the model of the mutual influence of all indicators of the social environment of the region. The model, presented in the form of a system of regression equations, allows to identify the vector and strength of the change in the state of the social environment when mediating one or more indicative indicators, taking into account the existing complexity of the system. As a result, the authors formed a goal-oriented graph-model of managing the state of the social environment of the region. This model is effective and prognostically effective pattern of development of the social environment of the region through the influence of such indicators of its state as the actual average monthly salary of the region's enterprise employees and the number of profitable businesses in the region. The developed model can be useful in the practical activities of representatives of state and municipal administration, as well as researchers in the field of social development.

Acknowledgements The research is financed as part of the project "Development of a method-ology for instrumental base formation for analysis and modeling of the spatial socio-economic development of systems based on internal reserves in the context of digitalization" (FSEG-2023-0008).

References

1. Jordan, D.T.L.: Human behavior and the social environment I (2000)
2. Fan, Y., Fang, C., Zhang, Q.: Coupling coordinated development between social economy and ecological environment in Chinese provincial capital cities-assessment and policy implications. J. Clean. Prod. **229**, 289–298 (2019)
3. Charles, V., D'Alessio, F.A.: An envelopment-based approach to measuring regional social progress. Socioecon. Plann. Sci.. Plann. Sci. **70**, 100713 (2019)
4. Becchetti, L., Bruni, L., Zamagni, S.: Happiness, relational goods, and social progress (2020). https://doi.org/10.1016/B978-0-12-816027-5.00012-4
5. Rodionov, D.G. et al.: Information environment quantifiers as investment analysis basis. Economies **10**(10), 232 (2022)
6. Cillo, V., Petruzzelli, A.M., Ardito, L., Del Giudice, M.: Understanding sustainable innovation: a systematic literature review. Corp. Soc. Responsib. Environ. Manag.Responsib. Environ. Manag. **26**(5), 1012–1025 (2019)
7. Arocena, R., Sutz, J.: Universities and social innovation for global sustainable development as seen from the south. Technol. Forecast. Soc. Chang. **162**, 120399 (2021)
8. Tanina, A. et al.: The tourist and recreational potential of cross-border regions of russia and kazakhstan during the covid-19 pandemic: estimation of the current state and possible risks. Economies **10**(8), 201 (2022)
9. Rodionov, D., Ivanova, A., Konnikova, O., Konnikov, E.: Impact of COVID-19 on the Russian Labor market: comparative analysis of the physical and informational spread of the coronavirus. Economies **10**(6), 136 (2022)
10. Woolliscroft, J.O.: Innovation in response to the COVID-19 pandemic crisis. Acad. Med. (2020)
11. Rodionov, D., Zaytsev, A., Konnikov, E., Dmitriev, N., Dubolazova, Y.: Modeling changes in the enterprise information capital in the digital economy. J. Open Innov. Technol. Market Complex. **7**(3), 166 (2021)
12. Benitez, G.B., Ayala, N.F., Frank, A.G.: Industry 4.0 innovation ecosystems: an evolutionary perspective on value cocreation. Int. J. Prod. Econ. **228**, 107735 (2020)
13. Kovalenko, I.I., Sokolitsyn, A.S., Semenov, V.P.: Industrial injuries in the socio-economic aspect. In: Proceedings of the 3rd International Conference Ergo-2018: Human Factors in Complex Technical Systems and Environments, Ergo 2018, pp. 205–208 (2018)
14. MacPherson, S.: St. Social development as an organizing concept for social progress (2019)
15. Skhvediani, A.E., Kudryavtseva, T.Y.: The socioeconomic development of Russia: some historical aspects. Eur. Res. Stud. J. **21**(4), 195–207 (2018)
16. Hickel, J.: The sustainable development index: measuring the ecological efficiency of human development in the anthropocene. Ecol. Econ. **167**, 106331 (2020)
17. UNDP: Human Development Report 1990: Concept and Measurement of Human Development. United nations Development Programme. Oxford University Press, New York, NY (1990)
18. Omar, H.C.: Human development index: a critique. Bangladesh Dev. Stud. **19**(3), 125–127 (1991)
19. Koronakos, G., Smirlis, Y., Sotiros, D., Despotis, D.K.: Assessment of OECD better life index by incorporating public opinion. Socioecon. Plann. Sci.. Plann. Sci. **70**, 100699 (2020)
20. OECD Better Life Index: Organisation for Economic Cooperation and Development. http://well-formed-data.net/archives/618/oecd-better-life-index/. Last accessed 15 Dec 2022
21. Krason, S.: A better life index that ignores what makes for a better life. Crisis (2014)

22. Kasparian, J., Antoine, R.: OECD's 'Better Life Index': can any country be well ranked? J. Appl. Stat. **39**(10), 2223–2230 (2012)
23. Marks, N.: The unhappy planet index: an index of human well-being and environmental impact. New Econ. Found. (2006)
24. NEF The Happy Planet index 2.0: The New Economics Foundation. https://static1.squ arespace.com/static/5735c421e321402778ee0ce9/t/578de9f729687f525e004f7b/146891827 2070/2009+Happy+Planet+Index/+report.pdf/. Last accessed 15 Dec 2022
25. Johns, H., Ormerod, P.: Happiness, economics and public policy. Inst. Econ. Aff. Res. Monogr. **62** (2007)
26. Campus, A., Porcu, M.: Reconsidering the well-being: the happy planet index and the issue of missing data. Contributi di Ricerca, CRENoS. C, **43**, E01 (2010)
27. Decancq, K., Schokkaert, E.: Beyond GDP: using equivalent incomes to measure well-being in Europe. Soc. Indic. Res. **126**, 21–55 (2016)
28. Costanza, R., et al.: Development: time to leave GDP behind. Nature **505**(7483), 283–285 (2014)
29. Kostina. E.Y.: Social well-being and social security in the context of the globalization of modern society. Universum: Soc. Sci. 1–9 (2015). (in Russian)
30. Nagaytsev. V.V., Pustovalova E.V.: The Level of Social Tension and Conflict in the System of Social Well-Being of the Population of the Region. News of the Altai State University, pp. 209–213. Scientific book, Barnaul (2010). (in Russian)
31. Vlasova. A.A.: Social well-being of a person and society as a strategic goal of social work. In: XX International Conference in memory of Professor L.N. Kogan Culture, Personality, Society in the Modern World: Methodology, Experience of Empirical Research (2015)
32. Kirkhmeer. L.V. On the issue of the concept of environmental and social development of the region in the concept of sustainable development. Collection of Materials of the II All-Russian Scientific and Practical Conference, pp. 8–12. TsNRS, S., Novosibirsk (2017). (in Russian)
33. Rodionov, D., et al.: Methodology for assessing the digital image of an enterprise with its industry specifics. Algorithms **15**(6), 177 (2022)
34. Jennings, V., Bamkole, O.: The relationship between social cohesion and urban green space: an avenue for health promotion. Int. J. Environ. Res. Public Health **16**(3), 452 (2019)
35. Anikina, E.A. et al.: Social, economic and emotional well-being of the older generation in Russia: social and pension policy. In: Barysheva, G.A. (ed) Dr. Economy Sciences. STT Publishing. (2018)
36. The concept of social protection: a multidimensional analysis. https://cyberleninka.ru/article/ n/ponyatie-sotsialnoy-zaschity-mnogoaspektnyy-analiz/. Last accessed 18 Dec 2022
37. Coogan, B.A.: Regional model of socio-pedagogical support of education in the Kurgan region. Bulletin of the South Ural State University. Series: Education. Pedagogical Sci. **38**(171), 27–32 (2009)
38. Chan, D.T.S.: Factors of increasing the social well-being of the population in the conditions of the knowledge economy. In: Continuous Well-Being in the World: A Collection of Scientific Papers of the International Scientific Symposium, Tomsk (2016)
39. Babayan, I.V., Pashinina, E.I.: Social well-being of youth in the socio-economic conditions of modernization. News of the Saratov University. New episode. Series Sociology. Polit. Sci. **16**(1), 15–22 (2016)
40. EMISS state statistics. https://fedstat.ru/. Last accessed 18 Dec 2022

Development of Digital Solutions for the Evaluation and Selection of R&D Projects as a Part of the Formation of a Scientific Budget

Svetlana Soldatova ⓘ and Elena Maslennikova ⓘ

Abstract Russian scientific budget as an integral system of organization, planning, implementation and control of state budget expenditures for scientific research and development is in its infancy. For its successful completion, a number of prerequisites are required. The formation of a methodological and information technology basis for the implementation of scientific budget processes, as well as optimization of the implementation of procedures for assessing and selecting scientific and scientific and technical projects for their financial support from the budget are among them. To solve these problems, a study was conducted aimed at developing proposals for using the methodology for assessing technology readiness levels in order to select scientific and scientific and technical projects and justify the volume of their budget financing, taking into account the possible expansion of the functionality of the integrated national information system for research, development and technological work. The article presents the results of the study, including an analysis of foreign experience in the application of the technology readiness level in the management of scientific and technological development. The paper presents a methodology for assessing the validity of the volumes of budgetary financing of research, development and technological work projects using the potential capabilities of integrated national information system for research, development and technological work, and proposes conceptual justifications for decisions aimed at automating the performance of a number of functions for the formation and implementation of the scientific budget.

Keywords Scientific budget · Technology readiness level · Information system

S. Soldatova (✉)
FSBI «STP Management», Moscow, Russia
e-mail: soldatova@fcntp.ru

E. Maslennikova
Russian Presidential Academy of National Economy and Public Administration, Moscow, Russia

21

M. Sari and A. Kulachinskaya (eds.), *Digital Transformation: What are the Smart Cities Today?*, Lecture Notes in Networks and Systems 846,
https://doi.org/10.1007/978-3-031-49390-4_2

1 Introduction

In modern world, the idea of financing scientific research and development within the framework of the scientific budget as an integral system of organization, planning, implementation and control of budget expenditures for research, development and technological work is gaining real features. Its full-fledged practical implementation requires the optimization of a number of key processes for the formation and implementation of the scientific budget, including the evaluation and selection of scientific and scientific and technical projects that have sufficient potential to ensure the development of domestic science within the framework of national strategic priorities. In this regard, the research was initiated, the purpose of which is to develop mechanisms for more effective integration of research and development planning with the budget process based on the methodology of technology readiness levels and expanding the functionality of the integrated national information system of research, development and technological work.

2 Research Methodology

The research methodology covers two groups of instruments. The tools of the first group were used for the conceptual substantiation of the developed proposals. These include the study of scientific publications about the technology readiness level methodology, its strengths and limitations, as well as the scope of useful application [1–4]. This group also includes an analysis of materials from US federal agencies, including the U.S. Government Accountability Office (GAO) [5], and European Commission documents [6, 7], which summarize experience and set out official recommendations regarding the use of technology readiness level scale for making decisions on financing scientific and scientific-technical projects from budgetary sources.

The tools of the second group are used as part of the justification of digital solutions aimed at ensuring the optimal implementation of the most important processes of the scientific budget—the assessment and selection of scientific and scientific and technical projects, as well as determining the amount of their funding. In this part of the study, both local digital solutions for automating the assessment of the level of technological maturity of projects, represented by various technology readiness level calculators [8–10], and system ones are analyzed. The latter cover the study and evaluation of the potential opportunities of national information system of research, development and technological work for closer linkage of research and development planning with the budget process. A methodology was developed for the evaluation and selection of scientific and scientific and technical projects in order to provide them with financial support from the federal budget based on the use of the technology readiness level scale. The technique was tested during the experiment.

3   Results

3.1   Experience in Applying the Technology Readiness Level Methodology in Budget Financing of Research, Development and Technological Work Expenditures

The technology readiness level characterizes the degree of maturity of the developed technology in the process of its advancement through the stages of the innovation cycle from idea to mass production. To measure technology readiness level, a scale is used, which, as a rule, has nine levels—from technology readiness level 1 (the initial level at which the technology exists in the form of a general idea) to technology readiness level 9 (the most mature level, reflecting the readiness of the technology for production use). The technology readiness level methodology is actively used in the public and private sectors in the US and the EU, including as a tool for financing scientific and scientific and technical projects [1, 6], as well as managing public funding of the scientific and technical sphere [2, 6]. US federal departments (NASA, ministries of defense, energy, transport, health care) rely on the UGT scale when funding research programs [5, 11, 12]. In the EU, the methodology has been applied as part of the Horizon 2020 program [7].

As noted above, a number of US departments use the technology readiness level scale, but the US Department of Defense pays special attention to its use in matters of research and development funding. This department carries out its program activities within the framework of planning and executing the state budget in the context of technology readiness level, distributing funds between scientific and scientific and technical projects depending on the level of their technological maturity. In the budgetary authority of the US Department of Defense for research, development and technological work expenditures, about 13% is allocated for financing basic and applied research, which correspond to technology readiness level 1–3. Financing of projects of the fourth and fifth levels of maturity is half of the total cost provided for these purposes. The rest of the volume falls on projects of the seventh level of maturity. More technologically mature projects are not subject to funding from the ministry [11].

In the European Union, the rules governing the science budget set the maximum thresholds for public participation in the financing of certain types of expenditures for research and development, as well as innovation. Its maximum size depends on the stage of innovation activity, the size of the capital of the enterprise, the organization of joint activities of large and small businesses. If state support for fundamental research provides 100% coverage of the expenses of organizations, regardless of their size, then applied research is funded taking into account the size of organizations, the existence of effective cooperation between them and the scale of dissemination of project results. For small enterprises, the threshold value of state funding for industrial research is 80%, for medium-sized enterprises—75%, for large organizations—65%. For experimental developments, taking into account the organization of joint activities or the wide dissemination of project results, the maximum share of state participation for small enterprises is 60%, for medium enterprises—50%, for large enterprises—40%. Support for the pre-production stage of innovation is a maximum of 50% for small and medium-sized enterprises and no more than 15% for large businesses.

The rules also state that thresholds can be raised if necessary. Under the condition of effective cooperation between organizations and wide dissemination of the results, the share of state funding for applied research can be 90% for small businesses, 80% for medium businesses, and 70% for large businesses [13].

Thus, the technology readiness level methodology has taken root in the practice of various countries as one of the main tools for determining the amount of public funding for research and development. The most common is the approach according to which the share of financial support for a particular project from the state decreases as the level of its technological maturity increases. However, the overall distribution of budget funds for research, development and technology work support looks different. The smallest share is accounted for by the total support of the least mature projects, and the largest share is for the support of more technologically mature projects.

It should be noted that the boundary of activities associated with the creation of new knowledge and defined, in accordance with the Frascati manual [14], as scientific research and experimental development, is the seventh level of the technology readiness level scale. Accordingly, research, development and technological work projects with technological maturity levels from the first to the seventh are supported by the scientific budget.

In Russia, separate steps are being taken to integrate the UGT methodology into the processes of managing public funding for research and development. Relevant decisions were made by the government of the Russian Federation and individual federal executive bodies. In particular, the order of the Ministry of Education and Science of Russia dated 06.02.2023 No. 107 prescribed to reflect in the information about the research, development and technological work project, which must be entered in the integrated national information system of research, development and technological work, an assessment of the level of technological maturity of scientific research or experimental development [15].

These steps undoubtedly contribute to the integration of the technology readiness level methodology into the processes of public funding of research and development. However, they can be characterized as local solutions that do not form a systemic basis that is adequate to the requirements of the budgetary process, within which the scientific budget is formed and implemented.

3.2 Methodological Approaches to the Implementation of the Concept of the Scientific Budget on the Basis of Integrated National Information System Research, Development and Technological Work

The solution to the problem described above is seen in the formation of unified mechanisms for the evaluation and selection of scientific and scientific and technical projects using the technology readiness level methodology, implemented on a common information technology platform, as proposed by integrated national information system of research, development and technological work. Implementation of the plan involves expanding the functionality of the information system.

At the moment, the functions of integrated national information system of research, development and technological work are limited to coordinating work on the formation of projects of scientific research topics, the examination of projects of scientific topics by the Russian Academy of Sciences and bringing the limits of budgetary funds to organizations performing research and development.

The regulatory resolutions of the Ministry of Education and Science of Russia, fixed in Order No. 107, provide for the placement in the integrated national information system of the research, development and technological work project, in addition to the previously required one. This is information about the current and target (planned) level of technology readiness level of scientific and scientific and technical projects. However, for what purposes this information should be used is not clear.

The solution proposed in this paper is aimed at developing mechanisms that make it possible to carry out a directed selection of research, development technological work projects and determine the amount of their budget financing using the technology readiness level methodology. As expected, the expansion of the integrated national information system of research, development and technological work functionality will ensure decision-making on the feasibility of financing new projects from the federal budget, on the feasibility of continuing budget financing of previously launched projects, as well as on the validity and sufficiency of the volumes of project financing announced by the applicants.

To implement the plan, a methodology was developed that includes not only evaluation, but also the selection of research, development technological work projects according to the criteria for compliance with the priorities of scientific and technological development, as well as technological maturity. The methodology provides for the implementation of three stages.

At the first stage, the applicant of a new project, who is obliged to post information about the research, development technological work project in the system, independently assesses the compliance of the topic with the current priorities of the country's scientific and technological development. The results of self-assessment are confirmed or refuted by experts. In the future, this step can be carried out automatically using technologies for analyzing big data, text mining and statistical information. Projects that have not passed this stage of selection cannot qualify for budget funding.

At the second stage, both for new and ongoing projects for which the deadlines for reporting on the results of the implementation of the corresponding stage have come, automatic generation of samples from the composition of projects available in the integrated national information system of research, development and technological work database should be provided. Samples should include projects grouped simultaneously by all key criteria (science and technology development priorities, thematic focus, technology readiness level, sectoral affiliation, specificity of results, and scale) to ensure comparability of information. The results of automated distribution should be verified by experts.

At the third stage, the system for each sample assumes the automatic calculation of the average amount of project financing required to move from the current level of technological maturity to the planned one. For each new research, development and technological work project, the amount of budget funding requested by the applicant is compared with the average value in the sample. In case of exceeding the requested amount of funding by 20 percent or more, the recommended amount of funding for the research, development and technological work project is determined based on the average value. For previously approved (implemented) projects, the requested amount of funding for the implementation of the next stage is also compared with the average of the sample. If the funding request exceeds the sample average by 20 percent or more, then an examination of financial documents is carried out.

If the examination confirms the reliability of financial justifications, then the state customer of the project approves the amount of financing. Otherwise, financial documents are returned to the applicant along with expert opinions to adjust the amount of funding for the next stage of the project. Budget financing of the project may be reduced or terminated if the contractor has not ensured the achievement of the planned (target) technology readiness level.

The proposed method has been tested in manual mode. On the example of two projects, it is possible to demonstrate the possibilities of using the integrated national information system of research, development and technological work in making decisions on the amount of funding for research, development and technological work projects from the budget. The first project relates to the field of nuclear energy, and the second—to the field of medicine.

The technology readiness level of the first project corresponds to technology readiness level 1, its financing from the federal budget is determined by the applicant in the amount of 68 million rubles. The project is financed on the basis of a government assignment. In the course of approbation of the methodology, a selection of projects close to the one evaluated according to the criteria described above was manually formed from the integrated national information system research, development and technological work database. All projects under consideration, including the one under evaluation, are of comparable scale. The requested amount of project financing was compared with the average in the sample (Table 1).

Since there is no excess of the requested amount of funding for the project being evaluated compared to the average in the sample, you should agree with the applicant's requirement and accept the amount of budget funding without changes. If, following the results of the implementation of the intermediate stage of the project,

Table 1 Sample of projects selected according to similarity criteria (nuclear power)

Name of the project	Technology readiness level of the project	Funding gap of the project, million rubles	Instrument of project financing
Estimated project			
Complex experimental and computational researches of radiation phenomena in reactor materials of nuclear and thermonuclear power engineering based on irradiation in an ion accelerator	1	68	State assignment
Selection of projects for comparison			
New radiation phenomena in gallium oxide and their application in devices	1	90	State assignment
Development of new methods, complexes of physical and mathematical models for solving the problems of safe use of atomic and thermonuclear energy, innovative nuclear energy technologies and nuclear fuel cycle, integrated disposal and decommissioning of nuclear and radiation hazardous facilities, rehabilitation of radiation-contaminated territories, radioactive waste management based on fundamental research of phenomena, workflows and material properties using probabilistic-deterministic methods	2	168	State assignment
Physical basis of the technology of ion-beam processing of gallium oxide for the creation of power electronics devices	1	21	State assignment
Research of the influence of convective and radiation phenomena and phase transitions on the thermal physics of building elements	1	18	State assignment
Average amount of project financing in the sample, million rubles	74,25		

the contractor presents the declared results and provides their documentary evidence, then it is necessary to make a decision to continue the implementation of the project while maintaining funding.

The technology readiness level of the second project corresponds to technology readiness level 2. Its financing should be carried out from the federal budget and funds for supporting scientific and scientific and technical activities in the total amount of 3 million rubles. By analogy with the first example, a sample of projects similar in a number of criteria was formed, and then a comparison was made of the amount of financing of the project being evaluated with the average in the sample (Table 2).

Since there is no excess of the requested amount of funding compared to the average in the sample, the decisions for this project are similar to those for the first one.

Table 2 Sample of projects selected according to similarity criteria (medicine)

Name of the project	Technology readiness level of the project	Funding gap of the project, million rubles	Source/instrument of project financing
Estimated project			
Research of genetic variants associated with non-specific resistance and intercellular regulation in various forms of the course of a new coronavirus infection COVID-19 for the development of panels for targeted gene diagnostics	2	3	Funds from the federal budget and funds from support scientific and scientific and technical activities funds
Selection of projects for comparison			
Molecular mechanisms of participation of mast cells in the formation of the immune and stromal landscape of a specific tissue microenvironment in normal and pathological conditions (oncogenesis, inflammation, fibrosis)	2	23	Organization's own funds
Quantitative characteristics of external and internal connections as an integral biomarker of the influence of pathogenetic factors on the regulation of microcirculatory blood flow in the human body with diabetes mellitus	2	21	RSF grant
The role of the expression profile of intercellular contact proteins in tumor resistance to therapeutic effects	2	1,2	RFBR grant
Characterization of innate and adaptive immunity and evaluation of the effectiveness of immunotropic therapy for socially significant diseases during the coronavirus pandemic	3	0,595	Private purchase
Average amount of project financing in the sample, million rubles	11,4		

If the implementation of the intermediate stage of the project will be accompanied by the presentation of the declared results and their documentary confirmation, then it is also necessary to make a decision to continue the implementation of the project while maintaining budget financing.

The implementation of the methodology based on digital solutions will reduce the requests for expert support for a number of functions in the process of evaluating and selecting research, development and technological work projects within the framework of the scientific budget.

3.3   Digital Solutions that Ensure the Evaluation and Selection of R&D Projects Within the Scientific Budget

This research considers two categories of digital solutions for the evaluation and selection of research, development and technological work projects for financial support from budgetary sources. Solutions from the first category provide process automation to determine the level of technological maturity of a scientific and technical reserve or the results of research, development and technological work project. This group includes the development of so-called technology readiness level calculators. Digital solutions from the second category are aimed at creating a unified digital environment for the implementation of a full cycle of evaluation and selection of research, development and technological work projects based on the technology readiness level methodology to determine the amount of financial support for scientific research and development at the expense of the federal budget.

The technology readiness level calculator is a tool that makes it easy for users to make an assessment of technology readiness level. Abroad, there is a practice of using technology readiness level calculators in both the private and public sectors.

The US Air Force Research Laboratory uses the technology readiness level calculator as a tool for managing technology development programs. This technology readiness level calculator is a Microsoft Excel spreadsheet in which the user answers a series of questions about an innovative project or technology. When the user provides complete answers to all the questions, the calculator displays on the interface the technology readiness level achieved by the innovative project or technology. The assessment of research, development and technological work results within a certain technology readiness level stage cannot be higher than the assessment of the previous stage, even if there are complete answers. Each question posed relates to a single stage of technology readiness level and does not affect the others. The questions are grouped according to the specifics, and not according to the stages of technological maturity, which allows the user to be more objective in the answers.

As it was noted by W. L. Note, B. C. Kennedy, R. J. Dziegiel, this calculator greatly simplifies the process of determining the stage of technological maturity of an innovative product or technology. As part of working with the calculator, each user gets access to a standard set of questions, which makes the result more repeatable. The standard format facilitates the comparison of technology readiness level of different technologies [8].

The US National Aeronautics and Space Administration (NASA) technology readiness level calculator acts as a tool for assessing the level of technological maturity of systems, subsystems and their components. The technology assessment provided by this calculator consists of two main parts: (1) determining the current level of maturity of the product or technology; (2) selection of mechanisms for increasing technological maturity, taking into account cost, timing and risks. The NASA Calculator includes questions to assess the technological maturity of technology hardware and software, as well as a set of questions to determine levels of readiness for other aspects of project maturity—production and systems. The NASA calculator also has a Microsoft Excel basis using macros. Its display bears a visual resemblance to that of a US Air Force Research Laboratory calculator. The user must answer a series of questions grouped by specification. The NASA version of the calculator has more functionality that allows you to expand the user's options during the assessment, in particular due to the function of comparing the results of the technology readiness level assessment with the indices of production and system readiness.

Russia also has experience in developing and using calculators to assess the technological maturity of innovative developments. An example is a calculator developed for the Russian Railways Open Joint Stock Company (RZD). Unlike the foreign analogues discussed above, Russian development is a specially developed program in the form of an application for a personal computer. During the assessment, the user checks the fact of solving certain tasks, tied to the levels of technology readiness and formulated taking into account the specifics of the activities of Russian Railways. Each of the tasks can have the status "solved", "not resolved" or "not applicable". After determining the status of all tasks, the user can let the program calculate the final result.

In general, the currently known technology readiness level calculators are tuned to the specifics of companies and departments using them in their activities, and they should be evaluated as local solutions that facilitate the work of experts in assessing the technological maturity of research, development and technological work projects or innovative projects due to its streamlining and standardization. For the purposes of this study, it is necessary to develop a calculator based on the basic technology readiness level scale and the basic rubricator of scientific and scientific and technical results. The first version of this calculator is ready and is being tested.

The use of information and telecommunication technologies in the field of state financing of research, development and technological work projects in Russia is currently predominantly local in nature, which greatly limits interdepartmental and interorganizational interaction between participants in the scientific budget processes. To solve this problem, the study developed the concept of an information system for the evaluation and selection of research, development and technological work projects in the framework of the formation of the scientific budget. The motivation for its creation is the idea of combining the evaluation and selection of research, development and technological work projects based on the UGT scale with the budget process. The proposed solution is based on the development and implementation in the integrated national information system of research, development and technological work of a number of additional functions for the evaluation and selection of research, development and technological work projects, as well as determining the validity of the amount of their financing from budgetary sources.

Figure 1 describes the implementation of the evaluating and selecting R&D projects processes and determining the amount of their funding from the budget as part of the proposed expansion of the functionality of the integrated national information system R&D.

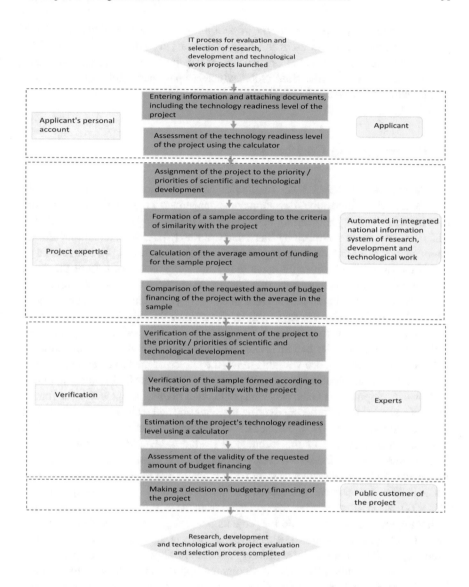

Fig. 1 The process of evaluation and selection of R&D projects within the scientific budget

The information system must comply with the principles of end-to-end digitalization using big data algorithms, machine learning and artificial intelligence. As part of expanding the functionality of an integrated national information system of research, development and technological work for the purpose of forming and implementing the scientific budget, it is proposed to expand the composition and refine a number of functional blocks of the system. In particular, in the "Applicant's Personal Account" block, it is necessary to provide for the possibility of self-assessment of the project's UGT using a calculator, as well as attaching documents to confirm the project's UGT.

It is proposed to develop the "Project Expertise" block, which operates in automatic mode, and the "Verification" block, which is a page accessible only to experts. The functions of these blocks are shown in Fig. 1.

4 Discussion and Conclusion

The technology readiness level methodology as an integrating concept within the framework of innovative development management is getting more and more convincing scientific justification. In this context, the use of the UGT scale in industrial organizations is evaluated by Bates and Clausen [3], A. Webster and J. Gardner apply similar assessments to the regenerative medicine industry [4], Markovska V. and Kabaivanov S. consider the technology readiness level scale from the point of view of the process approach to managing an organization [16]. A number of researchers support and develop the idea of extending the technology readiness level approach to wider areas. So, for example, S. Lokuge, D. Sedera, V. Grover and X. Dongming substantiate the concept of "organizational readiness for innovation" [17], and A. Webster and J. Gardner integrate UGT into a broader concept of institutional readiness [4]. In accordance with this trend, the choice of the CBT methodology as the basic tool for implementing scientific budgeting processes is absolutely justified.

However, approaches to the universal application of the concept within the framework of a single budget process covering the scientific and technical sphere have not yet been found. Moreover, a number of publications reflect a reserved or critical attitude towards the use of the technology readiness level scale in the activities of departments that allocate budgetary resources. A critical attitude to the use of the general technology readiness level scale in managing the financing of scientific and scientific and technical projects by the European Commission was expressed by Héder [2]. Y. Bar-Zeev justifies the exclusion from the selection procedures using the technology readiness level scale for projects belonging to the categories of fundamental and applied research [18]. However, there is also an opposite trend. For example, in the Technology Readiness Assessment Guidelines developed by the US Accounts Chamber, the lack of a common approach to assessing the technological maturity of projects, recognized in all areas, professional communities and departments, is assessed as a limitation [5]. However, the document does not contain specific proposals and recommendations on the desired approach, offering for review a generalization of the best practices for the application of technology readiness level in various departments.

The research contributes to overcoming the noted limitation by offering a conceptual development of an information system for the formation and implementation of the scientific budget, which is expected to optimize the processes of evaluation and selection of research, development and technological work projects, taking into account the use of the technology readiness level methodology.

The expected effects from the implementation of the information system can satisfy the needs of various parties. Representatives of the Russian Ministry of Education and Science and experts will be interested in reducing the time spent on the implementation of the processes of evaluation and selection of research, development and technological work projects within the framework of the scientific budget. Project implementers and experts will be able to appreciate the convenience of new ways of performing certain activities during the evaluation and selection of research, development and technological work projects. Improving the quality of the results of the evaluation and selection of research, development and technological work projects will be of interest to both representatives of the Russian Ministry of Education and Science and project implementers. Specialists of the Ministry of Education and Science of Russia are interested in reducing transaction costs for organizing the processes of evaluation and selection of research, development and technological work projects.

The overall effect of the development and use of the system is determined by the establishment of a single information technology framework that allows expanding the prerequisites for the implementation of scientific budget objects.

References

1. Artemis Innovation, http://www.artemisinnovation.com/images/TRL_White_Paper_2004-Edited.pdf. Accessed 27 Jun 2023
2. Héder, M.: From NASA to EU: the evolution of the TRL scale in public sector innovation. Innov. J.: Public Sect. Innov. J. **22**(2) (2017)
3. Bates, C.A., Clausen, C.: Engineering readiness: how the TRL figure of merit coordinates technology development. Eng. Stud. **12**(1), 9–38 (2020)
4. Webster, A., Gardner, J.: Aligning technology and institutional readiness: the adoption of innovation. Technol. Anal. Strat. Manag. **31**(10), 1229–1241 (2019)
5. GAO, https://www.gao.gov/assets/710/706680.pdf. Accessed 27 June 2023
6. EARTO, https://www.earto.eu/wp-content/uploads/The_TRL_Scale_as_a_R_I_Policy_Tool_-_EARTO_Recommendations_-_Final.pdf. Accessed 27 June 2023
7. European Commission, https://ec.europa.eu/programmes/horizon2020/sites/default/files/horizon_2020_first_results.pdf. Accessed 27 June 2023
8. Note, W.L., Kennedy, B.C., Dziegiel, R.J.: Technology readiness calculator. In: NDIA Systems Engineering Conference (2003)
9. Xavier, Jr. A., Veloso, A., Souza, J., Kaled Cás, P., Cappelletti, C.: AEB online calculator for assessing technology maturity: IMATEC. J. Aerosp. Technol. Manag., 12 (2020)
10. Summary 3. Main Menu. TRL Calculator. Release Notes, https://eclass.emt.ihu.gr/modules/document/file.php/MSC-TIE201/TRL%20Calc%20Ver%202_2.xls. Accessed 27 June 2023
11. Laws—Legal Research, https://law.onecle.com/uscode/10/2366b.html. Accessed 27 June 2023
12. DOE Directives, Guidance, and Delegations, https://www.directives.doe.gov/directives-documents/400-series/0413.3-EGuide-04a/@@images/file. Accessed 27 June 2023
13. EUR-Lex—Access to European Union law, https://eur-lex.europa.eu/legal-content/EN/TXT/PDF/?uri=CELEX:52014XC0627(01). Accessed 27 June 2023
14. OECD iLibrary, https://www.oecd-ilibrary.org/science-and-technology/frascati-manual-2015_9789264239012-en. Accessed 27 June 2023
15. Order of the Ministry of Education and Science of Russia dated 06.02.2023 N 107: On approval of the procedure for determining the levels of readiness of developed technologies, as well as

research and (or) scientific and technical results corresponding to each level of technologies readiness. Official Database "ConsultantPlus"
16. Markovska, V., Kabaivanov, S.: Process mining in support of technological readiness level assessment. In: IOP Conference Series: Materials Science and Engineering, vol. 878(1), pp. 012080. IOP Publishing (2020)
17. Lokuge, S., Sedera, D., Grover, V., Dongming, X.: Organizational readiness for digital innovation: development and empirical calibration of a construct. Inf. Manag. **56**(3), 445–461 (2019)
18. Bar-Zeev, Y.: TRL scale in horizon Europe and ERC—explained. Enspire Science (Jan. 2018)

Machine Learning Algorithms as a Tool for Improving Road Safety

Maria Rodionova⊙, Tatiana Kudryavtseva⊙, and Angi Skhvediani⊙

Abstract The chapter relates to the development of methods' comparative analysis. It investigates predictive machine learning techniques for the aim of the road safety improving. This study provides baseline analysis as a method for the evaluation of the model better quality. A baseline method allows us to compare different predictive models between each other and solve the problem of finding better way for an analysis. The analysis in this study is carried out on the basis of the machine learning techniques and the crash data in Saint-Petersburg. The results show a significant impact of oversampling in the slight class of severity level. Such peculiarity of the dataset influences on the reliability of models' results. The current analysis confirms that baseline is effective method of different methods' comparison, but also the crash data has the pattern of unbalancing that can cause of obtaining the proper results. Therefore, further direction of the research is applying different oversampling methods with machine learning predictive models.

Keywords Crash severity · Machine learning · ANN · Baseline · Classifiers

1 Introduction

Nowadays one of the global targets is reducing road traffic deaths and injuries by 50% by 2030. A huge amount of people is killed each year as a result of a road traffic crash. Millions more suffer non-fatal injuries including lifelong disability. Road traffic injuries cause considerable economic losses to victims, their families and to nations. Losses arise from the cost of treatment and reduced or lost productivity.

Road traffic injuries and deaths are largely preventable by developing and implementing national road safety programmes. Examples include social marketing and

M. Rodionova (✉) · T. Kudryavtseva · A. Skhvediani
Peter the Great St. Petersburg Polytechnic University, 29 Polytechnicheskaya St., St. Petersburg 195251, Russia
e-mail: rodionovamary98@gmail.com

© The Author(s), under exclusive license to Springer Nature Switzerland AG 2024
M. Sari and A. Kulachinskaya (eds.), *Digital Transformation: What are the Smart Cities Today?*, Lecture Notes in Networks and Systems 846,
https://doi.org/10.1007/978-3-031-49390-4_3

comprehensive legislation and enforcement for five major behavioural risk factors: speed, helmets, drink driving, seatbelts and child restraints.

In order to effectively establish a set of measures to reduce road traffic accidents, it is necessary to clearly understand and compute the impact of certain parameters on the number and severity of road accidents. That is why it is impossible for every territory to conduct research on the subject. The results of which will serve as the basis for the implementation of innovative transport infrastructure projects. In this regard, many authors consider this issue in terms of providing new algorithms for predicting crash severities, investigating the distribution of severity levels to decrease fatal and severe injuries after road crashes [1–8]. Several studies have examined the effectiveness of machine learning (ML) techniques in predicting crash severity and compared their results with statistical models. Various models, including the logit model and multivariate models developed by different researchers such as Kwon et al. [9], Zeng and Huang [10], Iranitalab and Khattak [11], Zhang et al. [12], and Wahab and Jiang [13] were used for this purpose. The findings showed that ML models outperformed the statistical models. Many papers describe analyses with the use of neural networks, mostly, it is artificial neural networks (ANN), as Shiran et al. [14] provide a comparison of ANN, multinomial logistic regression and decision tree techniques. Astarita et al. [15] compare ANN and hybrid Intelligent Genetic Algorithm, providing that we can find more complex algorithms, that can have better predictive abilities than simple ANN.

The major groups of techniques identified for road safety modeling were nearest neighbor classification, decision trees, evolutionary algorithms, support-vector machines, and artificial neural networks.

Therefore, the road safety problem can be solved by many methods of predictive analysis. But what the algorithm of comparison between different models results. In all provided researches the main method of comparison is metrics of confusion matrix. That will help everyone with such aim.

2 Materials and Methods

The research continues the previous study of the authors [16]. The data was used which affect the severity level and frequency of road traffic accidents in Saint-Petersburg. Such dataset is provided with the source "Karta DTP". The obtained data refers to 2015–2021 years and has 37,585 observations.

For our examined set "severity level" with 3 output classes (slight injury, severe injury and fatal outcome) was determined as dependent variable. 38 dummy variables and "participants count" that takes values from 1 to 26 were determined as independent variables. The data, which can be seen more detailed in the earlier published paper [16].

Based on the data and literature reviews from previous part of the work, artificial neural network was selected for the analysis.

Inspired by the capabilities of the human brain for its incredible processing capabilities due to interconnected neurons. An artificial neural network (ANN) is a highly complex, non-linear, parallel processor with a natural propensity for storing experimental knowledge and making it available afterward. A multi-layer perceptron ANN is typically made up of three kinds of layers: an input layer, an output layer, and one or more hidden layers. The input layer receives the values of the explanatory variables, i.e., the input data. The hidden layer, made up of m neurons, adds up the weights of the input values of the various explanatory variables, and calculates the complex association patterns. A single hidden layer is usually enough for crash analysis applications, but the definition of the number of neurons in it is generally the object of experimentation. ANN are designed by processing units known as perceptrons. These perceptrons consist of one layer, and they can solve linearly separable problems. However, to solve non-linear problems multilayer-perceptrons are used which usually contains three layers which are; an input layer, one or more hidden layers and an output layer. The concept of multilayer-perceptron is illustrated in Fig. 1.

The building block of an artificial neural network (ANN) is a neuron. These are simple computational units consisting of weighed input signals that produce an output signal by an activation function. Each neuron has a bias which can be thought of as an input that always has the value 1 or 0 and it must also be weighed. Weights are often initialized to small random values, ranging between 0 and 0.3, however, more complex initialization schemes could be used. The weighted inputs are added and passed through an activation function also known as the transfer function. The activation function is a mapping of the summed input to the output of the neuron. There are different types of activation functions used for specific purposes. Determination of such hyperparameters is so-called neural network architecture.

As in the previous study we examined ordered probit model for the crash severity prediction [16], we need to explore methods for the comparison of predictive abilities and models' quality.

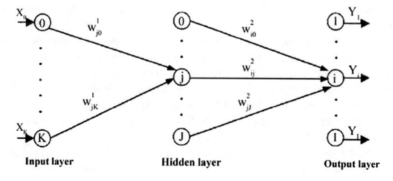

Fig. 1 Structure of multilayer perceptron [3]

This research provides the results of designing of the artificial neural network with multilayer perceptron to predict the severity of accidents by using Python language and its comparison with other methods by baseline method.

. Baseline method is a method that uses heuristics, simple summary statistics, randomness, or machine learning to create predictions for a dataset. Such predictions are used to measure the baseline's performance, in our case we use F1 score (1), this metric is computed to compare baseline with ANN. The aim is getting better performance of the developed model, than baseline.

$$\text{F1 score} = \frac{2 * (\text{precision} * \text{recall})}{(\text{precision} + \text{recall})}, \tag{1}$$

where precision—positive predictive value or the fraction of relevant instances among the retrieved instances;

recall—sensitivity or the fraction of relevant instances that were retrieved.

As baseline, it is considered dummy classifier with "straight" strategy and different meta—estimators, related to multiclass algorithms. Meta-estimators require a base estimator to be provided in their constructor. Meta-estimators extend the functionality of the base estimator to support multiclass problems, which is accomplished by transforming the multiclass problem into a set of simpler problems, then fitting one estimator per problem. Such estimators is implemented with classifiers (the one-vs-rest, the one-vs-one and error-correcting output code-based strategy) and pipeline with scaler and other classifiers (random forest, kneighbors, gradient boosting) or logistic regression.

The one-vs-rest strategy consists in fitting one classifier per class. For each classifier, the class is fitted against all the other classes. In addition to its computational efficiency (only n classes classifiers are needed), one advantage of this approach is its interpretability. Since each class is represented by one and only one classifier, it is possible to gain knowledge about the class by inspecting its corresponding classifier. This is the most commonly used strategy.

OnevsOne classifier constructs one classifier per pair of classes. At prediction time, the class which received the most votes is selected. In the event of a tie (among two classes with an equal number of votes), it selects the class with the highest aggregate classification confidence by summing over the pair-wise classification confidence levels computed by the underlying binary classifiers.

Error-Correcting Output Code-based strategies are fairly different from one-vs-the-rest and one-vs-one. With these strategies, each class is represented in a Euclidean space, where each dimension can only be 0 or 1. Another way to put it is that each class is represented by a binary code (an array of 0 and 1).

In the case, we provide some baselines for comparison that presented in Table 1.

Afterwards, we can conclude if our ANN architecture is optimal or not. And if some kinds of problems in the predictive analysis.

Table 1 Baselines

Method	Source	Specific hyperparameters
Dummy classifier	[17]	Strategy = 'stratified'
OutputCode & RandomForest classifiers	[18, 19]	Pipeline of OutputCode and RandomForest classifiers
OutputCode & GradientBoosting classifiers	[18, 20]	Pipeline of OutputCode and GradientBoosting Classifiers
OutputCode & KNeighbors classifiers	[18, 21]	Pipeline of OutputCode and KNeighbors classifiers
OutputCode Classifier & Logistic regression (multiclass = 'multinomial')	[18]	Pipeline of OutputCode and KNeighbors classifiers; multiclass = 'multinomial' for logistic regression
OneVsRest & RandomForest classifiers	[19, 22]	Pipeline of OneVsRest and RandomForest classifiers
OneVsRest & GradientBoosting classifiers	[20, 22]	Pipeline of OneVsRest and GradientBoosting Classifiers
OneVsRest & KNeighbors classifiers	[21, 22]	Pipeline of OneVsRest and KNeighbors classifiers
OneVsRest Classifier & Logistic regression (multiclass = 'multinomial')	[22]	Pipeline of OneVsRest and KNeighbors Classifiers; multiclass = 'multinomial' for logistic regression
OneVsOne & RandomForest classifiers	[19, 23]	Pipeline of OneVsOne and RandomForest classifiers
OneVsOne & GradientBoosting classifiers	[20, 23]	Pipeline of OneVsOne and GradientBoosting classifiers
OneVsOne & KNeighbors classifiers	[21, 23]	Pipeline of OneVsOne and KNeighbors Classifiers
OneVsOne Classifier & Logistic regression (multiclass = 'multinomial')	[23]	Pipeline of OneVsOne and KNeighbors Classifiers; multiclass = 'multinomial' for logistic regression
Ordered probit model	[16]	Evaluation and hyperparameters provides in the previous stage of research

3 Results

The designed ANN architecture consists of three layers: an input layer, a hidden layer and an output layer. The input layer and hidden one consist of the number of neurons that are equal to the number of input features of the dataset. The output layer contains three neurons as the network will predict three outcomes. The neurons in the input and the hidden layers will be activated using the relu function.

From the shown equation it is seen that this function gives an output of "x" given that "x" is positive and 0 otherwise. To carry out this experiment, relu is considered among other activation functions such as sigmoid and tanh functions because it is computationally less expensive as it involves less complicated calculations. Relu function is also useful in cases where a network has weights randomly initialized and about 50% of network results in 0 activation due to the features of relu. It means that the lower number of neurons would be activated resulting in a lighter network.

The Neurons of the output layer will be activated using the softmax function. The softmax function converts numbers (logits) into probabilities that add up to one. This function produces a vector showing the probability distributions of a list that are of a list of possible outcomes. Equation (2) shows that function where y_i is each element in the vector y. The ideal cost function to be used with the softmax function is the categorical cross-entropy loss which measures the similarity of the predictions to the actual value.

$$S(y_i) = \frac{e^{y_i}}{\sum_{j=1}^{N} e^{y_j}}, \tag{2}$$

where S—softmax function;

y_i—input vector;

e^{y}_i—standard exponential function for input vector;

N—number of classes in the multi-class classifier;

e^{y}_j—standard exponential function for output vector.

The choice of a loss function and optimizer could be vital for a neural network to generate better results. Optimization functions are used to modify the weights of a neural network. For this research, the adamax will be implemented. It is a variant of adam based on the infinity norm.

For Data Analysis and processing in python, pandas library provides high-level data structures and methods to ease the data analysis process. The matplotlib library is chosen as it provides quality plotting functionalities. For implementing the neural network, the keras library will be used. Keras is a high-level neural network API written in python capable of running on top of other libraries such as tensorflow, theano or CNTK and it focuses on fast experimentation.

The created network has the design with such hyperparameters as epochs (11), batch size (128).

For this ANN the accuracy (3), that performs predicting power of the network, equal to 62,3% and it exceeds the score for the training set that is 61,7%. The plot of losses values and accuracy with comparison of two datasets can be seen more detailed in the Fig. 2.

$$Accuracy = \frac{TP + TN}{TP + FP + FN + TN}, \tag{3}$$

where TP—true positive prediction in confusion matrix;

TN—true negative prediction;

FP—false positive prediction;

Fig. 2 Training and
validation sets

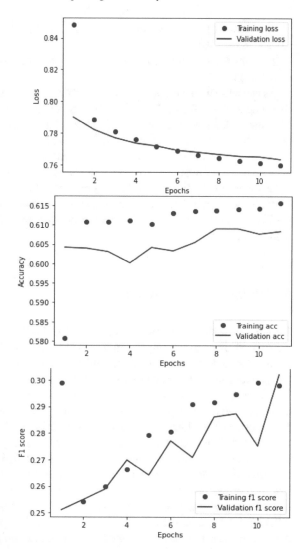

FN—false negative prediction.

As is seen in Fig. 2 training losses and losses of the validation dataset are quite similar. Whereas the validation accuracy is higher than training one. Hence, it can be concluded that the network is working properly.

But in our case to assess the quality of problems with multiple classes, we consider macro F1-score (short for macro-averaged F1 score).

The final calculations on the training set are presented in Fig. 2.

As can be seen, macro F1 score also presents higher value for validation set than for training one.

Final results for the test set of the model are following: loss—77%; accuracy—61%; f1 score—31%.

The obtained results with metrics are presented in Table 2 with baselines comparison.

As can be seen, the highest value of macro f1 score is obtained by OneVSRest and KNeighbors classifiers (37%). The OneVSOne and OutputCode with KNeighbors Classifiers also have high values of macro f1 score (36% and 34%, respectively). Our ANN has f1 score that equals to 30%. It means that this architecture of neural network is not good enough to predict severity level of road accidents on the basis of our dataset. However, if we compare f1 score of ANN for the first class and the same value of OneVSRest and KNeighbors classifiers, predicting of ANN is better (75% vs. 66%), while for the second class it is worse (13% vs. 40%); ANN doesn't predict the third class at all, whereas OneVSRest and KNeighbors classifiers do it with 5% f1 score.

4 Discussion and Conclusion

During the research one more benchmark for ANN performance has been implemented—ordered probit regression model [16]. Ordered probit models explain variation in an ordered categorical dependent variable as a function of one or more independent variables. Categories must only be ordered (for example, lowest to highest, weakest to strongest, strongly agree to strongly disagree)—the method does not require that the distance between the categories be equal. In our case we have 3 categories of road accidents severity level from the lowest (slight injury) to the highest one (fatal). Ordered probit models are typically used when the dependent variable has three to seven ordered categories. More than that, and researchers often turn to ordinary least squares regression, while if the dependent variable only has two categories, the ordered probit model reduces to simple probit [16].

Ordered probit is one example from the family of Generalized Linear Models (GLMs). GLMs connect a linear combination of independent variables and estimated parameters—often called the linear predictor—to a dependent variable using a link function. The link function typically involves some sort of non-linear transformation, which in the case of ordered probit means that the probabilities that a given observation in the dataset falls into each of the categories of the dependent variable are non-linear functions of the independent variables. The parameters of GLMs are typically estimated using Maximum Likelihood Estimation (MLE).

The results of comparison are depicted in Table 3.

As is seen, macro f1 score of the ordered probit is higher, than ANN and all methods that is observed earlier (39%). However, the first class is better predicted by ANN (75% vs. 65%), while the second class vice versa, ANN has f1 score for the second class—13%, ordered probit model—51%. The third class is not predicted at all with both methods.

Table 2 Baselines metrics

	Precision	Recall	F1-score
ANN			
1	0.62	0.95	0.75
2	0.43	0.08	0.13
3	0.00	0.00	0.00
Accuracy			0.61
Macro avg	0.35	0.34	**0.30**
Weighted avg	0.53	0.61	0.51
Dummy classifier (strategy = 'stratified')			
1	0.61	0.60	0.60
2	0.35	0.36	0.36
3	0.02	0.02	0.02
Accuracy			0.49
Macro avg	0.33	0.33	**0.33**
Weighted avg	0.50	0.49	0.50
OutputCode & RandomForest classifiers			
1	0.63	0.86	0.73
2	0.41	0.19	0.26
3	0.00	0.00	0.00
Accuracy			0.60
Macro avg	0.35	0.35	**0.33**
Weighted avg	0.53	0.60	0.54
OutputCode & GradientBoosting classifiers			
1	0.63	0.93	0.75
2	0.45	0.02	0.04
3	0.10	0.21	0.14
Accuracy			0.59
Macro avg	0.39	0.39	**0.31**
Weighted avg	0.55	0.59	0.48
OutputCode & KNeighbors classifiers			
1	0.63	0.74	0.68
2	0.38	0.30	0.33
3	0.25	0.01	0.02
Accuracy			0.56
Macro avg	0.42	0.35	**0.34**
Weighted avg	0.53	0.56	0.54
OutputCode Classifier & LogisticRegression (multiclass = 'multinomial')			
1	0.62	0.96	0.75

(continued)

Table 2 (continued)

	Precision	Recall	F1-score
2	0.46	0.07	0.12
3	0.00	0.00	0.00
Accuracy			0.61
Macro avg	0.36	0.34	**0.29**
Weighted avg	0.54	0.61	0.50
OneVsRest & RandomForest classifiers			
1	0.63	0.84	0.72
2	0.39	0.20	0.27
3	0.09	0.02	0.03
Accuracy			0.59
Macro avg	0.37	0.35	**0.34**
Weighted avg	0.53	0.59	0.54
OneVsRest & GradientBoosting classifiers			
1	0.62	0.97	0.76
2	0.48	0.06	0.10
3	0.60	0.01	0.02
Accuracy			0.62
Macro avg	0.57	0.35	**0.30**
Weighted avg	0.57	0.62	0.50
OneVsRest & KNeighbors classifiers			
1	0.64	0.69	0.66
2	0.40	0.39	0.40
3	0.13	0.03	0.05
Accuracy			0.56
Macro avg	0.39	0.37	**0.37**
Weighted avg	0.54	0.56	0.55
OneVsRest Classifier & Logistic regression (multiclass = 'ovr')			
1	0.62	0.96	0.75
2	0.46	0.07	0.12
3	0.00	0.00	0.00
Accuracy			0.61
Macro avg	0.36	0.34	**0.29**
Weighted avg	0.54	0.61	0.50
OneVsOne & RandomForest classifiers			
1	0.64	0.84	0.72
2	0.40	0.21	0.28
3	0.14	0.02	0.03

(continued)

Table 2 (continued)

	Precision	Recall	F1-score
Accuracy			0.59
Macro avg	0.39	0.36	**0.34**
Weighted avg	0.54	0.59	0.54
OneVsOne & GradientBoosting classifiers			
1	0.62	0.97	0.76
2	0.47	0.06	0.11
3	0.60	0.01	0.02
Accuracy			0.62
Macro avg	0.56	0.35	**0.30**
Weighted avg	0.57	0.62	0.50
OneVsOne & KNeighbors classifiers			
1	0.63	0.80	0.70
2	0.42	0.27	0.33
3	0.24	0.02	0.03
Accuracy			0.58
Macro avg	0.43	0.36	**0.36**
Weighted avg	0.54	0.58	0.55
Onevsone Classifier & Logistic regression (Multiclass = 'Multinomial')			
1	0.62	0.96	0.75
2	0.47	0.07	0.13
3	0.40	0.01	0.02
Accuracy			0.61
Macro avg	0.49	0.35	**0.30**
Weighted avg	0.55	0.61	0.50

The results of the research allow us to conclude that the current ANN architecture doesn't precisely predict the severity level of road accidents. Fatal accidents are not predicted at all by ANN as well as ordered probit regression. Hence, this neural network should be improved or changed. Also, it can be useful to improve our dataset for the analysis.

Dataset—Our study shows that the data has majority class that accounts for 61% and minority one, which equals to 4%. This fact can be cause of prediction impossibility of the fatal accidents (minority class) [24, 25]. Possibly, for training set can be used such methods as oversampling and undersampling [25, 26]. When we increase by duplicating examples in the minority class and deleting examples in the majority class [24–27].

Table 3 ANN and ordered probit model comparison

	Precision	Recall	F1-score
ANN			
1	0.62	0.95	0.75
2	0.43	0.08	0.13
3	0.00	0.00	0.00
Accuracy			0.61
Macro avg	0.35	0.34	0.30
Weighted avg	0.53	0.61	0.51
Ordered probit regression model [16]			
1	0.62	0.70	0.65
2	0.53	0.48	0.51
3	0.00	0.00	0.00
Accuracy			0.58
Macro avg	0.38	0.39	0.39
Weighted avg	0.56	0.58	0.57

Changing hyperparameters of ANN doesn't show significant changes in performance of network, only range within plus or minus 2%. Also, can be considered other types of neural networks or unsupervised learning.

Acknowledgements This research was funded by the Russian Science Foundation (project No. 23-78-10176, https://rscf.ru/en/project/23-78-10176/).

References

1. Papadimitriou, E., Theofilatos, A.: Meta-analysis of crash-risk factors in freeway entrance and exit areas. J. Transp. Eng. Part A Syst. **143**, 04017050 (2017)
2. Lyu, N., Cao, Y., Wu, C., Xu, J., Xie, L.: The effect of gender, occupation and experience on behavior while driving on a Freeway Deceleration Lane based on Field Operational Test Data. Accid. Anal. Prev.. Anal. Prev. **121**, 82–93 (2018)
3. Shi, X., Wong, Y.D., Li, M.Z.-F., Palanisamy, C., Chai, C.: A feature learning approach based on XGBoost for driving assessment and risk prediction. Accid. Anal. Prev.. Anal. Prev. **129**, 170–179 (2019)
4. Gianfranchi, E., Spoto, A., Tagliabue, M.: Risk profiles in novice road users: relation between moped riding simulator performance, on-road aberrant behaviors and dangerous driving. Transp. Res. Part F. Traffic Psychol. Behav. **49**, 132–144 (2017)
5. Orsini, F., Gecchele, G., Gastaldi, M., Rossi, R.: Collision prediction in roundabouts: a comparative study of extreme value theory approaches. Transp. A Transp. Sci. **15**, 556–572 (2019)
6. Barua, S., El-Basyouny, K., Islam, M.T.: Multivariate random parameters collision count data models with spatial heterogeneity. Anal. Methods Accid. Res. **9**, 1–15 (2016)
7. Dong, C., Clarke, D.B., Nambisan, S.S., Huang, B.: Analyzing injury crashes using random-parameter bivariate regression models. Transp. A Transp. Sci. **12**, 794–810 (2016)

8. Zeng, Q., Wen, H., Huang, H., Pei, X., Wong, S.: A Multivariate random-parameters Tobit model for analyzing highway crash rates by injury severity. Accid. Anal. Prev.. Anal. Prev. **99**, 184–191 (2017)
9. Kwon, O.H., Rhee, W., Yoon, Y.: Application of classification algorithms for analysis of road safety risk factor dependencies. Accid. Anal. Prev.. Anal. Prev. **75**, 1–15 (2015)
10. Zeng, Q., Huang, H.: A stable and optimized neural network model for crash injury severity prediction. Accid. Anal. Prev.. Anal. Prev. **73**, 351–358 (2014)
11. Iranitalab, A., Khattak, A.: Comparison of four statistical and machine learning methods for crash severity prediction. Accid. Anal. Prev.. Anal. Prev. **108**, 27–36 (2017)
12. Zhang, J., Li, Z., Pu, Z.: Comparing prediction performance for crash injury severity among various machine learning and statistical methods. IEEE Access **6**, 60079–60087 (2018)
13. Wahab, L., Jiang, H.: A comparative study on machine learning based algorithms for prediction of motorcycle crash severity. PLoS ONE **14**(4), 1–17 (2019)
14. Shiran, G., Imaninasab, R., Khayamim, R.: Crash severity analysis of highways based on multinomial logistic regression model, decision tree techniques, and artificial neural network: a modeling comparison. Sustainability **13**, 5670 (2021). https://doi.org/10.3390/su13105670
15. Astarita, V., Haghshenas, S.S., Guido, G., Vitale, A.: Developing new hybrid grey wolf optimization-based artificial neural network for predicting road crash severity. Transp. Eng. **12**, 100164 (2023)
16. Rodionova, M., Skhvediani, A., Kudryavtseva, T.: Prediction of crash severity as a way of road safety improvement: the case of saint Petersburg, Russia. Sustainability **14**, 9840 (2022)
17. Scikit-learn documentation, sklearn.dummy.DummyClassifier—scikit-learn 1.3.0 documentation. Accessed 21 Sep 2023
18. Scikit-learn documentation, sklearn.multiclass.OutputCodeClassifier—scikit-learn 1.3.0 documentation. Accessed 21 Sep 2023
19. Scikit-learn documentation, sklearn.ensemble.RandomForestClassifier—scikit-learn 1.3.0 documentation. Accessed 21 Sep 2023
20. Scikit-learn documentation, sklearn.ensemble.GradientBoostingClassifier—scikit-learn 1.3.0 documentation. Accessed 21 Sep 2023
21. Scikit-learn documentation, sklearn.neighbors.KNeighborsClassifier—scikit-learn 1.3.0 documentation. Accessed 21 Sep 2023
22. Scikit-learn documentation, sklearn.multiclass.OneVsRestClassifier—scikit-learn 1.3.0 documentation. Accessed 21 Sep 2023
23. Scikit-learn documentation, sklearn.multiclass.OneVsOneClassifier—scikit-learn 1.3.1 documentation. Accessed 21 Sep 2023
24. Hussain, S.F., Ashraf, M.M.: A novel one-vs-rest consensus learning method for crash severity prediction. Expert Syst. Appl. **228**, 120443 (2023)
25. Mohammadpour, S.I., Khedmati, M., Zada, M.J.H.: Classification of truck-involved crash severity: dealing with missing, imbalanced, and high dimensional safety data. PLoS ONE **18**(3), e0281901 (2023)
26. Wei, Z., Zhang, Y., Das, S.: Applying explainable machine learning techniques in daily crash occurrence and severity modeling for rural interstates. Transp. Res. Rec. **12**, 03611981221134629 (2023)
27. Wan, J., Zhu, S.: Cross-city crash severity analysis with cost-sensitive transfer learning algorithm. Expert Syst. Appl. **208**, 118129 (2022)

Smart Destinations for Advanced Development of Tourism

Artur Kuchumov, Galina Karpova, and Yana Testina

Abstract Smart destinations are a modern form of existence of tourist areas. The emergence of smart devices and the development of digitalization facilitate carrying out destination management processes within the framework of sustainable development, accounting for the opinions of all interested parties. The goal of the chapter is to identify the opportunities and prospects for the development of Russian tourist destinations during application of smart technologies and to develop indicators for monitoring their transformation. The article describes the prerequisites for application of smart technologies in tourism: smart cities, now becoming increasingly technologically advanced and widespread, are the foundation for the smart destinations. Based on the analysis of various methods for assessing the effectiveness of smart territories, a rating for evaluating smart destinations was proposed, with the following indicators: smart technology, smart innovation, smart sustainability, smart accessibility, smart inclusivity, as well as a system of sub-indicators. It is proposed to use this methodology both in conjunction with the methodology for assessing smart cities, and independently. This assessment method has been tested on five Russian cities with a population of over one million: Moscow, St. Petersburg, Nizhny Novgorod, Ufa, and Kazan, leaders among the country's smart cities. The calculated correlation between the results of the analysis of smart destinations and the tourist flow in terms of the number of tourists suggests a direct strong relationship. The application of this technique will make it possible to calculate the prospective tourist flows, depending on the investment in the development of smart technologies in tourism.

Keywords Smart destinations · Smart tourism · Smart city · Index of smart destinations · Digitalization of tourism

A. Kuchumov (✉) · G. Karpova
Saint-Petersburg State University of Economics, 191023 Saint Petersburg, Russia
e-mail: arturspb1@yandex.ru

Y. Testina
Saint-Petersburg State University, 199034 Saint Petersburg, Russia

© The Author(s), under exclusive license to Springer Nature Switzerland AG 2024
M. Sari and A. Kulachinskaya (eds.), *Digital Transformation: What are the Smart Cities Today?*, Lecture Notes in Networks and Systems 846,
https://doi.org/10.1007/978-3-031-49390-4_4

1 Introduction

In the modern world, the role of technology involvement in human life is significant [1]. The development of the global Internet, together with the growing dependence of the population on technology leads to an increase in the number of users of the World Wide Web (Fig. 1).

An exponential trend line with an approximation confidence of 0.9971, which is sufficient to confirm the hypothesis [3], indicates a further increase in the number of Internet users in the world in 2023.

In the Russian Federation, according to the Global Digital 2023 Meltwater report [4], there are currently 127.6 million Internet users, which is 88.2% of the total population of the country. The share of Internet users in Russia is significantly higher than the global average of 64.4% [4]

At the same time, the shares of both urban and rural users are high, and they are constantly increasing, as shown in Fig. 2.

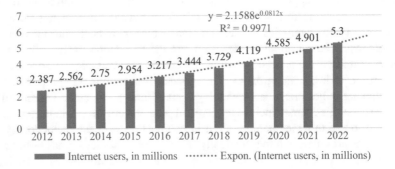

Fig. 1 The number of Internet users in the world, mln. [Compiled by the authors on the basis of data in [2]]

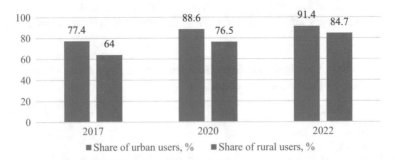

Fig. 2 Dynamics of the share of urban and rural Internet users among the population aged 15–74 in Russia for 2017–2022 [Compiled by the authors on the basis of data in [5]]

Such an extensive use of the Internet in the world and in Russia has led to the emergence of smart technologies and their widespread use [6], among other things, in the support of the sustainable development of regions [7].

The use of the term *smart* refers to technological, economic and social changes fueled by these technologies [8], their synchronization and interconnection [9] to increase the efficiency of resource use, create favorable conditions for life, protect the environment, increase competitiveness [10] with the aid of various digital technologies: Big Data [11], Internet of Things (IoT) [12], Artificial Intelligence (AI) [13], Blockchain [14], Sustainable Infrastructure [15], etc.

Smart technologies, as a rule, indicate a qualitative improvement in existing technologies, for example:

Big Data → Smart Data [16]
Internet of Things → Smart Things [17]
Blockchain → Smart Contracts [18].

Smart technologies have proven to be effective in the process of sustainable development of territories, implemented in the Smart City concept.

Smart cities are a technologically congruent environment that allows to optimize interactions of all urban systems with the needs of their residents [19]. In its optimal state, the stay of citizens in smart cities should lead to Smart Life [20].

According to Lim et al. [21], smart cities is a union between city management, local businesses, value creation for the citizen, urban big data, development and application of technological innovations, economy, and other areas. Vukovic, N. A., Larionova, V. A., Morganti, P. believe that Smart Cities should primarily be based on the principles of sustainability and benefit both the society and the environment, based on the Smart Sustainable City concept [22]. Pięta-Kanurska and Małgorzata [23] emphasizes the importance of a relational approach in the study of the essence of a smart city, which makes it possible to construct its functionality involving the residents in the processes of interaction between different loci [24].

Doubtlessly, smart cities attract tourists. According to the IESE Business School ranking, IESE Cities in Motion Index 2022 [25], the top five cities in the ranking are among the most popular destinations in the world (Table 1): four of them are in the top ten in terms of the number of tourists in 2022.

Table 1 Relationship between the ranking position of smart cities and attractiveness for tourists in 2022

City	Rank in the IESE cities in motion Index [25]	Rank in the TOP 100 city destinations Index [26]
London	1	6
New York	2	10
Paris	3	1
Tokyo	4	20
Berlin	5	8

Therefore, within this study, it is important to apply the concept of smart city in the activities of tourist destinations, since at present tourism is closely related to information and communication technologies (ICT) in association with changes in the consumer behavior of tourists [7], while the process is equidirectional, which is confirmed by the fact that the Internet and Information Communication Technologies (ICT) also radically transform the habits of travelers [27]. Smart destinations are key to the transformation of the tourism sector.

The United Nations World Tourism Organization (UNWTO), noting the increase in the share of digital processes in the organization of travel, considers that a smart destination is one with a strategy for technology, innovation, sustainability, accessibility and inclusivity along the entire tourism cycle: before, during and after the trip [28].

2 Materials and Methods

Scientists have long been trying to evaluate the effectiveness of smart cities in an attempt to calculate and prove the effectiveness of their transformation [29, 30], since smart cities are the foundation of smart destinations [31].

The study of smart destinations should be comprehensive, multifactorial in nature [32]

Bosch et al. [33] note that the use of smart city ranking indices is a rather laborious process, which is accompanied by a constant change in indicators, taking into account the dynamic development of the objects of study.

The smart city rating published by the IMD World Competitiveness Center [34] in 2023, the IMD Smart City Index (SCI), evaluated 141 world cities according to four groups of indicators: Economic Performance, Government Efficiency, Business Efficiency, and Infrastructure. This methodology is not acceptable due to the lack of indicators that can be used to assess the tourism industry in cities.

The IESE Cities in Motion Index [25] is based on 114 indicators that characterize the effectiveness of urban development, while the following indicators are used to reflect tourism development: Number of passengers per airport, Hotels, Restaurant Price Index, McDonald's, Number of congresses and meetings. These indicators are absolute, not relative, which negatively affects their reliability and significance, since, for example, the number of hotels cannot be equated with the Occupancy indicator, and high prices in restaurants do not attract a large number of tourists.

The authors of the article in their paper titled "Smart Cities as Drivers of Digitalization and Environmentalization of the Arctic" [19] suggest the use of a Balanced Score Card adapted for smart cities, including ten indicators of the degree of digitalization of a smart city: Smart Grid, Smart business environment, Smart government, Smart retail, Smart tourism, Smart mobility, Smart home, Smart healthcare, Smart industry, Smart society. At the same time, it is proposed to evaluate Smart tourism using indicators (1 and 2): Tourist destination digitalization index (I_{TDD}) and Tourist infrastructure digitalization index (I_{TID})

$$I_{TDD} = \frac{\text{Gross tourist product of destination}}{\text{Implementation costs of digital technologies}} \quad (1)$$

$$I_{TID} = \frac{\text{Number of tourist infrastructure facilities using digital technologies}}{\text{Total number of tourist infrastructure facilities in the smartcity}} \quad (2)$$

These relative indicators can be used for a comparatively shallow analysis of the urban environment, however, for a more detailed analysis from the point of view of the effectiveness of tourism development, they are not enough.

In the "Methodology for assessing the progress and effectiveness of the digital transformation of the urban economy in the Russian Federation (IQ of cities)", presented by the Ministry of Construction of Russia [29], the assessment of the tourist component of smart cities is proposed to be conducted according to the following criteria [35]:

1. Availability of digital cards of a city resident and of a city guest (binary);
2. Number of unique active users of digital cards of a city resident (who have performed at least 1 action) per 10 thousand people of the city's population;
3. Availability of a comprehensive system for informing tourists and residents of the city (binary);
4. Number of views of the city's online portal per 10 thousand people of the city's population.

In 2022, Moscow, St. Petersburg and Nizhny Novgorod were in the lead in the ranking [6].

Figure 3 presents the map of smart cities in Russia in 2021. On it, the size of the circle depends on the value of the index: the large circle indicates the index value of 60, the medium circle is 59–30, and the small circle is 30. Cities vary in color depending on the size of the population: green—1 million inhabitants, blue—1 million to 250 thousand, blue—250 to 100 thousand, purple—less than 100 thousand.

Fig. 3 Smart cities of the Russian Federation in 2021 [35]

We should note that the proposed methodology evaluates the smart city as a whole in a multifaceted way, however, the assessment of tourism is not effective enough, since half of the indicators are binary in nature, that is, they do not allow us to assess the dynamics of changes: improvement or deterioration of indicators. The dynamics of the indicator "The number of views of the city's online portal" depends not only on the quality of the content on the portal, but also on its SEO promotion and the presence of similar websites that duplicate certain sections, for example, the process of booking hotels for tourists and residents of the city.

In the opinion of the authors, when assessing a smart destination, it is possible to use all the indicators applicable to smart cities described in the Methodology of the Ministry of Construction of Russia [29], and "enhance" the tourist part with additional relative parameters. In fact, it is proposed to use both a cumulative methodology that combines indicators for assessing smart cities and smart destinations for a more accurate study, as well as separately apply the methodology for evaluating smart destinations in order to reduce the cost of the process and increase the frequency of rating calculations, which will allow systematic monitoring of digitalization.

It is proposed to use the UNWTO approach to the definition of smart destinations to identify the factors that have the strongest impact on tourism [28]: technology, innovation, sustainability, accessibility, inclusivity.

3   Results

Within the Russian statistical practice of providing information on the activities of tourist destinations, it is possible to propose the use of the following indicators presented in Table 2, which will allow us to trace the dynamics of the transformation of smart destinations.

We will test the methodology for evaluating smart destinations in the 5 largest smart cities in Russia [37] with a population of more than 1 million people [35]. The data are presented in Table 3.

Based on the conducted point-rating assessment, it was concluded that St. Petersburg is the destination most adapted for smart tourism, while the highest number of points was obtained for the availability of information for disabled people at the city's tourism portal. Table 4 shows the rating of the assessed Russian smart cities.

According to the Table 4, with the exception of the data on the tourist flow of Moscow and its position in the ranking of smart destinations, the value of the tourist flow directly correlates with the scores received by the destinations in the course of the study listed in the Table 3. The linear correlation coefficient is 0.93, which indicates a strong relationship between the indicators (see Fig. 4).

The reliability of the approximation of 0.9927 indicates that it is possible to conclude that there is a relationship between the number of tourists in the destination and its smart index. Consequently, further improvement of destinations in the context of smart technologies will increase the tourist flow and attract new tourists.

Table 2 Indicators of the effectiveness of digitalization of tourism of smart destinations

Parameter	Indicators	Sub-indicators	The value of the coefficient
Tourism technology	Payment methods on the territory of the destination	Only cash payment and bank transfers—0 points Use of bank cards—1 point Use of NFC technologies—1 point Use of digital currency—1 point Payment using biometrics—1 point	0–4 points
	Digital guest card	Possibility of obtaining discounts when visiting the destination—1 point Possibility of making payments—1 point	0–2 points
	Use of digital tourism technologies	Application of specialized software by the enterprises of the tourism industry—1 point Use of the "Electronic voucher" service—1 point Participation in tourist arrivals by E-visa—1 point	0–3 points
Tourism innovation	Use of biometric parameters	For payment in the enterprises of the tourism industry—2 points At airports at check-in—1 point For personal identification at the enterprises of the tourism industry—2 points	0–5 points

(continued)

Table 2 (continued)

Parameter	Indicators	Sub-indicators	The value of the coefficient
	Application of AR and VR technologies	Use of AR technologies in tourist attractions within the territory of the destination—1 point Use of AR technologies outside of attractions in the territory of the destination—1 point Use of VR technologies in tourist attractions within the territory of the destination—1 point Use of VR technologies outside of attractions in the territory of the destination—1 point	0–4 points
	Use of sharing economy elements	Car sharing in the territory of the destination—1 point Road sharing—1 point Flat sharing—1 point Kick sharing or bike sharing—1 point	0–4 points
Tourism sustainability	Websites dedicated to specially protected natural areas of the destination	0–30% of protected areas of destination have a website—0 point 31–70%—1 point 71–100%—2 points	0–2 points
	City's tourism portal provides information on centers for the protection of the environment	On separate waste collection points—1 point On battery collection points—1 point On ecofriendly organizations—1 point	0–3 points
	City's tourism portal provides information on eco-tourism	On eco-trails—1 point On specially protected natural areas—1 point On rare and endangered plants and animals living within the territory of the destination—1 point	0–3 points

(continued)

Table 2 (continued)

Parameter	Indicators	Sub-indicators	The value of the coefficient
Tourism accessibility	Multilingualism of the city's tourism portal	Only Russian language—0 points 2–4—1 point 5–7—2 points 8–10—3 points More than 10—5 points	0–5 points
	Adaptability of the city's tourism portal	Availability of the site—0 points Adaptive mobile version—1 point Application for mobile devices—1 point	0–2 points
	Availability of an inclusive version of the city's tourism portal	Availability of a version for the visually impaired—1 point Availability of a voice interface—1 point Availability of a chat for people with deafness or speech limitations—1 point Absence of flashes [36] on the screen to prevent epitheptic seizures—1 point	0–4 points
Tourism inclusivity	Availability of information for disabled people at the city's tourism portal	On available accommodation facilities—1 point On accessible places of public catering—1 point On available attractions—1 point On available tourist routes—1 point On transport companies—1 point On specialized tourist enterprises—1 point	0–6 points

(continued)

Table 2 (continued)

Parameter	Indicators	Sub-indicators	The value of the coefficient
	Availability of information for religious people at the city's tourism portal	On places for ceremonies and prayers—1 point On religious buildings (churches, mosques, synagogues, temples, etc.)—1 point On specialized accommodation and catering facilities—1 point	0–3 points
	Availability of information for the people with small budget at the city's tourism portal	On places providing free meals—1 point On places providing free accommodation—1 point On places providing assistance in a critical situation—1 point	0–3 points

4 Discussion

The authors' conclusions regarding the prospects of using smart technologies are confirmed by the dynamics of their operation, including "Application of smart technologies for the development of a region: A case study on tourism industry in St. Petersburg" [6], written in 2020. The processes described in it were further developed after the end of the pandemic and led to the rapid involvement of smart technologies in the tourism industry.

Platov A. V., Petrash E. V., Silaeva A. A. in the article "The role of smart technologies in shaping the experience of visitors to sustainable tourist destinations" [7] note that the correct application of smart technologies can contribute to the formation of attractive and memorable impressions. At the same time, the authors note that the growing involvement of digital technologies will increase the attendance of smart destinations.

Kazandzhieva V. in the article "Could smart technologies make tourist destinations more sustainable?" [44] notes the effectiveness of the development of smart destinations in supporting the sustainable development of the territory. We should note that most of the indicators of the rating of smart destinations proposed by the authors do not contradict the vectors of sustainable development of destinations.

Table 3 Point-rating assessment of smart destinations

Indicators	Moscow [38, 39]	Saint Petersburg [40]	Nizhny Novgorod [41]	Ufa [42]	Kazan [43]
Payment methods on the territory of the destination	3	3	2	3	3
Digital guest card	1	1	2	0	2
Use of digital tourism technologies	2	2	1	1	1
Tourism technology	6	6	5	4	6
Use of biometric parameters	0	0	0	0	0
Application of AR and VR technologies	4	4	4	2	4
Use of sharing economy elements	4	4	4	4	4
Tourism innovation	8	8	8	6	8
Websites dedicated to specially protected natural areas of the destination	0	1	0	1	1
City's tourism portal provides information on centers for the protection of the environment	0	0	0	0	0
City's tourism portal provides information on eco-tourism	0	1	0	1	0
Tourism sustainability	0	2	0	2	1
Multilingualism of the city's tourism portal	1	1	0	0	3
Adaptability of the city's tourism portal	2	1	1	1	2
Availability of an inclusive version of the city's tourism portal	1	4	3	2	2
Tourism accessibility	4	6	4	3	7
Availability of information for disabled people at the city's tourism portal	0	6	0	0	0
Availability of information for religious people at the city's tourism portal	0	0	0	1	1
Availability of information for the people with small budget at the city's tourism portal	0	1	1	0	1
Tourism inclusivity	0	7	1	1	2
Total	18	29	18	16	24

Table 4 Analysis of the rating of smart destinations

City	Position in the ranking of smart cities	Position in the ranking of smart destinations	Tourist flow by number of trips [43]
Moscow	1	3–4 (18)	16,908,169
Saint Petersburg	2	1 (29)	9,772,334
Nizhny Novgorod	3	3–4 (18)	2,397,809
Ufa	4	5 (16)	2,136,314
Kazan	5	2 (24)	3,908,170

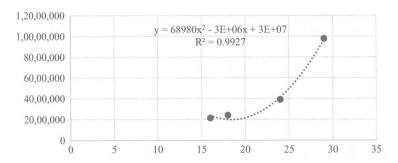

Fig. 4 Scatter diagram of the Smart Destination Rating and tourist flow by the number of trips

5   Conclusion

Smart destinations are a promising direction for the development of tourist areas. The involvement of digital technologies in tourism is happening everywhere, aiding in the creation of a new type of tourism product. Smart destinations in Russia have not yet become widespread, however, the situation will improve with the increase in the spread of smart technologies.

The research conducted in the article made it possible to conclude that there is a direct positive relationship between the level of development of a smart destination and the tourist flow in terms of the number of tourist arrivals, which makes it possible to conclude that it is necessary to further improve smart destinations to attract new tourists. Competent management of the processes of digitalization of smart destinations, cooperation with tourist portals, and accounting for the diversity of tourists, will increase the attractiveness of a region, including within the framework of sustainable development.

References

1. Weisburg, A.V.: The influence of dependence on information technology on the socialization of the individual. Bull. Econ. Law Sociol. **1**, 178–180 (2018)

2. Global number of internet users 2005–2022 Published by Ani Petrosyan, Feb 23, 2023, https://www.statista.com/statistics/273018/number-of-internet-users-worldwide/. Accessed 15 Sep 2023
3. Nikolaeva, I.V.: Specifics of the application of absolute and relative approximation error in regression analysis. Forum Young Sci. **6**(10), 1282–1286 (2017)
4. Churanov E. All statistics of the Internet and social networks for 2023—figures and trends from the Global Digital 2023 report, https://www.web-canape.ru/business/statistika-interneta-i-socsetej-na-2023-god-cifry-i-trendy-v-mire-i-v-rossii/. Accessed 15 Sep 2023
5. Federal State Statistics Service Selective Federal Statistical Observation on the Use of Information Technologies and Information and Telecommunication Networks by the Population, https://rosstat.gov.ru/free_doc/new_site/business/it/ikt22/index.html. Accessed 15 Sep 2023
6. Kuchumov, A.V., Karpova, G.A., Testina, Y.S.: Application of smart technologies for the development of a region: a case study on tourism industry in St. Petersburg. In: ACM International Conference Proceeding Series, pp. 1–4 (2020)
7. Platov, A.V., Petrash, E.V., Silaeva, A.A.: The role of smart technologies in shaping the experience of visitors to sustainable tourist destinations. Res. Result. Bus. Serv. Technol. **8**(1), 12–24 (2022)
8. Gretzel, U., Sigala, M., Xiang, Z., Koo, C.: Smart tourism: foundations and developments. Electron. Mark. **25** (2015)
9. Höjer, M., Wangel, J.: Smart sustainable cities—definition and challenges. In: Hilty, L., Aebischer, B. (eds.) ICT Innovations for Sustainability. Springer Series Advances in Intelligent Systems and Computing, pp. 333–349 (2016)
10. Kandpal, V., Manoj, K.N., Chandra, D.: Call for chapters for edited book titled: smart technology for a smart city and industry 4.0. LNCS, vol 13378, pp. 473–484 (2023)
11. Schramm, M., Shafaghi, M.: Moving from Big Data to Smart Data for Enhanced Performance, Business Efficiency, and New Business Models, pp. 1–17 (2020)
12. Rock, L.Y., Tajudeen, F.P., Chung, Y.W.: Usage and impact of the internet-of-things-based smart home technology: a quality-of-life perspective. Univers. Access Inf. Soc. (2022)
13. Marr, B.: The key definitions of artificial intelligence (AI) that explain its importance. https://www.forbes.com/sites/bernardmarr/2018/02/14/the-key-definitions-of-artificial-intelligence-ai-that-explain-its-importance/. Accessed 16 Sep 2023
14. Mohanta, B., Panda, S., Jena, D.: An Overview of Smart Contract and Use Cases in Blockchain Technology, pp. 1–4 (2018)
15. Bibri, S.E.: On the sustainability of smart and smarter cities in the era of big data: an interdisciplinary and transdisciplinary literature review. J Big Data **6**, 25 (2019)
16. Iafrate, F.: Other Titles from ISTE in Information Systems, Web and Pervasive Computing, pp. G1–G9 (2015)
17. Madakam, S.: Internet of things: smart things. Int. J. Futur. Comput. Commun., 250–253 (2015)
18. Stancu, A., Dragan, M.: Logic-based smart contracts. In: Rocha, Á., Adeli, H., Reis, L., Costanzo, S., Orovic, I., Moreira, F. (eds.) Trends and Innovations in Information Systems and Technologies. WorldCIST 2020. Advances in Intelligent Systems and Computing, vol. 1159. Springer, Cham (2020)
19. Testina, Y., Kuchumov, A., Boykova, J., Voloshinova, M.: SMART CITIES as drivers of digitalization and environmentalization of the arctic. In: Proceedings—International Scientific Conference on Innovations in Digital Economy, SPBPU IDE 2021. Association for Computing Machinery, pp. 220–227 (2021)
20. Khalifa, N., Taha, M.: Empowering AI Applications in Smart Life and Environment. Springer Nature (2024)
21. Lim, C., Kim, K.-J., Maglio, P.: Smart cities with big data: reference models, challenges, and considerations (2018)
22. Vukovic, N.A., Larionova, V.A., Morganti, P.: Smart sustainable cities: smart approaches and analysis. Economy of region **17**(3), 1004–1013 (2021)
23. Pięta-Kanurska, M., Małgorzata, P.: Smart City and Inclusive Growth. Polska akademia nauk Komitet przestrzennego zagospodarowania kraju pan biuletyn zeszyt. City and Regional Dynamics, pp. 273–274 (2020)

24. Agafonova, A.G.: Relational approach in urban studies. J. Sociol. Soc. Anthropol. **18**(4), 96–110 (2015)
25. IESE Business School—IESE Cities in Motion Index 2022, https://media.iese.edu/research/pdfs/ST-0633-E.pdf. Accessed 17 Sep 2023
26. The most popular cities in the world in 2022 are named, https://www.atorus.ru/node/50752. Accessed 17 Sep 2023
27. Paliwal, M., Nistala, N., Singh, A., Dikkatwar, R.: Smart tourism: antecedents to Indian traveller's decision. Eur. J. Innov. Manag. (2022)
28. Digital Transformation I UNWTO, https://www.unwto.org/digital-transformation. Accessed 17 Sep 2023
29. Order of the Ministry of Construction of Russia dated December 31, 2019 No. 924/pr "On approval of the methodology for assessing the progress and effectiveness of the digital transformation of the urban economy in the Russian Federation (IQ of cities)" https://www.minstroyrf.gov.ru/docs/120502/. Accessed 17 Sep 2023
30. Gutman, S., Teslya, A.: Selecting and assessing indicators for monitoring environmental safety in the Russian Arctic IOP Conference Series: Earth and Environmental Science, vol. 302, p. 012137 (2019)
31. Molchanova, V.: Tendencies of innovative development of tourist destinations: "smart destination", pp. 715–720 (2017)
32. Gahr, D., Rodríguez Rodríguez, Y., Hernández-Martín, R.: Smart Destinations: The optimization of Tourism Destination Management, pp. 63–110 (2014)
33. Bosch, P., Jongeneel, S., Neumann, H., Iglár, Br., Huovila, A., Airaksinen, M., Seppä, I.: Recommendations for a Smart City Index, pp. 1–33 (2016)
34. Smart City Observatory—IMD business school for management and leadership courses https://www.imd.org/smart-city-observatory/home/. Accessed 17 Sep 2023
35. Smart City https://russiasmartcity.ru/documents. Accessed 17 Sep 2023
36. GOST R 52872-2019 Internet resources and other information presented in electronic digital form. Applications for stationary and mobile devices, other user interfaces. Accessibility requirements for people with disabilities and other persons with disabilities, https://slabovid.ru/info/requirements/. Accessed 18 Sep 2023
37. The Ministry of Construction named the most "smart" cities in Russia—Vedomosti, https://www.vedomosti.ru/society/news/2022/07/15/931561-minstroi-nazval-samie-umnie-goroda-rossii. Accessed 18 Sep 2023
38. Tourist portal of the Moscow City Tourism Committee, https://discovermoscow.com/. Accessed 18 Sep 2023
39. Tourist portal of St. Petersburg, https://www.visit-petersburg.ru/. Accessed 18 Sep 2023
40. Tourist portal of Nizhny Novgorod, https://nn-tourist.ru/. Accessed 18 Sep 2023
41. Tourist portal of Ufa, https://www.visit-ufa.ru/. Accessed 18 Sep 2023
42. Tourist portal of Kazan, https://visit-tatarstan.com/. Accessed 18 Sep 2023
43. Rosstat—Tourism, https://rosstat.gov.ru/statistics/turizm. Accessed 18 Sep 2023
44. Kazandzhieva, V.: Could Smart Technologies Make Tourist Destinations More Sustainable? Modern Tourism. Smart Solutions for the Development of Tourism in Bulgaria in the Context of the COVID-19 Pandemic, Varna. Bulgaria, pp. 1–12 (2021)

Green Financing for Sustainable ESG Development of Smart City Industrial Ecosystems in the Circular Economy

Ekaterina Malevskaia-Malevich⬤

Abstract The development of industrial ecosystems, which are an interconnected network of companies and organizations in the region that use the products, waste and energy generated along the way in accordance with ESG principles, promotes the smart cities development. The monograph chapter describes the problem of financing the development of cyber-social industrial ecosystems built on smart digital technologies in terms of the cost of capital used. The study substantiates that traditional investment analysis methods are not suitable for projects that implement the concept of sustainable development, since they do not include a number of features. These features are highlighted and substantiated in the work. Thus, the authors have developed a cost of capital model of an enterprise implementing green projects by attracting green financing. In addition, within the framework of the study, the concept of a green premium to the cost of equity capital of such enterprise is defined. In addition to the capital cost model, the authors formulated a cost model for a manufacturing enterprise producing green products or implementing green projects. It was found that in green production, the volume of production losses is reduced, which allows reducing reserves in case of production losses. Based on the totality of the two models presented in the study, the overall positive nature of the influence of the implementation of green projects through green financing on the company's market value was determined.

Keywords Sustainable development · Capital investment financing · Green finance · Industrial ecosystems · Green capital cost · Smart city · Green premium

E. Malevskaia-Malevich (✉)
Peter the Great St. Petersburg Polytechnic University, Saint-Petersburg, Russia
e-mail: Mmed11@yandex.ru

The North-West Institute of management - Branch of the Russian Presidential Academy of National Economy and Public Administration (RANEPA), Saint-Petersburg, Russia

1   Introduction

It is widely acknowledged, that the concept of a smart city is very broad and can be interpreted in different ways, but it can be unequivocally stated that this concept is consonant with the concept of Sustainable Development, as well as the SDGs adopted by the UN. Smart urban transport system, its environmental safety, green building, machine learning systems, use of renewable energy sources, digitalization and development of bioenergy, solid waste management systems, eco-industrial parks, water and energy supply of large cities, public safety, smart government—all these terms combine the directions of the transition of the country's economy in general to the concept of sustainable development.

The industrial complex, which is the first in the value chain, and, a priori, the most "dirty" of the types of economic activity, bears additional responsibility in connection with its leading role in the transition to a sustainable development trajectory.

In order to make the transition to the concept of sustainable development, industrial ecosystems should pay special attention to their impact on the environment by implementing green production projects. The peculiarity of such projects is that, despite the high social and environmental significance, they can often be inefficient in terms of obtaining a cost result. Taking into account the fact that obtaining a financial result is the main goal of commercial activity, such a contradiction creates a stalemate. To solve it, a new type of capital was created—"green" capital, that is, endowment capital aimed at financing only those projects that, in addition to economic, also have environmental efficiency.

To finance environmental projects, ecosystems can attract so-called "green" financing. Green financing involves a set of measures, institutions, tools to stimulate the activities of ecosystems aimed at achieving SDG, that is, the implementation of environmental and social projects [1, 2].

A feature of investing in such projects is that the result is often non-financial, but expressed in other categories of value. Thus, the traditional methods of assessing the return on invested capital are losing their relevance; in addition, the classical investor is also not interested in investments of a social or environmental nature, since he is focused specifically on the value result [3, 4].

Investments aimed at implementing environmental projects are highly costly at the initial stages and are characterized by a long payback period. To stimulate the financing of such projects, the state needs to introduce tax incentives in order to increase the attractiveness for private investors.

1.1   Purpose of the Study

The purpose of this study is to identify the most promising methods for financing industrial ecosystems, as the first links in the value chain, on the path to sustainable development an integrated smart city system [5].

1.2 Research Objectives

To achieve this goal, the following research tasks were formulated and solved during the study:

- formulate the green finance concepts in the condition of the smart city economy.
- develop a green capital cost model for an enterprise operating in the smart city economy.
- formulate a cost model for a manufacturing enterprise producing green products/ implementing green projects.
- determine how the implementation of green projects using green financing affects the company's market value.

1.3 Research Methods

Generally accepted methods were used as research methods, such as the method of synthesis and generalization, analogies, methods of content and expert analysis. Econometric modeling methods were used to set optimization problems in the study.

The theoretical basis of the study was the provisions of neoclassical and neo-institutional economic theory. Scientific groundwork in the field of features of the functioning of a circular economic system, research in the field of pricing the value of green capital.

Determining the cost of capital of a green enterprise is based on the classical model of the cost of capital assets CAPM [6].

The work is based on studies previously conducted by the team of authors [4, 5, 7–12].

In article [5], the authors analyzed the cost of equity capital of a green enterprise and introduced the concept of a "green premium". In [4], the authors substantiate the existence of a "green premium" to the cost of equity capital when issuing green securities.

1.4 Literature Review

Like other terms related to the green economy, green finance does not have a single definition.

Porfiriev B. N. writes that, according to a number of international organizations of economists, the formation of "green" finance is one of the main changes in the existing global financial system. In a narrow sense, B. N. Porfiriev believes that "green" finance is "a set of financial products and services, the development, production and use of which is focused on reducing environmental and climate development risks."

In a broad sense, B. N. Porfiriev understands "green" finance as financial mechanisms to stimulate the implementation of alternative energy projects that help reduce greenhouse gas emissions and adapt to climate change. For example, special quotas and grid tariffs for the use of electricity from renewable sources by grid companies. Financial institutions, such as institutions and specific legal and administrative support for their activities, specialized in the aforementioned green investments, financing or hedging such investments, for example, through green bonds [13, 14].

According to the definition of the Group of 20 (G20), "green" finance can be understood as "investment that ensures environmental protection in the context of sustainable development" [15]. Green finance promotes internalization of environmental externalities.

In their study, Zhou X. and Cui Yu prove that empirical results show that the green bond issue factor has a positive effect on not only company stock prices, their profitability and operating performance, but on innovation potential and can improve corporate social responsibility [16].

Currently, the dynamics of the global growth of "green" financing can be traced. However, a number of Russian researchers have identified factors hindering green growth, among which external and internal factors of the investment environment can be distinguished. External factors of the investment environment include the imperfection of the environmental regulation market (the presence of external effects or externalities in the economy due to imperfect regulation of the relationship between economic entities); conservative type of thinking of business entities; insufficiently correct measurement of indicators of "green" growth [9]. Specialists refer to insufficient development of competencies in the financial sector as internal factors of the investment environment. Evaluation of green projects requires technical knowledge, for which financial specialists need to improve their skills. Experts also note the ineffectiveness of the monitoring and control system for the green finance policy used by financial institutions and the weak diversification of the green finance market [17].

2 Research Results

Traditional criteria for assessing the investment attractiveness of shares on the stock market are financial results and potential growth prospects of the company. The concept of sustainable development presupposes an orientation towards other values characterized by the SDGs. The concept of a smart city, through the integration of information systems into the management of city property, is designed precisely to implement the SDGs at the city level to improve the living comfort of the population. It is obvious that investments at the smart city level are aimed primarily at achieving a social effect rather than an economic one.

Methods for measuring social return on investment are described by the authors in [12]. It is clear that achieving social efficiency alone is not enough. If a project is not economically efficient, it brings losses, which indirectly also falls on the shoulders of the population.

One of the objectives of this study is to find an answer to the question of whether it is possible to harmonize the achievement of social efficiency together with economic efficiency.

As was shown in [5, 12], if "green" capital is used to finance investment activities, the so-called "green halo effect" arises, which increases the liquidity of the asset in the stock market. Increased liquidity allows you to offer a smaller premium to the cost of capital, thereby reducing its cost to the enterprise.

An attempt to calculate the cost of capital for "green" projects was made by the authors in [5]. The calculation of the cost of capital for green projects/enterprises $\left(r_{green\ equity}\right)$ will have a number of features. Based on previously analyzed studies, we can conclude that in some cases, raising capital to finance "green" projects will be cheaper than financing conventional projects due to the increased interest of investors. As various studies show [2–4], a modern investor can be characterized as socially responsible, this is manifested in the fact that when choosing an investment object, he tends to favor projects with a high ESG rating aimed at to achieve the SDGs, rather than projects, the implementation of which only leads to an economic effect, often to the detriment of other types of efficiency.

"Green premium" is a relatively new concept in finance, associated primarily with the increased attention of investors to "green" projects, and creating an obvious paradox of traditional ideas about the investment attractiveness of a company, how it was shown on an example in [4]. The "green premium" that an investor is willing to pay for the opportunity to purchase a "green" security is compensated by the increased demand for it. The high liquidity of the paper allows it to be sold on the secondary market at a higher price.

The "green" premium can also be interpreted as a reduction in the risk premium to the cost of capital of the enterprise. According to the CAPM model, any return on the market other than the risk-free one consists of the risk-free return and the risk premium. In order to obtain a "green" status, a project must undergo additional verification by independent rating agencies. In addition, it is necessary to disclose non-financial reporting. Greater transparency of activities naturally reduces the risk of capital investment. Thus, the risk premium becomes lower and capital is cheaper.

Let us present a model for calculating the cost of equity for financing green projects/enterprises. (based on Dorofeev, 2020 [18], previously presented by the authors in [5] in Russian).

$$r_{green\ equity} = r_f + \beta \times r_{market\ premium} + r_{UR} + r_{greenium} \qquad (1)$$

where $r_{green\ equity}$—the company's cost of equity (calculated based on the CAPM model); r_f—risk-free rate of return, such as T-Bills yield; β—stock systematic risk beta; r_{UR}—asset idiosyncratic risk premium; $r_{greenium}$—the premium that an investor is willing to pay for investing in green projects, or a high ESG rating or a discount for a low rating.

To get such "green premium" ($r_{greenium}$) an enterprises need to care additional expenses. For example, new capital ($\Delta Capex$) has to be invested, when replacing technologies with more environmentally friendly ones, as well as operational costs

(ΔNWCI) associated with continuous provision of preventive measures in order to minimize the risk of accidents, as well as to minimize climate risks. This leads to a decrease in the FCFE, however if these activities simultaneously increase the amount of the "green" premium for investors, the cost of equity for the enterprise will decrease, if we consider the change equal, then the value of the company (Value) will not change.

$$FCFE_{green} = NI + D - \Delta NWCI - \Delta Capex + ND \tag{3}$$

$$Value_{green} = \frac{FCFE_{green}}{r_{green\,equity}} = Value \tag{4}$$

However, based on studies [17, 19], where the authors show a smaller spread for green portfolios in comparison with comparable classical ones, it can be assumed that the "green premium" will grow at a faster pace than the growth of an enterprise's costs for ensuring environmental friendliness. Then the value of the company's equity capital will decrease faster.

The conditional level of green value growth can be expressed as an elasticity coefficient:

$$Value_{green} = \frac{decline\,rate\,FCFE_{green}(\%)}{decline\,rate\,r_{green\,equity}(\%)} \tag{5}$$

If the rate of decline in the cost of capital is faster than the rate of growth in the costs of the enterprise:

$$Value_{green} > Value \tag{6}$$

The market value of a green company for owners will be higher, despite the increase in the total costs of the enterprise [5].

As part of our study, the following cost–benefit model was adopted: all production costs as a result of the production process are transformed into sales or into losses, depending on the degree of probability of losses. Therefore, it is considered that all costs are transformed into losses if the probability of losses is equal to one and it is equal to zero if the probability of losses is equal to zero. Thus, the loss probability varies from zero to one; all intermediate values of the loss probability in this interval are possible and reflect the real degree of controllability of the production process [10, 11]:

$$Production\ costs = costs + losses$$
$$E[C_i] = P_i \times C_i$$
$$0 \le P_i \le 1 \tag{7}$$

$E[C_i]$—mathematical expectation of the value of production costs; P_i—loss probability; C_i—production costs.

It follows that if the costs of production are «C_i», then, assuming the conditions of a perfect competitive market (all costs are transformed into sales in the absence of losses, product prices are equal to production costs), sales will also amount to «C_i» if no loss. Otherwise, the costs completely turn into losses.

Production losses under the conditions considered are:

$$\Pi\Pi = (1 - P_i) \times (-C_i) \tag{8}$$

The mathematical expectation of the value of production costs/sales (Π) in this case will be:

$$E[\Pi] = C_i \times P_i + (1 - P_i) \times (-C_i) = C_i \times (2 \times P_i - 1) \tag{9}$$

The sales value of green enterprise products under the accepted conditions is a function of two independent variables«C_i» and «P_i».

The value of «C_i» is determined by the costs and production capabilities of the enterprise for the implementation of green projects: what products are intended to be produced, by what technologies, etc. The factors that determine the variable «C_i» are resource and balance restrictions and market demand for products. All these factors refer to products.

The variable «P_i» is determined by the presence or absence of losses in production. If there are no losses, then all production costs/costs are transformed into products, if not, then part of the production costs goes into losses, the enterprise is forced to create reserves of various kinds to overcome the impact of losses, in the form of production reserves, additional control operations during the production process, others preventive measures. The variable «P_i» therefore relates primarily to the production process.

Green production in the circular economy—waste-free lean production, involves the reduction of production losses. Then production costs are transformed into sales of the enterprise, and there is no need to additionally reserve capital, which releases capital from the company's operating cycle, so that FCF increases, which in turn maximizes EV.

It has been established that the factors determined by the variable «P_i» are most characteristic of green products or projects with incomplete control over the production process, which is generally also true for the products of "green enterprises". The advantages of "green technologies" are, in particular, that they allow in many cases to reduce the costs of production of products and thereby increase the interest of the manufacturer (for example, by saving on the costs of the manufacturer to preserve the environment, create reserves and safety stocks, etc.). The absence of the need to create additional reserves frees up capital, allowing the payment of additional dividends, which, in turn, will provide large revenues to the budget through personal income tax from owners.

Thus, a cost model is formulated for an enterprise operating within the framework of a smart city economy, which is, implementing any "green" projects aimed at achieving the goals of the SDGs—reducing resource consumption, negative impact on the environment, etc. The model clearly demonstrates that, despite the high cost of introducing new technologies, the final effect of their implementation will be positive, both at the level of the enterprise itself and at the level of a smart city.

3 Discussion

The presented models together form a kind of circular model of the cost of green capital. On the one hand, any technological innovation is expensive for enterprises and can have a long payback period. However, taking into account the fact that it is the "green" innovations necessary for the development of an integrated system of a smart city that have a number of additional positive effects beyond the scope of a single enterprise, the cumulative effect will be achieved much faster.

4 Conclusion

As a result of the study, the concept of green finance was formulated in the context of the circular economy, as a set of measures, institutions, tools to stimulate the activities of enterprises aimed at achieving sustainable development goals, that is, the implementation of environmental and social projects, green products or projects.

A model of the cost of capital of an enterprise implementing green projects by attracting green financing was developed. The concept of a green premium to the cost of equity of such an enterprise is defined.

The authors formulated a cost model for a manufacturing enterprise that produces green products or implements green projects. It was revealed that in the case of green production, the volume of production losses is reduced, which makes it possible to reduce reserves in case of production losses. Based on this model, the positive nature of the impact of the implementation of green projects through green financing on the indicator of the company's market value is determined.

5 Directions for Further Research

Of course, the proposals formulated in this study are theoretical, but in order to test the model on stock market data, it is recommended to wait for a more stable stage. At the moment, the Russian stock market has "survived" the shock consequences of the geopolitical crisis and is again showing interest in the green finance market, which can form the basis for empirical testing of the model.

Acknowledgements The study was funded by the Russian Science Foundation grant No. 23-28-01316 "Strategic management of effective sustainable ESG development of a multi-level cluster-type cyber-social industrial ecosystem in a circular economy based on the Industry 5.0 concept: methodology, tools, practice", https//rscf.ru/project/23-28-01316.

References

1. Wang, K.-H., et al.: Does green finance inspire sustainable development? Evidence from a global perspective. Econ. Anal. Policy 75, 412–426 (2022)
2. Dikau, S., Volz, U.: Central bank mandates, sustainability objectives and the promotion of green finance. Ecol. Econ. **184**, 107022 (2021)
3. Heinkel, R., Kraus, A., Zechner, J.: The effect of green investment on corporate behavior. J. Financ. Quant. Anal. **36**(4), 431–449 (2001)
4. Babkin, A., et al.: The relationship between socially responsible investment and the market value of an enterprise. In: E3S Web of Conferences, vol. 291. EDP Sciences (2021)
5. Babkin, A.V., Malevskaya-Malevich, E.D.: Impact of socially responsible investment on the value of innovatively active industrial enterprises. Sci. Tech. J. SPbSPU. Econ. Sci. **14**(4), 82–94 (2021). https://doi.org/10.18721/JE.14406
6. Fama, E.F., French, K.R.: The CAPM is wanted, dead or alive. J. Financ.Financ. **51**(5), 1947–1958 (1996)
7. Babkin, A., et al.: Methodology for economic analysis of highly uncertain innovative projects of improbability type. Risks **11**(1), 3 (2022)
8. Babkin, A., et al.: Framework for assessing the sustainability of ESG performance in industrial cluster ecosystems in a circular economy. J. Open Innov.: Technol., Mark., Complex. **9**(2), 100071 (2023)
9. Kvasha, N., Malevskaia-Malevich, E.: Intelligent quasi-rent as a value-creating factor for modern industrial enterprises. In: European Conference on Knowledge Management. Academic Conferences International Limited (2021)
10. Babkin, A.V., Kurcheeva, G.I., Aprelova, L.A.: Problems of green construction in the context of the implementation of the concept of a healthy city. π-Economy **15**(2), 59–78 (2022)
11. Demidenko, D.S., et al.: Assessing the efficiency of enterprises based on the ESG concept. π-Economy **15**(4), 82–95 (2022)
12. Malevskaia-Malevich, E., et al.: Adaptation of investment analysis to the features of socially oriented investments. In: E3S Web of Conferences, Vol. 402. EDP Sciences (2023).
13. Porfiriev, B.N.: "Green" trends in the global financial system. World Economy and International Relations **60**(9), 5–16 (2016)
14. Porfiriev, B.N.: "Green" Economy: Realities, Prospects and Limits of Growth. Publisher (2013)
15. Based on https://g20sfwg.org/wp-content/uploads/2021/07/2016_Synthesis_Report_Summary_RU.pdf. Accessed 23 Jan 2022
16. Zhou, X., Cui, Y.: Green bonds, corporate performance, and corporate social responsibility. Sustainability **11**(23), 6881 (2019)
17. Based on https://ru.valdaiclub.com/a/highlights/printsipy-esg-faktory-mirovoy-ekonomiki/. Accessed 13 Apr 2022
18. Dorofeev, M.L.: Features of the cost of capital in the green bond market. All-Russ. Econ. J. ECO **5**(551), 62–76 (2020)
19. Henning, S.L., Lewis, B.L., Shaw, W.H.: Valuation of the components of purchased goodwill. J. Account. Res. **38**(2), 375–386 (2000)
20. Martin, J.D., William Petty, J.: Value based management. Bayl. Bus. Rev. **19**(1), 2 (2001)
21. Based on https://natwest.us/insights/articles/green-halo-20/. Accessed 05 Sep 2023

22. Kleyner, G., Babkin, A.: Forming a telecommunication cluster based on a virtual enterprise. In: Internet of Things, Smart Spaces, and Next Generation Networks and Systems: 15th International Conference, NEW2AN 2015, and 8th Conference, ruSMART 2015, St. Petersburg, Russia, August 26–28, 2015, Proceedings 15. Springer International Publishing (2015)
23. Kurcheeva, G.I., Kopylov, V.B.: Approaches to the development of the concept of "digital city": the role of the population in management. π-Economy **14**(1), 21–33 (2021)

The Legal Regime of Digital Platforms at the National Level (on the Example of the Russian Federation)

Oxana Zhevnyak⊙

Abstract The purpose of the study is to comprehensively analyze the legal regime of digital platforms that has currently developed in Russia. The study proceeded from the hypothesis that the content of the legal regime of digital platforms is influenced by their essential characteristics and is built at different levels. In this regard, the essential characteristics of digital platforms, as well as levels of legal regulation, have been identified. In addition, the author distinguishes the positive and negative impact of digital platforms on the economy and gives some data on the use of digital platforms for urban management. The main part of the study is an analysis of the legal regime of digital platforms, built at the national level on the example of the Russian Federation. When analyzing the legal regime, interpreting the rules of law governing digital platforms, the essential characteristics of digital platforms are used. The choice of laws is determined by the main types of relations emerging on a digital platform. These relationships are highlighted by the author. The author determines that the digital platform is a multidimensional concept, highlights certain aspects of understanding of a digital platform and various aspects of its economic understanding, identifies the levels of legal regulation of digital platforms, the main relations subject to legal regulation, and the main problems requiring a legal solution. The author carries out own legal analysis of the key federal laws that determine the legal regime of digital platforms at the Russian national level and makes a number of specific conclusions on the results of such an analysis.

Keywords Digital Platform · Legal Regime of Digital Platform · Laws Governing Digital Platforms

O. Zhevnyak (✉)
Ural Federal University Named After the First President of Russia B.N.Yeltsin, 19 Mira Street, 620002 Ekaterinburg, Russia
e-mail: o.v.zhevnyak@urfu.ru; zevnyak@mail.ru

© The Author(s), under exclusive license to Springer Nature Switzerland AG 2024
M. Sari and A. Kulachinskaya (eds.), *Digital Transformation: What are the Smart Cities Today?*, Lecture Notes in Networks and Systems 846,
https://doi.org/10.1007/978-3-031-49390-4_6

73

1   Introduction

1.1   The Concept, Main Characteristics and Types of Digital Platforms, Their Significance for the Development of the Economy, Including the Economy of a City

The topic is devoted to the legal regulation of digital platforms considering that they can be used, among other things, for the development of the city's economy. We will start by analyzing the economic essence of digital platforms in order to understand which economic features of digital platforms should be taken into account when developing their legal regime.

Digital platforms are massively used as a business model by economic entities. They have penetrated largely into the lives of consumers. We are surrounded by digital platforms, the world is experiencing a "platform revolution" (this term was used in the book by Parker, Alstin, Chaudary "Platform revolution: how networked markets are changing the economy – and how to make them work on you" [2]).

The term "digital platform" is quite common in theory and practice. However, other names of the same phenomenon are found in the literature such as "online platform", "electronic platform", "Internet platform".

The main features of platform relations that distinguish them from other types of economic relations are as follows:

(1)   two-sided or multi-sided nature of the digital platform;
(2)   the economies of scale provided by the digital platform;
(3)   network effects of the digital platform;
(4)   information component of the digital platform;
(5)   linking clients to the digital platform;
(6)   the digital nature of the platform;
(7)   organizational and regulatory nature of platform activities.

Types and examples of digital platforms are listed in Table 1.

The economic importance of digital platforms cannot be overestimated. The positive impact of digital platforms on the development of economic relations is as follows:

(1)   the use of platform business organization reduces costs and increases productivity for economic entities;
(2)   platforms provide new opportunities for business development, which consist not only in providing a new model for organizing activities, but also in a set of digital technologies embedded in the platform;
(3)   platforms eliminate geographical and language barriers, allow national producers to enter world markets, enable small and medium-sized businesses to reach out to a large number of buyers;
(4)   platform owners have great opportunities to develop innovations;

Table 1 Types of digital platforms with examples [1]

No	Type of platforms	Examples
1	Services that publish content (video, music, etc.)	Spotify, Apple Music, Youtube ads
2	Cartographic services	Google Maps, Yandex Maps, 2GIS
3	C2C platforms (based on the sharing economy)	eBay, Avito, Airbnb, Blablacar
4	Dating services	Tinder
5	Travel booking services	Booking.com, TripAdvisor, Airbnb
6	Food delivery services	Yandex Food, Delivery Club
7	Online games	Twitch
8	Job search services	HeadHunter
9	Service platforms	Profi.ru
10	Online trading services with consumers (B2C) (marketplaces)	Amazon Marketplace, OZON, Wildberries
11	Rental services	Airbnb, CYAN
12	Financial services (payment systems)	Paypal, QiWi, Yumoney, Apple Pay
13	Social networks	Facebook, Vkontakte, Instagram
14	Search engines with contextual advertising	Google search, Yandex
15	Passenger transportation services	Blablacar, Uber, Yandex Taxi
16	App stores	Apple App Store, Google Play
17	News aggregators	Google news, Yandex news
18	Communication platforms	Skype, Whatsapp

(5) platform services improve the level and quality of life of platform clients and the population in general;

(6) platforms save time for customers to find the information they need;

(7) platform ecosystems allow customers to receive services based on the principle of single sign-on, while at the same time providing individualization of offers, considering the needs of a particular client [3].

However, it is worth noting that in addition to a positive impact, platforms can have a negative impact on economic relations. The negative effects of digital platforms are as follows:

(1) platforms are crowding out other business models, becoming disruptive technologies that cannot be competed with. This leads to the bankruptcy of entrepreneurs using traditional business models. In addition, there is a risk of a sudden curtailment of activities by the platform owner in a market where there are no other players;

(2) the network effects inherent in digital platforms entail an objective process of monopolization of the platform market. Platforms abuse their dominant position, for example, to impose contract terms on customers, to use unfair pricing. Suppliers of goods and services are dependent on the rules and tariffs of the platform, there is a high risk of their discrimination by the platform operator;

(3) the digital platform market is dominated by several giant Internet platforms with which it is difficult to compete for national platform businesses;

(4) the ubiquitous use of the platform business model results in a loss of direct contact between the business and the consumer;

(5) consumers are tied to the platform, the ability to choose other services for them is reduced. The platform defines a model of consumer behavior, forms a model of endless consumption;

(6) the platform business is associated with the collection of excessive information about customers, which can be used by the platform operator in bad faith, for example, to manipulate consumer behavior, disconnect successful businesses from the platform, etc. In addition, there are high risks of leakage of customer personal data [3].

The negative characteristics of digital platforms are an objective manifestation of their features. The task of the state is to minimize this negative impact, to find a balance in the regulation of digital platforms in order not to slow down their development, but to reduce their harmful impact to the maximum.

Digital platforms can certainly be used for urban management and economic development of a city. They can become the basis or an integral part of smart city programs.

The data of the site "Digital Software Marketplace - a project of ANO "Digital Platforms"" can be cited as an example. Information about 156 companies from Yekaterinburg that are software developers and offer licenses for the use of digital platforms (vendor companies) is posted on this site.

In addition, 287 digital platforms and programs designed for the management of municipalities are registered on the site. Among them, 32 platforms are designed for urban healthcare management, 9 platforms for public safety, 43 programs are designed for local government, 1 taxi program, 11 platforms for the implementation of the public address system, 59 for video surveillance management, 16 for parking management, 109 for transport management and 16 platforms in other areas [4].

1.2   The Purpose and the Methodology of the Study

Digital platforms are subject to legal regulation. Efforts to regulate them are made by different actors. Among them are international organizations, states that act independently or in alliance with other ones, digital platform operators (platform business owners) who can independently regulate certain issues of the platform's activities or through agreements with platform users.

When developing rules governing the activities of platforms, many questions arise. At present, it cannot be said that an ideal legal regime of digital platforms has formed, the existing rules give rise to many problems. However, it can be noted that a certain legal regime of the platforms exists. There are also scientific studies of certain legal issues concerning digital platforms.

However, we believe that a comprehensive study of the legal regime of digital platforms is necessary. It should include the identification of levels and subjects of regulation, the clarification of features of digital platforms that require reflection in the content of the legal regime, issues that require legal regulation, problems that require a legal solution. In addition, the study of the legal regime should include an analysis of its current state, the history of its development, the formulation of prospects and forecasts for improvement. It is necessary to analyze the regulatory framework, law enforcement practice, empirical data on the legal experience of digital platforms. The present study attempts to address some of these questions.

The purpose of this study is to comprehensively analyze the national legal regime of digital platforms that has currently developed in Russia.

The study proceeded from the hypothesis that the content of the legal regime of digital platforms is influenced by the essential characteristics of digital platforms and that the legal regime is built at different levels. In this regard, the essential characteristics of digital platforms, as well as levels of legal regulation, have been identified. After that, an analysis is given of the legal regime of digital platforms, built at the national level on the example of the Russian Federation. When analyzing the legal regime, interpreting the rules of law governing digital platforms, the essential characteristics of digital platforms were used.

This analysis is the author's legal analysis of specific norms of key federal laws that determine the legal regime of digital platforms in the Russian Federation. The choice of laws is dictated by the main types of relations emerging on the digital platform. These relationships are highlighted by the author.

2 Results

2.1 Digital Platforms as a Multidimensional Concept

The legal regime of digital platforms is influenced by their features. In this regard, it is necessary to understand that the digital platform can be considered from different aspects. Each aspect can be reflected in the legal regime of digital platforms.

We propose to single out the following aspects of understanding the digital platform, which can potentially be reflected in the legal regime of digital platforms. So, the digital platform can be viewed from various perspectives, including:

(1) a technological aspect. It allows us to consider a digital platform as a set of technologies, computer programs that can be accessed from client devices via the Internet and other communication networks;

(2) an economic aspect. Digital platforms from an economic point of view can be considered as a type of economic market relations, a type of economic activity of a platform operator, a form of business organization, etc.;

(3) a legal aspect. From this point of view, the digital platform is the subject of legal regulation. In addition, the platform itself participates in the regulation of relations between the platform operator and its clients and between clients themselves. The regulation is carried out through the rules developed by the platform itself (user agreements, etc.) and contracts;

(4) a sociological aspect. A digital platform can be understood as a form of social interaction, as well as a social community.

(5) a political aspect. Digital platforms control their customers and can have a huge impact on shaping their will. This influence may extend beyond economic life. The owners of giant platforms have enormous power and are able to participate in political life. The platform can shape the political consciousness of clients through the dissemination of certain information. It is no coincidence that the owners of large digital ecosystems are viewed as private governments;

(6) a biological aspect. This approach is less obvious, but a digital platform can be viewed as something similar to a biological organism that has its own life cycle, develops in accordance with certain objective laws, has an ecosystem, a "habitat" in which platform clients exist.

The consideration of the digital platform as an economic phenomenon seems to be the most important direction of the scientific study of platforms, which is the basis for understanding the other aspects mentioned above. It was economic research that laid the foundation for the theory of digital platforms. Among the most famous authors are Rochet and Tirole, Parker and Alstyne, Evans, Armstrong, Caillaud and Jullien.

It is the economic understanding of the digital platform that is most significant for the formation of the legal regime of the digital platforms.

The economic understanding of the digital platform also allows it to be considered in different aspects. We propose to highlight the following aspects of the economic understanding of the digital platform:

(1) the digital platform as a type of economic market relations. The approach to the digital platform as a bilateral or multilateral market prevails in the economic literature. As a rule, two groups of market actors interact in this market, for example, those who offer their goods and services and those who purchase goods and services. In addition, other groups of actors can operate on the platform market, for example, advertisers in the search engine market. There are no intermediaries between these groups of market participants, which distinguishes the platform market from traditional conveyor markets. The value of this market for participants is determined by the number of participants on the same side of the market and (or) on the other side of the market;

(2) the digital platform as a type of economic activity. The essence of this activity is to provide services for organizing interaction between a large number of clients of two or more sides of the platform and for regulating this interaction. The

digital platform from this point of view is the organizer of trade, while the term "trade" should be considered in a broad sense;

(3) the digital platform as a form of business organization. A platform business model is a form of business in which a product or service is offered by posting information in a special application to a generous number of potential consumers. On this platform, business contacts are established with consumers of goods and services, and in some cases (for example, when providing information services and content) activities are directly implemented.

2.2 Levels of Legal Regulation of Digital Platforms

We propose to single out the following levels of regulation, which generally form the legal regime of digital platforms. There are several levels at which rules for regulating digital platforms can be formed:

(1) international level;
(2) national level;
(3) local level;
(4) contractual level.

International treaties can play an important role in regulating global digital platforms. Currently, there are no international acts with the participation of Russia that have a special focus on the regulation of digital platforms. There are examples of international regulation of relations developing on the platform in the world, for example, documents adopted within the framework of the European Union. At the international level, there are also examples of the adoption of advisory rules, such as model laws.

Currently, in Russia, the solution of legal issues related to digital platforms is mainly carried out at the level of national law, namely at the level of regulatory legal acts. Examples of the main federal laws in this area will be considered below.

In addition, policy and strategic documents are being adopted that lay the foundation for the formation of a legal framework for regulating digital platforms. For example, the Passport of the national project "National Program "Digital Economy of the Russian Federation"" [5], the Strategy of the State Policy of the Russian Federation in the field of consumer protection for the period up to 2030 [6], the Concept of general regulation of the activities of groups of companies developing various digital services based on one "Ecosystem" [7].

Each digital platform, based on the specifics of its activities, develops rules for using the platform. This is the so-called local level of regulation. These rules become binding for a specific user of the platform, provided that this user joins them and, thereby, the terms of these rules are included in the content of the agreement with the user. An example is the branched structure and content of local acts developed and operating for the Yandex ecosystem [8].

The regulation is also carried out at the level of agreements concluded by the platform operator with each of its users. Typically, such an agreement is an adhesion contract (Article 428 of the Civil Code of the Russian Federation [9]), that is, it is concluded by joining in general to the developed standard rules of the platform. The contract may also contain individual conditions.

2.3   The Main Types of Relations Developing on the Digital Platform that Require Legal Regulation

Relationships that develop on a digital platform have features that should be considered in the legal regulation of digital platforms or in the practice of applying the rules of law. The specifics of digital platforms require consideration in many areas of regulation.

We propose to single out the main types of relations arising on digital platforms that are the subject of legal regulation. They are listed below. When considering them, examples of issues requiring a legal solution in each area of relations will be given. It should be noted that the list of such relationships and issues is not exhaustive.

1. Contractual relationships between platform users and the platform operator. The issues that require a legal solution are the significance of the local rules developed by the platform for regulating these relations, the procedure for concluding an agreement between the parties to the relationship, the application of the rules on the adhesion agreement, the right of the platform to unilaterally change user agreements, etc.
2. Conclusion of agreements between users of the platform through the platform. Issues that require a legal solution are the use of electronic documents and digital signatures when concluding contracts, the qualification of an offer posted on the platform as a public offer, the qualification of various actions and inactions of the client as an acceptance, etc.
3. Protection of the rights of consumers of platform services. The possibility of applying consumer protection legislation in connection with the gratuitous nature of relations with the platform should be legally resolved, the problems of distance selling should be properly regulated, as well as the responsibility of information aggregators to consumers, etc.
4. Protection of personal data of platform users. The problems that need to be addressed in the law are the collection by the platform operator of excessive information about users, the automatic processing of personal data, the use of information to manipulate consumer behavior, the threat of information leakage, etc.
5. Labor relations on the platform. The issues subject to legal regulation are the legal status of the platform employee, whether he is an employee, a self-employed person or an individual entrepreneur, social protection of such an employee

regardless of legal status, remote work, considering the characteristics of the platform, irregular working hours, etc.

6. Protection of intellectual rights on platforms. Issues that arise before the law are the rights to the software of the platform itself, liability for the sale and other use of counterfeit products through the platform, etc.

7. Protection of competition from the market power of the platform. The issues requiring a legal solution are the definition of the product market and market boundaries for the platform business and the business carried out through the platform, market dominance criteria for the digital platform, pricing on the platform, abuse of dominance by the platform, etc.

8. Tax relations with the participation of the platform. Issues that need to be addressed are taxation, considering the global nature of many platforms, determining the place of provision of services and the state of taxation for platform business and business carried out through the platform, etc.

2.4 Analysis of the Main Federal Laws of Russia Regulating Digital Platforms

Next, a legal analysis of the provisions of the key federal laws governing digital platforms in the Russian Federation will be given. Thus, the national level of the legal regime of digital platforms will be analyzed using the example of one specific country. The main criterion for choosing federal laws for analysis is the main types of relations that require legal regulation, discussed above.

Civil Code of the Russian Federation [9, 10]

Article 128 stipulates that the provision of services is an object of civil rights. Therefore, they should include the services provided by the operator of the digital platform, that is, services for organizing interaction between users of the platform. It follows that civil legal relations arise in connection with these services. These relationships are contractual obligations. Chapter 39 governs the relationship arising from the contract for the provision of paid services. However, services are often provided by a platform to users of one of its sides at no cost. From an economic point of view, these users pay for the platform services by participating in the platform, which increases its value for other users, as well as by providing their data and information about transactions that help for running platform business. In this case, the question arises about the legal qualification of the contract. A contract for the provision of services free of charge should be recognized as an unnamed agreement, the possibility of concluding which is allowed by Article 421.

Article 160 states the observance of a simple written form of the transaction if it is made using electronic or other technical means that allow the content of the transaction to be reproduced on a tangible medium in an unchanged form. Contracts on the digital platform are concluded by electronic means. The content of the agreement concluded between the platform operator and the user, as a rule, is determined by the rules of the platform, therefore these rules or the text of the user agreement itself must be available for downloading to a tangible medium in an unchanged form. This also applies to agreements concluded between the users themselves through the platform. In addition, Article 160 states that when concluding contracts by electronic means, the requirement for a signature is considered fulfilled if any method is used that allows one to reliably identify the person who expressed the will. This rule allows using different ways of expressing the will of the party to complete the transaction, including codes received via SMS messages, e-mail, expressing consent through a personal account, which is entered using a password known to the user. However, Article 160 indicates that a law or an agreement may provide for a special method for determining the person who has expressed the will. Thus, the terms of the user agreement, which can determine specific ways of expressing will, are of particular importance in this case.

Article 428 helps to qualify the agreement concluded between the platform operator and a user of the platform as an adhesion agreement, since its terms are set out in the standard text of the user agreement to which the user joins as a whole. Joining the contract can occur, as already noted, in different ways, namely by ticking the appropriate box, pressing the appropriate button ("click"), etc. At the same time, pre-checked boxes, in fact, deprive the possibility of free expression of one's will, therefore they are unacceptable. It is important that the text of the user agreement is provided to the user with simple and convenient access, of which the user must be explicitly informed. The rules on the adhesion agreement can also be applied when concluding agreements between users. It is worth noting that Article 178 allows to recognize transactions as invalid if at the time of the transaction the user did not have some information about the subject of the transaction. However, paragraph 5 of Article 178 significantly reduces this possibility by pointing out the need for "ordinary prudence" when making a transaction. It is worth recognizing that reading the text of the user agreement is a matter of ordinary prudence for a user and the user should bear the risk of not reading the agreement.

Article 432 establishes that the contract is concluded by sending an offer by one of the parties and its acceptance by the other party. Articles 437 and 494 makes it possible to qualify as a public offer the placement of information about the product on a digital platform (on a website, in a mobile application). So, the contract is concluded when consumer accepts the terms of the offer. If information about a product with different terms of sale (for example, different terms of delivery and payment) is posted on the platform, when the consumer selects certain order parameters provided for in the offer, these actions of the consumer cannot be qualified as a counteroffer (Article 443). It seems that in this case there is a plurality of offers, one of which is accepted by the consumer. Article 438 considers actions to fulfill the conditions specified in the offer as an acceptance. In relation to the conclusion of an agreement with the platform operator, these actions can be expressed in the beginning of the client to use the platform.

Article 497 regulates the remote method of sale. Unless otherwise provided by law, the buyer has the right to refuse to perform such an agreement before the goods are transferred to him but is obliged to reimburse the seller for the costs of executing the agreement.

Article 310 is relevant for resolving the issue of a unilateral change by the platform operator of the terms of the user agreement. In relations between persons engaged in commercial activities, such a change is possible if it is provided for by the agreement. The article prohibits unilaterally changing the terms of an agreement with an individual who is not an entrepreneur, unless this is provided for by law, but not by the contract. Therefore, to obtain the right to change the user agreement, the platform operator must take certain actions. We can assume possible solutions to this issue for the platform operator. It is worth providing in the contract for a certain period of its validity, so that when concluding a new contract, the operator would have the opportunity to update the text of the user agreement. It is possible to describe the relationship with the user in the user agreement in such a way as to try to ensure that each user entry on the platform can be qualified as a new contract. None of these options is an unequivocal legal solution, especially considering Article 16 of the Law "On Protection of Consumer Rights", which will be discussed below.

Article 1253.1 defines the concept and types of information intermediary, from which it follows that in many cases the digital platform operator is an information intermediary. Thus, the operator of the digital platform is not responsible for the violation of intellectual rights when information is posted on the platform by a third party, if the operator did not know and should not have known that intellectual rights were violated and took measures to stop their violation at the request of the copyright holder. This may include selling counterfeit products through the platform, posting unlicensed content, etc.

Law of the Russian Federation "On Protection of Consumer Rights" [11]

Preamble of the law, Articles 9 and 12 give the concept of an aggregator of information about goods and services, which include many digital platforms. These norms determine that the owner of the aggregator is responsible for providing incomplete or inaccurate information to the consumer, but not for breach of the contract by the seller (executor). However, it is possible to hold the owner of the aggregator liable for non- or improper performance of the contract by the seller (executor), if a conscientious citizen-consumer could have the opinion that the contract is concluded directly with this aggregator. There is judicial practice on bringing to responsibility aggregators of information about taxi services for harm caused to the life and health of a consumer by a taxi driver.

Article 26.1 regulates the remote method of selling goods when the buyer is an individual, that is, a consumer. This article gives the consumer the right to refuse the goods at any time before they have been transferred, and within the period after the date of the transfer. The consumer does not have the right to refuse the goods that have individual properties and can be used exclusively by this consumer. If the consumer refuses the goods, the seller must return the money paid to him, except for the costs of delivering the goods from the consumer.

Article 16 provides for a ban on the inclusion in the contract with the consumer of certain unacceptable conditions that infringe on the rights of the consumer. When they are included in the contract, they are recognized as null and void. The right of a person interacting with the consumer to unilaterally withdraw from the contract or change its terms, if this right is not provided for by regulatory legal acts, is named among other unacceptable terms of the contract. This has already been mentioned above in relation to the right of digital platform operators to unilaterally change user agreements. According to this law, such a condition of the contract with the consumer is void.

Federal Law "On Electronic Signature" [12]

Article 5 distinguishes between types of the electronic signatures. A simple electronic signature, according to this article, is an electronic signature that, using codes, passwords, and other means, confirms the fact that a signature was formed by a certain person. Registration of a personal account and sending messages from this account, SMS codes sent to the phone and then entered in a certain field of the platform, sending a message from a certain e-mail address can be considered as a simple electronic signature of a digital platform user.

Article 6 provides that an electronic document signed with simple electronic signature is equivalent to a regular document signed with a handwritten signature, only if it is provided for by regulatory legal acts or an agreement between the participants in electronic interaction. Thus, the possibility of using a simple electronic signature when users of a digital platform interact with the platform operator or between themselves should be initially provided for in the user agreement.

Federal Law "On Information, Information Technologies and Information Protection" [13]

Articles 2, 10, 10.1, 10.3–10.7 give the concept of an information system, search engine, news aggregator, audiovisual service, social network, ad placement service and their operators (owners), the organizer of the dissemination of information on the Internet and regulate the obligations of these persons related to the dissemination and provision of information.

Federal Law "On Personal Data" [14]

Article 5 provides that the content and volume of the processed personal data must correspond to the purposes of their processing, that the processed personal data should not be redundant. Thus, in fact, a ban has been established on the collection of personal data that is excessive, which is relevant when collecting and processing personal data of users on digital platforms.

Article 16 establishes a ban on making legally significant decisions affecting the rights and legitimate interests of the owner of personal data based on exclusively automated processing of personal data. Exceptions may be provided by federal laws. In addition, the owners of personal data themselves can express written consent to this. These rules are relevant for digital platforms on which information is processed automatically, while the results of processing become the basis for making legally significant decisions, for example, to terminate relations, lower a rating, apply other negative consequences.

Labor Code of the Russian Federation [15]

Chapter 49.1 defines the concept of remote work and remote workers, regulates the conclusion of an employment contract with such workers, interaction with the employer, working hours, labor organization and protection, termination of the contract, as well as additional wages guarantees. These rules may be relevant for digital platforms. Many employees who have an employment relationship with a digital platform operator work remotely.

Federal Law "On Conducting an Experiment to Establish a Special Tax Regime "Tax on Professional Income"" [16]

This law is relevant for digital platforms whose activities are related to the provision of taxi services, personal services, professional services, and others, where clients of the platform providing services are often payers of professional income tax (the so-called "self-employed").

In some cases, for example in the taxi industry, the economic essence of the relationship between the platform operator and the individuals directly providing the services is that the individuals are subordinate to the platform. The platform not only organizes the interaction between providers and consumers of services, but also determines the parameters of such interaction (price and other conditions) and regulates and controls this interaction. Thus, such relations between the digital platform operator and service providers are labor relations in essence, although these relations are formalized through civil law contracts. This problem requires its solution. Moreover, in this case, the consumer of services may be in good faith mistaken about who is the service provider, namely, a person who is an independent entity (a self-employed or an individual entrepreneur), or a digital platform operator, whose employees this person is. The answer to this question may affect the holding of a digital platform operator liable for the actions of direct providers of services. This issue has been discussed above.

Federal Law "On Protection of Competition" [17]

There are numerous and varied violations of antitrust laws in the practice of digital platforms. Many digital platforms dominate their markets and abuse their dominance [18]. Specific manifestations of the anti-competitive practices of digital platforms can be qualified under this law as various types of its violations. Thus, the law copes with the regulation of digital platforms.

However, some categories of law (commodity market, monopoly low price, market dominance) require new approaches to their understanding in relation to digital platforms. On September 1, 2023, the Federal law of July 10, 2023 N 301-FZ "On Amendments to the Federal Law "On Protection of Competition"" came into force, which has the goal of improving the antimonopoly regulation of digital markets. This law introduces the concept of network effect as a special criterion for determining the dominant position in the market [19].

Tax Code of the Russian Federation [20]

Article 174.2 defines the concept of the provision of services in electronic form (this is the provision of services through the information and telecommunications network, including Internet, automatically using information technologies), determines the types of such services for tax purposes, and also regulates the specifics of calculating and paying value added tax when foreign organizations provide services in electronic form, the place of sale of which is the territory of the Russian Federation (Article 148). These rules are relevant for digital platforms whose operators are foreign organizations.

3 Discussion

The first idea, announced as a result, speaks of the multidimensionality of digital platforms. The consideration of the platform as a multidimensional phenomenon is not completely new (see, for example [21]), however, this study highlights a wide range of aspects, including several economic aspects. We assume that there may be studies in the future, adding aspects. The author is planning to consider the stated aspects.

The second idea, concerning the levels of regulation, is found in studies devoted to the legal regulation of other relations, for example, artificial intelligence (see, for example, [22]). However, the author did not come across the idea of a level of regulation in relation to the legal regime of digital platforms. Here, it is especially important to single out the local level, since the essence of the activity of a digital platform lies in the organization and regulation of relations between users of the platform, which is noted by many researchers (see, for example, [2]). Further consideration should be given to splitting the levels of regulation and adding non-legal forms of regulation, such as soft law forms, recommendations, model contracts, and the like.

The third idea of identifying the main relationships that require legal regulation is based on an analysis of the relationships that are developing on platforms and the most striking problems associated with these relationships. In fact, it is a survey idea based on the ideas of many scientists considering mostly separate relationships.

Finally, the analysis of the existing legal regime of digital platforms on the example of the national level of regulation is the author's analysis, the application of legal analysis of the rules, in which the problems of legal regulation of many issues are identified. The considered federal laws are key, but they do not exhaust the list of sources of legal regulation of digital platforms. A discussion about the main relations that require legal regulation and, accordingly, about the key sources of law can take place. The nuances of the legal analysis carried out by other authors may further complement this analysis.

Further, the specific conclusions of the author, drawn from the analysis of key federal laws regulating digital platforms in the Russian Federation, will be discussed.

The conclusions regarding the legal qualification of the contract between the operator of the digital platform and the user give rise to further discussion about the need to change Chapter 39 of the Civil Code of the Russian Federation to cover gratuitous relations for the provision of services by the norms of this chapter.

The conclusion about the options for a simple electronic signature used by the user of the platform can be supplemented with new options used in practice.

In theory and practice, there is a discussion about the form in which an agreement on the use of a simple electronic signature should be concluded—in paper or electronic. Obviously, for most digital platforms, the paper form of agreements with users is not used at all, although there may be exceptions: some banks, to avoid the risks of not recognizing a simple electronic signature as equivalent to a handwritten one, conclude an agreement in paper form with a client, which provides for the possibility of signing electronic documents using certain types of simple electronic signature.

The conclusion that the familiarization of the user with the text of the user agreement before the conclusion of the contract is included in the user's ordinary due diligence seems to be quite reasonable. Further discussion may relate to the definition of "ordinary due diligence" in the case of a contract concluded on a digital platform.

The conclusion about the plurality of offers is debatable when several options for the terms of the contract are posted on the digital platform.

The conclusion that the buyer, in the event of a distance sale of goods, has the right to refuse to fulfill the contract, subject to reimbursement of the necessary expenses to the seller, should be compared with the norms of the Law "On Protection of Consumer Rights". It is necessary to resolve the issue of the correlation of norms: whether they correlate as general and special or general and concretizing.

It is necessary to consider the issue of granting the platform operator the right to change user agreements unilaterally, for example, giving him the opportunity to do this with a certain frequency (infrequently, for example, once a year). The options for resolving the issue proposed in the work are not an unambiguous legal solution, especially considering Article 16 of the Law "On Protection of Consumer Rights".

The conclusion that the conscientiousness and delusion of the consumer as to who he concludes the contract with (with the seller (executor) or the owner of the information aggregator) affects the possibility of holding the owner of the information aggregator liable for violation of the contract by the seller (executor) may be the subject of discussion. Perhaps it should be used only in certain areas of activity, for example, the provision of taxi services.

Further research is needed aimed at identifying the qualification criteria for different types of digital platforms, building a system of legal types of digital platforms.

It is necessary to resolve the issue of what data collected by the platform operator about its customers is redundant, inconsistent with the purposes of their processing.

A discussion is needed about the relationship between the concepts of "personal data" and information about platform users and their transactions collected by the platform. Accordingly, it is necessary to resolve the issue of the extent to which the legislation on the protection of personal data is applied to the activities of the platforms.

The problem of legal qualification of the relationship between the platform operator and individuals providing services through the platform requires discussion and solution. The proposed conclusion that in many cases these relations are labor relations, although they are formalized by civil law contracts, can lead to a violation of the freedom of contract. A possible solution is to extend the same social guarantees to such individuals, regardless of their legal status.

It is necessary to develop criteria for dominance in the market of digital platform operators, the legal concept of network effects, the development of the concept of a commodity market for the digital platform market.

It is useful to develop the research of a scientific and legal concept of an electronic service, which has a universal character. It is necessary to develop qualifying features of such a service, to distinguish it from related categories.

4 Conclusion

1. It is determined that the digital platform is a multidimensional concept. The following aspects of understanding the digital platform are highlighted: technological, economic, legal, political, social, and biological. Various aspects of the economic understanding of the digital platform have been identified: as a type of economic market relations, as a type of economic activity, as a form of business organization.

2. The levels of legal regulation of digital platforms are identified: international, national, local, contractual.

3. The main relations subject to legal regulation are singled out. They are contractual relationships between platform users and the platform operator, conclusion of agreements between users of the platform through the platform, protection of the rights of consumers of platform services, protection of personal data of platform users, labor relations on the platform, protection of intellectual rights on platforms, protection of competition from the market power of the platform, tax relations with the participation of the platform. The main problems requiring a legal solution are identified.

4. An analysis of the national legal regime of digital platforms allows us to draw several conclusions:

(1) the services provided by the operator of the digital platform are subject to civil rights. Their provision is subject to the norms of the Civil Code of the Russian Federation on the contract for the provision of services for a fee, however, services are often provided on a legally free basis, which leads to the recognition of the contract concluded in this case as an unnamed contract;

(2) contracts on a digital platform are concluded using electronic means. At the same time, the will of the platform user to make a transaction is expressed by entering codes received via SMS, email and other messages, registering

a personal account, sending a message from a specific email address. Such actions of the platform user should be qualified as signing an electronic document using a simple electronic signature;

(3) the possibility of using a simple electronic signature in the above options when interacting by digital platform users with the platform operator or between themselves should be initially provided for in the user agreement;

(4) the terms of the agreement between the user and the platform are determined by the rules of the platform, to which the user joins as a whole, that is, agreements with platform operators are agreements of adhesion. Joining to the rules occurs by ticking the appropriate box, "clicking" and the like. The fact that the client starts using the platform should be recognized as an acceptance to conclude an agreement with the platform operator. The rules on the adhesion agreement can also be applied when concluding agreements between users;

(5) the familiarization with the text of the user agreement should be included in the ordinary due diligence of the user and put on him the risk of negative consequences from not reading the agreement;

(6) the placement of information about the product on the digital platform must be qualified as a public offer, which makes it possible to state the fact of the conclusion of the contract when the other side of the platform (consumer) accepts the terms of the offer. A situation of multiple offers is possible, one of which is accepted by the consumer, thereby concluding a contract;

(7) the platforms use a remote method of selling goods, which, in accordance with the Civil Code of the Russian Federation, gives the buyer the right to refuse to perform such an agreement, subject to reimbursement to the seller of the incurred expenses. Somewhat different rules are provided for by the Law of the Russian Federation "On Protection of Consumer Rights";

(8) the platform operator has the right to unilaterally change user agreements only in relation to clients engaged in entrepreneurial activities, if this is provided for in the agreement. The law contains a ban on changing the terms of an agreement with an individual who is not an entrepreneur, until this is provided for by law, but not by an agreement. At the same time, the Law of the Russian Federation "On Protection of Consumer Rights" provides for a ban on including in a contract with a consumer the right of the other party to unilaterally withdraw from the contract or change its terms, if this right is not provided for by regulatory legal acts, and recognizes such a condition as null and void;

(9) the operator of the digital platform should be recognized as an information intermediary and, accordingly, his liability for infringement of intellectual rights when a third party places information on the platform should be excluded if he did not know and should not have known that intellectual rights are infringed as a result of the placement of this information;

(10) many digital platforms should be recognized as aggregators of information about goods and services. The owner of the aggregator is responsible for providing incomplete or inaccurate information to the consumer but is not responsible for the violation of the contract by the seller (executor). However, it is possible to hold the owner of the aggregator liable for such a violation of

the contract if a conscientious citizen-consumer could have the opinion that the contract is concluded directly with the owner of the aggregator;

(11) different types of digital platforms may have different legal qualifications in accordance with information legislation: an information system, a search engine, a news aggregator, an audiovisual service, a social network, an ad placement service. The public duties of their operators (owners) in accordance with the information legislation depend on this;

(12) digital platform operators are not entitled to collect excessive personal data about their customers;

(13) operators of digital platforms are not entitled to make legally significant decisions affecting the rights and legitimate interests of the subject of personal data, based solely on automated processing of personal data;

(14) many platform users who sell their services through it are payers of professional income tax ("self-employed"). In some cases, the economic essence of the relationship between the platform operator and such persons is that they are subject to the digital platform, which not only organizes the interaction between them and service consumers, but also determines the parameters of such interaction, regulates, and controls this interaction. Such relations between the digital platform operator and service providers are labor relations;

(15) many digital platform workers should be subject to remote labor laws;

(16) numerous and varied violations of antitrust laws are encountered in the practice of digital platforms. Specific manifestations of the anti-competitive practices of digital platforms can be qualified under antitrust law as various types of its violations, that is, the law "copes" with the regulation of digital platforms. However, some categories of law (commodity market, monopoly low price, market dominance) require new approaches to their understanding in relation to the activities of digital platforms;

(17) the civil law governing contractual relations for the provision of services does not provide for the concept of an electronic service. Such a concept (provision of services in electronic form) is given in the tax legislation. This definition should be used to define the concept of an electronic service for the purpose of their civil law regulation or to make it universal.

References

1. Digital Ecosystems in Russia: Evolution, Typology, Approaches to Regulation. Institute of Economic Policy named after E. T. Gaidar (Gaidar Institute) (2022). https://www.iep.ru/files/news/Issledovanie_jekosistem_Otchet.pdf. Last accessed 20 May 2023
2. Parker, G., Choudary, S., Van Alstyne, M.: Platform Revolution: How Networked Markets are Transforming the Economy—and How to Make Them Work for You, 1st edn. W.W. Norton & Company Inc., New York (2016)
3. Ecosystems: Regulatory Approaches: Public Consultation Report. Bank of Russia (2021). https://cbr.ru/Content/Document/File/119960/Consultation_Paper_02042021.pdf. Last accessed 20 May 2023

4. Digital marketplace of software—project of ANO "Digital Platforms". https://platforms.su. Last accessed 20 May 2023
5. Passport of the national project "National Program "Digital Economy of the Russian Federation", approved by the Presidium of the Council under the President of the Russian Federation for Strategic Development and National Projects, protocol dated 04.06.2019 No. 7. https://www.consultant.ru/document/cons_doc_LAW_328854. Last accessed 20 May 2023
6. Decree of the Government of the Russian Federation of August 28, 2017 N 1837-r "On approval of the Strategy of the state policy of the Russian Federation in the field of consumer protection for the period until 2030". https://www.consultant.ru/document/cons_doc_LAW_256217/. Last accessed 20 May 2023
7. The concept of general regulation of the activities of groups of companies developing various digital services based on one "Ecosystem", posted on the website of the Ministry of Economic Development of Russia. https://www.economy.gov.ru/material/departments/d31/koncepciya_gos_regulirovaniya_cifrovyh_platform_i_ekosistem/. Last accessed 20 May 2023
8. Yandex. Legal Documents. https://yandex.ru/legal/rules/. Last accessed 20 May 2023
9. Civil Code of the Russian Federation. Part one of November 30, 1994, No. 51-FZ (as amended). https://www.consultant.ru/document/cons_doc_LAW_5142/. Last accessed 20 May 2023
10. Civil Code of the Russian Federation. Part four of December 18, 2006, No. 230-FZ (as amended). https://www.consultant.ru/document/cons_doc_LAW_64629/. Last accessed 20 May 2023
11. Law of the Russian Federation of February 7, 1992, No. 2300–1 "On Protection of Consumer Rights" (as amended). https://www.consultant.ru/document/cons_doc_LAW_305/. Last accessed 20 May 2023
12. Federal Law of April 06, 2011, No. 63-FZ "On Electronic Signature" (as amended). https://www.consultant.ru/document/cons_doc_LAW_112701/. Last accessed 20 May 2023
13. Federal Law of July 27, 2006, No. 149-FZ "On Information, Information Technologies and Information Protection" (as amended). https://www.consultant.ru/document/cons_doc_LAW_61798/. Last accessed 20 May 2023
14. Federal Law of July 27, 2006. https://www.consultant.ru/document/cons_doc_LAW_61801/. No. 152-FZ "On Personal Data" (as amended). Last accessed 20 May 2023
15. Labor Code of the Russian Federation of December 30, 2001, No. 197-FZ (as amended). https://www.consultant.ru/document/cons_doc_LAW_34683/. Last accessed 20 May 2023
16. Federal Law of November 27, 2018, No. 422-FZ "On the Experiment to Establish a Special Tax Regime "Tax on Professional Income"" (as amended). https://www.consultant.ru/document/cons_doc_LAW_311977/. Last accessed 20 May 2023
17. Federal Law of July 26, 2006, No. 135-FZ "On Protection of Competition" (as amended). https://www.consultant.ru/document/cons_doc_LAW_61763/. Last accessed 20 May 2023
18. Zhevnyak, O.V.: Anti-competitive practice of digital platforms and response measures under Russian and foreign legislation. Law Polit. **5**, 14–41 (2021)
19. Federal Law of July 10, 2023 No. 301-FZ "On Amendments to the Federal Law "On Protection of Competition". https://www.consultant.ru/document/cons_doc_LAW_451662/. Last accessed 23 July 2023
20. Tax Code of the Russian Federation. Part two of August 05, 2000, No. 117-FZ (as amended). https://www.consultant.ru/document/cons_doc_LAW_28165/. Last accessed 20 May 2023
21. Dunaev, O.N., Kulakova, T.V.: Translogistic platform: development of logistics information platforms. Transp. Russ. Fed. **1**(62), 36–39 (2016)
22. Neznamov A. Law of robotics. How to regulate artificial intelligence. In: Forbes, January 16 (2018). https://www.forbes.ru/tehnologii/355757-zakony-robototehniki-kak-regulirovat-iskusstvennyy-intellekt. Last accessed 20 May 2023

Features of the Introduction of Digital Currency into the National Payment System in the Context of Global Socio-Economic Instability

Maria Alexandrovna Plakhotnikova⬤, Alexander Yurievich Anisimov⬤, Anna Yurievna Kashirtseva⬤, Alexey Sergeevich Kashirsky, and Alexander Adolfovich Grabsky

Abstract At the present stage, many well-established systems are being rethought, including due to the fact that the sanctions policy of the leadership of the European Union and the United States has become a trigger for rethinking elements of the Russian monetary and payment and settlement systems. The traditional financial system around the world has begun to experience the challenges and risks of a period of socio-economic instability. Development, testing, implementation of digital settlement tools in national payment systems, as well as clarification of the essence of the concept of "digital currency" is an urgent direction in the near future of the development of national financial systems. The definition of the main features in the world practice of the introduction and use of digital means of payment in national payment systems today is a popular scientific and practical direction. The purpose of the article is to identify the true essence of the concept of "digital currency", its interpretation from the standpoint of practical use as a tool of monetary and payment and settlement policy in Russia. The comparative analysis of domestic and foreign sources, the synthesis of scientific knowledge, the method of expert evaluation, as well as the method of system analysis were used in the study. As a result, a roadmap for the introduction of the "national digital currency" into the national payment

M. A. Plakhotnikova (✉)
PJSC Rosseti-Center, Malaya Ordynka Str., 15, Moscow, Russia
e-mail: erelda@mail.ru

A. Y. Anisimov
The Synergy University, Meshchanskaya Str., 9/14 P. 1, Moscow, Russia

A. Y. Kashirtseva
Southwest State University, 50Th Anniversary of October Street, 94, Kursk, Russia

A. S. Kashirsky
LLC "Gormash Global" Commercial Director, Mytishchi, Novoslobodskaya Str. 1, Moscow, Russia

A. A. Grabsky
RGGU Named After. Sergo Ordzhonikidze, Miklukho Maklaya Str., 23, Moscow, Russia

© The Author(s), under exclusive license to Springer Nature Switzerland AG 2024
M. Sari and A. Kulachinskaya (eds.), *Digital Transformation: What are the Smart Cities Today?*, Lecture Notes in Networks and Systems 846,
https://doi.org/10.1007/978-3-031-49390-4_7

system of Russia has been developed and presented, taking into account the forecast opportunities and risks.

Keywords Digital currency · Digital ruble · National payment system ·
Cryptocurrency · National digital currency

1 Introduction

The actions of central banks to issue digital currencies are rapidly moving from the plane of theoretical research to a substantive discussion of ways to implement and test options. The emergence of the technical possibility of introducing a national digital currency in the countries of the world in the last decade has created the prerequisites for a number of scientific research in this area. The study showed that the concepts of "national digital currency" and "cryptocurrency" differ significantly, their interpretations by theorists and practitioners are contradictory, sometimes clearly incorrect [1].

The first one is completely in the legal field and is a digital analogue of cash, and the second one has legal restrictions. On the one hand, the provision by Russian legal entities of services for the exchange of cryptocurrencies for rubles and foreign currency, as well as for goods (works, services) is considered as a potential involvement in the implementation of questionable transactions in accordance with the legislation on countering the legalization (laundering) of proceeds from crime and the financing of terrorism [1].

In 2022, according to this position, the Bank of Russia has taken actions to clarify the current situation with cryptocurrency. Namely, a report has been published that examines the risks of the spread of cryptocurrencies and the proposals of the Bank of Russia to limit operations with cryptocurrencies in Russia [2].

At the moment, legislative studies and additions continue to be carried out for the introduction of digital payment instruments into the Russian payment system and the formation of a national "digital currency" [3], which will establish the regulatory legal framework for regulating both cryptocurrencies and the national "digital currency". The improvement of legislation in this direction was initiated in 2022.

The purpose of the work is to identify the true essence of the concept of "digital currency", its interpretation from the standpoint of practical use as a tool of monetary and payment and settlement policy in Russia. The disclosure of the goal involves the solution of a number of tasks set in the study:

- to study the domestic and successful foreign experience in the development and features of the use of national digital calculation tools and experience in the development of "digital currency" at the state level;
- to identify the positive aspects and disadvantages of using digital currencies in national payment systems, as well as the limitations and opportunities for their development;

– to consider the experience of development and digital settlement tools, as well as their impact on the implementation of "digital currency" in national payment systems in developed and developing countries [4];
– to study theoretical approaches to the study of the concept of "digital currency" and the process of using it in the national payment system [5];
– to analyze statistical indicators on the development and experience of the introduction of "digital currency" in national payment systems of different countries of the world;
– to assess the prospects for further development of national payment systems in the context of the transition to the digital economy.

The novelty of the study is that a forecast roadmap has been developed for the introduction of a "digital currency" in conditions of global socio-economic instability. In 2020, there were systemic changes in the entire global economy, the demand for digital calculation tools increased. As a result, the processes of developing mechanisms for the introduction of "digital currency" into national payment systems have intensified.

2 Materials and Methods

A number of modern methods were used in the study of the problem identified in the article. Theoretical methods, analysis and synthesis were applied: analysis of theoretical provisions on digital currency in the national payment system, innovations in monetary policy. The article focuses on identifying gaps in theory and practice, due to the acute crisis conditions that have arisen in the world today. Using the method of concretization and generalization, problems and gaps in research and theory were identified. The limitations and possibilities of the development of digital calculation tools and the prerequisites for the emergence of a "digital currency" in the national payment system were investigated [3]. The domestic and successful foreign experience in the development and features of the use of the national digital currency are analyzed. The main positive aspects and disadvantages of using digital currencies in national payment systems are revealed. The experience of developing digital calculation tools and their transformation into a "digital currency" and the mechanisms of introducing a "digital currency" in national payment systems in developed and developing countries are considered. The article critically evaluates the opinions of scientists and practitioners about the essence of the concept of "digital currency", shows their inconsistency. The prospects of further development of national payment systems in the conditions of transition to the digital economy are assessed. There has been a significant increase in interest in the national "digital currency" in many countries of the world over the past two years and the factors that influenced this trend [6].

The degree of reliability of the research results and the validity of the conclusions is provided by the use of official statistics and publications of reputable domestic and foreign analysts. The author of the article studied: the report of the Central Bank of Russia for public consultations "Digital Ruble" [3, 5], Strategy for the development of the national payment system for 2021–2023 (2021) [7], materials of the European Central Bank, materials and reports of specialists of the Bank of England [4, 8].

The study of the problem was carried out in several stages:

At the first stage, the analysis of theoretical concepts, existing methodological approaches in the economic scientific literature on the problem of developing and implementing a "digital currency" in the national payment system was carried out. The purpose, objectives, methods, research are defined and the main hypothesis is put forward, which is that the current state of the national financial system requires the introduction of a "digital currency" to adapt it to the requirements of the external environment [6]. The interest of countries in the development of a "digital currency" of national banks has been studied.

At the second stage, based on the study of information sources and analysis of data reflecting changes and demand in the development and implementation of digital currency in different countries, it was concluded that it was necessary to form a forecast roadmap for the introduction of "digital currency" in conditions of global socio-economic instability.

The third stage is the formulation of conclusions. The experience of developing and implementing a "digital currency" in various countries of the world is systematized and generalized [2]. The features of the use of digital currency in national payment systems are revealed. The results of the study can be used by employees of state and financial management bodies, specialists in the management of banking structures.

3   Results

In many countries of the world, central banks are considering the possibility of introducing a "digital currency" into the national payment system as a logical development of digital settlement tools. Financial experts and economic scientists assess the prospects of such a process and its consequences for the economies of individual countries, in particular, and the global financial system as a whole. At the moment, many questions have not received reliable answers. An important aspect of such studies is forecasting the impact of the processes of introducing a national "digital currency" in different countries on economic growth, the stability of the financial system and monetary policy [9].

Many scientists have investigated the impact of the national "digital currency" on the US economy, namely, the activities of the central bank digital currency (CBDC). In their opinion, the issue of the so-called "digital currency": CBDC and CFR can be considered as new instruments in the government bond market. The researchers used the stochastic dynamic general equilibrium model (DSGE model) in their work, they showed that the digital currency will contribute to an increase in gross domestic product by 3% [10]. DSGE models are widely used by central banks and other financial institutions for forecasting and developing economic policy.

Another team of authors assessed the impact of the type of "digital currency" being developed on the possibilities of monetary policy. The authors suggested that the "digital currency" will provide greater price stability by ensuring that the real value of the currency can be maintained at a stable level. The use of "digital currency" in the national payment system will cause significant changes in the liquidity of assets in financial markets. By applying a non-monetary approach to the formation of monetary policy, researchers have shown that, despite the benefits in the form of increasing the efficiency of settlements in the economy, the introduction of digital currency can reduce the volume of deposits placed by the population in banks [11]. As a result of a reduction in the supply of deposits, banking costs and financial disintermediation may increase.

The problems of the development and implementation of national currencies are the subject of research by many domestic and foreign authors: Bailey (2021), Wilkins (2021), Kozhevina, Egorova (2020), Kuznetsov (2020), Soldatkin, Sigov(2021), Sitnik (2020), Zhang, Tian, Cao (2021).

In a number of works that relate more to the engineering and instrument-making industries, various options for implementing the national digital currency in practice are considered. Bech and Garratt (2017) in their work systematized the main options for implementing the digital currency, and also highlighted their advantages and disadvantages.

The analysis of research by domestic and foreign authors on the problems of the introduction and use of "digital currency" in the national payment system revealed an insufficiently clear understanding of the stages of transformation of digital payment instruments into a full-fledged "digital currency" for a period of socio-economic instability, which served as the basis for highlighting the main areas of scientific research in this area.

Despite all the rapidly developing innovations of our time (digitalization, modernization of elements of the management system), cash has still not lost its priority and makes up a significant part of the money supply in developed and developing countries. Six years ago (in 2015), the cash–to–GDP ratio in the US was 8%, in Switzerland—11%, in Japan—20% [12]. Nevertheless, some countries express their willingness to switch to a "digital currency", which can be traced by changing their legislation in this area [12]. In addition, appropriate financial mechanisms and infrastructure are currently being developed. In Ecuador, for example, in 2015, a means of payment in digital form was already issued, which has dollar collateral [13].

The supposed advantages of "digital currencies", which served as the reasons for the initialization of projects to develop national systems for the introduction of "digital currencies", are the speed of transactions and the ability to control their flows [13], in contrast to the anonymity of the cryptocurrency and the inability to control its movement. Reliability, speed and control are the main factors contributing to the development of national digital currency systems worldwide.

The European Central Bank interprets the concept of "digital currency" in the following aspects—these are records on bank accounts, electronic money and virtual currencies (which do not have a material equivalent with the same name, but are legal tender).

In the Russian Federation, "digital currency" is "a set of electronic data (digital code or designation) contained in an information system that is offered and can be accepted as a payment that is not a monetary unit of the Russian Federation, a monetary unit of a foreign state, an international monetary unit and a unit of account, and as investments and in respect of which there is no person obligated to each owner of such electronic data, with the exception of the operator and (or) nodes of the information system, obliged only to ensure compliance with the procedure for the release of these electronic data and the implementation of actions in relation to them to make (change) records in such an information system with its rules" [14]. This definition is broad enough to cover all the main aspects of the concept of "digital currency", however, this approach is not the end point of the development of the concept under study. Current trends show that Russia is following the path of introducing a state-owned "digital currency", the digital ruble project is at the final stage of development.

The main purpose of the launch of the "digital currency" is to provide a simple and secure payment method, as well as to create an internal corporate "money circulation" in the form of coding of the ruble denomination, allowing settlements between companies under the control of the Bank of Russia. National money "digital currency" will not replace, but will only become an additional payment option. An additional advantage of "digital currencies" is the speed of transactions for customers of the digital financial system [1], i.e. "digital currency" can be calculated faster than conventional means of payment, since bank transfers go through a greater number of transactions between buyers' banks, intermediaries and sellers and settlements can take several hours or days.

In general, the concept of "national digital currency" is currently defined as "a digital analogue of national fiat currencies that are issued, regulated and guaranteed by central banks (CB)" [14]. At the same time, the "digital currency" should be distinguished from the cryptocurrency, which, in turn, is a "virtual digital currency that has no material analogue, created and used on the basis of cryptographic methods designed to ensure maximum confidentiality and protection of electronic data" [12]. The main difference between the national digital currency and the cryptocurrency is that the second is a decentralized element for which no regulatory authority has been established. The dominant approach to the definition of a "national digital currency", presented above, has, in our opinion, a number of disadvantages:

(1) the definition indicates only one basic difference from other means of payment, which does not reveal the essence of the concept as a whole;
(2) the legal status of the payment instrument is not specified;
(3) the purpose of the implementation of this tool is not specified.

Thus, we propose the author's definition of the concept of "national digital currency", namely: an analogue of the national fiat currency, which is issued by the central bank in digital form in addition to existing cash and non-cash means of payment, in order to maintain the competitiveness of the national payment system and adapt it to the digital transformation of the economy and has the status of a legal means of payment [15].

The national "digital currency" developed in different countries has a number of common differences from cash and from reserves placed by commercial banks in the central bank. Interest can also be accrued on them, currency transactions can be separated from other operations of the central bank. But when using digital currency stored in centralized accounts, economic agents lose the ability to perform transactions anonymously [16].

Increased interest in the development of national "digital currencies", which was established during the survey of the Official Monetary and Financial Institutions Forum (OMFIF, which studies the activities of central banks and economic policy) together with IBM in July–September 2019, was facilitated by the emergence of such private projects as Libra. More than twenty-three representatives of central banks expressed considerable concern about the appearance of Libra, which could create competition for central banks and undermine the monetary sovereignty of countries. And already more than two-thirds called the main reason for the development of national "digital currencies" the creation of an alternative to cash and other (namely, Libra) payment instruments. This was also noted by the head of the central bank of Sweden Stefan Ingves: "Libra has become a global catalyst for reforms: it has shown that central banks can no longer ignore digital reality" [15]. Libra is the digital currency of social networks, in particular Facebook, which can be paid on other platforms—WhatsApp and Instagram. The main difference between Libra (for example, from bitcoin with its free floating exchange rate) is the binding to a basket of several currencies and government securities. This should make it stable, with low volatility and inflation. Hence the concerns listed above about Libra's competition with central banks and its catalyzing effect on the implementation of reforms. Over the past four years, the situation with the introduction of "digital currency" in national payment systems has changed quite a lot. According to a survey conducted by the Bank for International Settlements, in 2017, two out of every three national banks showed interest in studying digital currency. And at the beginning of 2020, 80% were already involved in the work on the "digital currency" of central banks—four out of every five central banks in countries that together account for three quarters of the world's population and 90% of the global economy.

The difference between non-cash transactions in commercial banks and the national "digital currency", according to the authors of the article, is the organization of a register of data on transactions and the corresponding privacy policy, which ensures absolute transparency of payments, which, in turn, reduces the level of corruption and reduces the scale of the "shadow" economy [17].

Currently, to carry out non-cash transactions, each credit institution either has its own data processing center (data center), or uses the production facilities of companies providing data center services on a fee basis. At the same time, the credit institution is responsible for maintaining the register, the privacy policy of the commercial bank ensures the safety of data from hacking, but at the same time, in case of illegal actions, it allows you to identify real participants in the money exchange and bring them to justice. This procedure for maintaining the data register has a number of advantages for commercial enterprises, however, from the point of view of the state financial system, it is not optimal [9], since it is necessary to jointly operate these data centers based on a centralized privacy policy established by the regulator and agreed by all participants maintaining the register, which can be implemented due to the fact that many functions Data centers of various credit organizations are similar.

It is possible that a distributed registry, which is already used by the Central Bank of Russia for other purposes, can be used as the basis for combining physically different data centers to maintain one registry. However, in order to exclude illegal actions, to prevent the anonymity of operations, encryption algorithms and methods must comply with the privacy policy established by the regulator.

Thus, the use of a national "digital currency" can reduce transaction costs compared to classic non-cash transactions, and make participants in monetary circulation less susceptible to the risk of bankruptcy of credit institutions, as well as provide transparent intercorporate settlements controlled by the central regulatory authority. However, at the moment there are organizational and technical difficulties in implementing the national "digital currency", as well as the risk of a conflict of interest between infrastructure participants. The following steps are proposed within this research area:

(1) accelerated policy of import substitution of software in the financial sector;
(2) adaptation of the best foreign practices for the introduction of "digital currency" for Russian conditions and taking into account negative experience;
(3) formation of state loyalty and incentive programs for all interested parties of the transition to the "digital currency";
(4) using the experience of implementing a Fast Payment system in the Russian Federation.

As part of the implementation of the second of the steps proposed above, it is necessary to analyze the Chinese experience of introducing a "digital currency", because despite the presence of several payment systems in China, it occupies one of the leading positions in the processes of issuing a national "digital currency". Financial experts of the world community believe that China will be the first country to fully convert its currency to a digital analogue of the national currency based on a standardized blockchain, i.e. a blockchain based on a national standard. The development of the standard is carried out by the State Institute for Standardization of Electronic Products of China, an accredited institution specializing in the development of various standards. The importance of the Chinese experience lies in the fact that the Chinese digital yuan has been successfully scaled in some regions of China and has successfully begun to fulfill its intended purpose of replacing the US dollar as the main settlement currency. The analysis of the world experience in the development and testing of the national digital currency showed the following territorial differentiation of this process. China, the Caribbean, the European Union, the United States, and Russia are the countries that potentially should take a leading position in the process of creating and implementing digital currency in national payment systems. China's experience is one of the earliest and most consistent. The People's Bank of China started working on the digital yuan back in 2014. Over the past seven years, a digital currency concept has been developed in China and 10 system tests have been conducted. At the same time, the last test of the digital yuan was launched in April 2021, where 11 regions of the country participate. Participants of two types of citizens were selected in the testing: those who permanently reside in the regions and those who come for a while. The access of Chinese banks to the data of visitors is quite minimal and the types of their wallets differ from the wallets of local residents—in this case, only small transactions are supported.

In 2020, during the coronavirus pandemic, interest in the national digital currency increased significantly, as many processes began to go online and the socio-economic crisis began to take over more and more areas. And in the middle of 2020, several countries began actively experimenting with a digital national currency. Apart from China, South Korea and Sweden were the first to start this work. And by the end of 2020, the European Central Bank has decided on the possible issuance of a national digital currency and the beginning of experiments with the digital euro. The European Central Bank is the issuer of the second most important currency in the world, and such a step confirms the importance, relevance and prospects of such a process as the issuance of a national digital currency [9]. The report of the Bank of England published studies of the central bank's "digital currency" and the possibility of its issuance. In the United States, several federal reserve banks, together with the expert and scientific community, are studying and developing technologies that allow the issuance of a digital version of cash dollars.

PricewaterhouseCoopers (PwC), a 160–year-old network of auditing companies headquartered in London, conducted a study to identify leaders in the market for the development of national digital currencies (CBDC) [11]. The list of financial instruments (CBDC) in the course of the study was divided into two groups:

(1) tools for retail purposes
(2) tools for wholesale purposes

The leaders in the market of developing "digital currencies" of central banks (CBDC) for retail use were the Bahamas, Cambodia and China (Fig. 1) [18]. In the Bahamas, the national "digital currency" (the so-called "sand dollar") was launched in October 2020, the high predictive potential of this digital currency, pegged to the US dollar, became the basis for cooperation with Mastercard, which led to the issuance of a card for the national currency of the Bahamas. This is the world's first national "digital currency", which has its own bank plastic card from the largest financial corporation.

This ensured the initial leadership of the Bahamas in the top 10 countries that were the first to introduce a national digital currency (CBDC) for retail payments. The top-10 rating was compiled based on the level of readiness of CBDC projects and analysis of technical capabilities. As of the end of April 2021, more than sixty Central Banks around the world participated in the launch of national digital currencies. However, at the moment, the rate of spread of the "sand dollar" has decreased, since the negative events that were caused by the lockdown and the growing national tension have negatively affected the development of the country's economy. The leaders of the Bahamas, however, believe that the main tool for the growth of the use of the "sand dollar" is an information policy aimed at ensuring that as many people as possible

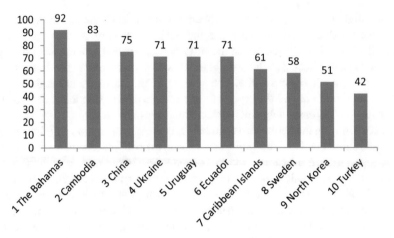

Fig. 1 Top 10 regions in the rating of the best national currencies (CBDC) for retail

understand the advantages of using the national "digital currency" and its difference from the cryptocurrency.

The Russian Federation is not in the top 10 regions of the rating of the best national currencies (CBDC). This can be explained by the fact that the issue of creating a national digital currency in the country was approached with great caution. The growing popularity of cryptocurrencies contributed to the awareness of the need for accelerated research, testing, and then the introduction of a national digital currency. Its development will reduce the spread of the popularity of cryptocurrencies, especially with the parallel displacement (full or partial) of such from the legal field of the state. As a result, this will contribute to increasing control over digital cash flows in the country. In the ranking of the top 10 CBDC for wholesale use, the top three were: Thailand, Hong Kong and Singapore (Fig. 2) [19], while in Thailand testing of the national digital currency was organized in the summer of 2020.

In the Caribbean, the digital currency was launched on March 31, 2021. The Eastern Caribbean Bank launched an analogue of the Eastern Caribbean dollar (DCash). In Antigua and Barbuda, Saint Lucia, Saint Kitts and Nevis, Grenada is currently undergoing a test regime. Citizens of the listed countries accept transfers and make payments in DCash, pay in digital currency for purchases, pay in banks.

The experience of the European Union demonstrates fully formed prerequisites for the development, testing and launch of the digital euro. In particular, the European Central Bank has begun to evaluate the pros and cons of launching a digital currency. At the current stage, the European Central Bank (European Central Bank) is analyzing the experience of other states and collecting the opinions of European citizens about the launch of an analogue of the euro. According to the report of the European Central Bank, first of all, citizens want to be sure of the security and confidentiality of the system, its usability throughout the European Union.

Five pilot projects of the digital dollar were launched in the United States in 2021. Representatives of the Digital Dollar Foundation and Accenture companies,

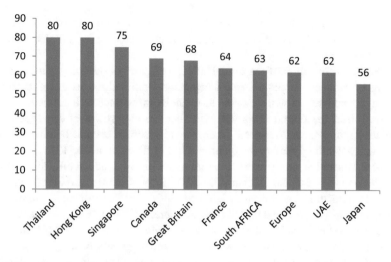

Fig. 2 Top 10 regions in the ranking of the best national currencies (CBDC) for wholesale settlements

who are responsible for launching these projects, explain that the digital dollar is an absolutely new form of money, which was created specifically for online payments and complements existing electronic money. Project experts at this stage analyze and determine the technical capabilities of the digital American currency, as well as identify the advantages and weaknesses of its development.

In the UK, the chairman of the Bank of England made a statement that the issue of creating a "national digital currency" is also being discussed in his department. "We are studying the issue of the need to issue a digital currency from the Bank of England. We will continue to study this issue, as it is of great importance both for the essence of money transfers and for society," said Andrew Bailey [5, 17, 20].

The Bank of Russia has announced the start of work on the creation of a digital ruble, which will be similar to the CBDC. The decision on the feasibility and form of issuing the national digital currency will be made after the end of the pilot project [21]. Now many countries are following this path.

In October 2021, for consultations with experts and the public, the Bank of Russia presented a report on the possibilities and prospects of issuing a digital ruble [14, 18]. The report outlines the main definitions and possibilities of issuing a digital currency of the central bank—the digital ruble. The digital ruble will be an additional form of the national Russian currency, which will be issued by the Bank of Russia in digital form. At its core, the digital ruble will combine the properties of cash and non-cash rubles. The priority property of the future digital ruble is its accessibility to all economic entities, including in remote, hard–to-reach, sparsely populated territories, i.e. citizens, businesses, the state, and financial market participants. The functions of the digital ruble are similar to cash and non-cash rubles: a means of saving, a measure of value and a means of payment. And another important parameter is the equivalence of the forms of the Russian ruble. One digital ruble will be equivalent to one cash ruble and one non-cash ruble [3]. The expected results of the introduction of digital currency in the world and Russia are the creation of safer, simpler and faster payment systems.

With the introduction of the digital ruble, a number of advantages may appear. The digital ruble can become a new convenient additional means of settlement for both buyers and sellers where access to financial infrastructure is limited. And the introduction of the digital ruble will contribute to the growth of coverage of the population with financial services, access to which should become better and easier, which will accordingly affect the quality of life of the population [13].

Global innovative trends in the financial sector have spread not only to developed countries and the European Alliance, the BRICS countries are also interested in the process of transition to a "national digital currency". While most banks are studying various concepts of issuing their "digital currencies", China has already made a definite decision—the digital yuan will not be associated with modern cryptocurrencies and blockchain technology that is not standardized by government agencies. Digital currency experts believe that in the future, the digital yuan can seriously compete with the cryptocurrency and the US dollar [2]. In Russia, the development of the digital ruble continues and the search for the most effective way to implement it.

In India, as well as other BRICS countries, the launch of the digital rupee, the Indian digital currency, is being prepared. In India, from December 1, 2022, a pilot project for the retail use of the national cryptocurrency e-Rupee (digital rupee) was launched [22]. Initially, it will be used in four cities: New Delhi, Mumbai, Bangalore and Bhubaneswar. The Government of India today considers the project a success and has embarked on a new stage of introducing the digital rupee. South Africa is also creating its own national digital currency, the digital rand, in accordance with global innovative financial trends. The Central bank of Brazil, represented by its head, Roberto Campos Neto, announced its readiness to switch to a digital currency in 2022. Roberto Campos Neto also noted that a technological base will be created in Brazil for the release of digital real [18]. At the moment, Brazil is modernizing the financial system based on the results of testing the digital real.

The adaptation of all the studied experience makes it possible for Russia to avoid mistakes already made on the path of digital transformation of the national financial system. Thus, it is proposed to develop a mechanism for transforming the conceptual essence of the concept of "national digital currency" into a working model for managing financial business processes at the state level and transnational relationships.

The roadmap mechanism involves the allocation and quantification of the expected results from the introduction of the "digital currency" into the Russian financial system. The expected results are formulated based on the identified problems that are characteristic of the current state of the Russian financial system, namely:

the presence of a significant share of the shadow economy and corrupt financial schemes [13];

- stagnation of the development of the national payment system, lagging behind its effectiveness from other countries [13];

Table 1 Expert assessment of expected results, %

Financial policy assessment blocks	Expert assessments					Average rating	Weighted assessment, %
Increasing transparency and speed of financial transactions throughout Russia	90	80	81	95	80	85.2	37
Growing loyalty to the national digital currency	30	25	30	25	39	29.8	13
Simplification of international financial settlements	15	25	22	30	35	25.4	11
Reducing the level of cybercrime	15	11	13	15	25	15.8	7
Increasing the efficiency of the national payment system	80	90	70	65	78	76.6	33
Total						232.8	100

– the complexity of international financial settlements due to the high volatility of currencies and the lack of a single means of payment recognized by all interested parties;
– high level of cybercrime in the financial sector [20];
– low loyalty to the national currency [13];

Thus, the list of expected results derived from the need to solve the problems listed above, as well as their expert assessment, are presented in Table 1 [17].

The expert assessment was carried out anonymously, by e-mail, as experts were representatives of the Bank of Russia and private credit organizations, a total of five experts. The scale is from 1 to 100%. As a result, the following roadmap was formed (Table 2) [17].

The roadmap presented in Table 1 takes into account the conditions of global socio-economic instability, and also shows the projected results from each of the stages of the introduction of the "national digital currency" or "digital ruble" into the country's economy [5]. The direction of the following scientific research is the development of a program to minimize the identified risks.

4 Discussion

The studied domestic and international experience in the development of digital currency in national payment systems allowed us to identify prospects for further development of digital settlement tools, measures to minimize the negative manifestations of the transition to a national digital currency and increase the efficiency of

Table 2 Roadmap for the introduction of "digital currency" into the Russian financial system

Stage name	Stage term	Expected results	Risks
1. Scaling of the Digital Ruble pilot project to the entire territory of Russia	End of 2023	Increasing transparency and speed of financial transactions throughout Russia up to 80%	The rejection of the "digital ruble" as a full-fledged means of payment by the population, as a result, a massive refusal to use it
2. Development of infrastructure for the use of the digital ruble	2024–2025	Increase in the efficiency of the national payment system by 35%	The growth of budget expenditures, which may lead to a reduction in other budget items and thereby cause social tension in society
3. Formation of a system of interaction with other national digital currencies	2026–2027	Simplification of international financial settlements by 25%	Increase in the volatility of the national digital currency during the period of regulation of the mutual exchange rate
4. Revision of the legislative and legal framework governing the use of the "digital ruble", taking into account the results of the previous stages	2028	Decrease in the level of cybercrime by 15% The growth of loyalty to the national digital currency 25%	The growth of contradictions in the current legislation, the increase in the timing of establishing the legal status of the digital ruble
5. Complete transformation of the national payment system (NPS), taking into account the legislative and infrastructural framework for the introduction of the "digital ruble"	2029–2030	Increasing the efficiency of the national payment system to 75%	Technical lag behind other countries in the field of fintech, and as a result, the low relevance of the final version of the NPS

the entire new emerging financial ecosystem. The application of such measures in the future, during periods of uncertainty, will increase the financial stability of the national payment system.

5 Conclusions

The topic of "national digital currency" is becoming more and more relevant for all countries that are active players in the global economic arena. The management of Central banks around the world is aware of the modern opportunities that a centralized "digital currency" can provide, namely:

(1) reliability and transparency of financial transactions;

(2) speed of transactions;
(3) availability of financial interactions in remote and hard-to-reach territories.

The prerequisites for the active creation of digital national currencies in the world have been created and socio-economic crises, including those caused by the 2020 pandemic, have only added acceleration to these processes, so the use of "digital currency" in national payment systems will develop much more dynamically than previously expected (until 2020), this is proved by the number of countries that have started to test your "national digital currency" [12].

The study shows that the introduction of "national digital currencies" contributes to partial opposition to the development and use of private cryptocurrencies. This is due to the fact that "national digital currencies" are standardized and are the digital expression of ordinary paper money, i.e. they have state security guarantees, therefore, they will be more attractive if the infrastructure for their use is developed.

An important aspect is that despite the development in commercial projects, the technology of cryptocurrencies and anonymous registry holders is not considered by state regulators as a basis for creating a national digital currency, while the use of digital currency in the national payment system will contribute to the development of digital technologies in general. Therefore, this problem requires a primary solution [2].

During the period of active progression of the coronavirus pandemic in the world, interest in creating a "national digital currency" has increased significantly. This is due to the departure of many processes online and the fact that the socio-economic crisis has begun to take over more and more areas. As a result, in the middle of 2020, several countries began to actively conduct experiments with the digital national currency. This work was started by countries—China, South Korea, Sweden. By the end of 2020, the European Central Bank has decided on the possible issuance of a national digital currency and the beginning of experiments with the digital euro. In the Russian Federation, there is a phase of active preparation for the launch of testing of the digital ruble in some regions of the country. At the same time, taking into account the experience of other countries, it is necessary to adhere to the strategy of phased introduction of the digital ruble into the national payment system, while it is proposed to rely on the roadmap developed and given above [7], which shows both the forecast results and possible risks.

References

1. On the use of virtual currencies when making transactions, in particular, Bitcoin (2014). http://www.cbr.ru/press/pr/?file=27012014_1825052.htm. Last accessed 11 Nov 2023
2. Cryptocurrencies: trends, risks, measures. Report of the Bank of Russia for public consultations (2022). http://cbr.ru/Content/Document/File/132241/Consultation_Paper_20012022.pdf. Last accessed 11 June 2023
3. Sitnik, A.: Digital currencies of central banks. Courier Kutafin Moscow State Law Univ. **9**, 180–186. (2020). https://doi.org/10.17803/2311-5998.2020.73.9.180-186. Last accessed 11 June 2023

4. Barrdear, J., Kumhof, M.: The macroeconomics of central bank issued digital currencies. Bank England Staff Working Paper **605**, 36 (2016)
5. Digital ruble: Public consultation report (2021). http://www.cbr.ru/analytics/d_ok/dig_ruble. Last accessed 11 June 2023
6. Bordo, M., Levin, A.: Central bank digital currency and the future of monetary policy. Natl. Bureau Econ. Res. **237**, 64 (2017)
7. Strategy for the development of the national payment system for 2021–2023 (2021). http://www.cbr.ru/content/document/file/124363/strategy_nps_2021-2023e.pdf. Last accessed 11 June 2023
8. Bech, M., Garratt, R.: Central bank cryptocurrencies. BIS Quart. Rev. (2017). https://ideas.repec.org/a/bis/bisqtr/1709f.html. Last accessed 11 June 2023
9. PwC analysts named leaders in the national digital currency development market (2021). https://bioomchain.ru/newsfeed/analitiki-pwc-nazvali-liderov-na-rynke-razrabotki-natsional nyh-tsifrovyh-valjut. Last accessed 11 June 2023
10. Jobst, C., Stix, H.: Doomed to disappear? The surprising return of cash across time and across countries (2017). https://ssrn.com/abstract=3042640. Last accessed 14 June 2023
11. Bailey, A.: Innovation in the public interest presentation (2021). http://www.bankofengland.co.uk/speech/2021/june/Andrew-bailey-cityuk-annual-conference. Last accessed 11 June 2023
12. Digital currency: what is the difference from cryptocurrency and which countries are planning to issue (2021). http://www.immigrantinvest.com/insider/digital-money-euro-dollar-ruble-2021. Last accessed 11 June 2023
13. Grigoriev, V. V.: National digital currency as a factor in the revival of the Russian economy, Economy. Taxes. Right. No.1. (2019). https://cyberleninka.ru/article/n/natsionalnaya-tsifro vaya-valyuta-kak-faktor-ozhivleniya-ekonomiki-rossii. Last accessed 14 June 2023
14. Kozhevina, O., Egorova, M.: Issues of regulation of national digital currencies in the context of global global challenges. Legal Sci China Russia **3**, 30–33 (2020). https://doi.org/10.17803/2587-9723.2020.3.030-033. Last accessed 11 June 2023
15. Digital currency of central banks: world experience (2020). https://econs.online/articles/reguli rovanie/tcifrovaya-valyta-tsentralnykh-bankov-mirovoy-opyt/. Last accessed 11 June 2023
16. Expert: China's Digital yuan Will Target the Dollar, Not Bitcoin (2020). https://cointelegraph.com/news/expert-chinas-digital-yuan-will-target-the-dollar-not-bitcoin. Last accessed 11 June 2023
17. Central Banks Digital Currency: World Experience. (2021). http://www.econs.online/articles/regulirovanie/tsifrovaya-valyuta-tsentralnykh-bankov-mirovoy-opyt. Last accessed 11 June 2023
18. The BRICS countries are interested in the transition to digital currencies (2021). http://www.tvbrics.com/bricslife/strany-brics-zainteresovalis-perekhodom-na-tsifrovye-valyuty. Last accessed 11 June 2023
19. Wilkins, C.A.: Under the western sky: the crypto frontier-speech (2021). https://www.bankofengland.co.uk.speech/2021/november/carolyn-a-wilkins-keynote-speaker-at-autorite-des-mar ches-financiers-annual-meeting. Last accessed 11 June 2023
20. Soldatkin, S., Sigov, V.: Digital ruble: features of the Russian model for the creation and operation of the national digital currency. Bull. Khabarovsk State Univ. Econ. Law **2**(106), 62–69 (2021). https://doi.org/10.38161/2618-9526-2021-2-62-69. Last accessed 11 June 2023
21. China is developing a government blockchain standard (2021). https://www.cnews.ru/news/top/2021-10-29_kitaj_razrabatyvaet_gosstandart. Last accessed 11 June 2023
22. Zhang, J., Tian, R., Cao, Y.A.: Hybrid model for central bank digital currency based on blockchain. IEEE Access **9**, 53589–53601 (2021). https://doi.org/10.1109/ACCESS.2021.307 1033,lastaccessed2023/06/11

Tax Support Tools for the Implementation of National Government Projects with the Digital Functioning of "Smart Government"

Irina A. Zhuravleva⬤

Abstract Russia is a country with different regional and municipal economic potential, budgetary security, tax component, relevant for the development of the country as a whole is the task of determining the direction vector for sustainable development in order to use financial, budgetary, tax, administrative and managerial resources, and increase the efficiency of the implementation and application of "Smart Government". All of them are aimed at the implementation of national goals and national projects based on digitalization tools. In modern conditions, the public authorities are particularly acute task to ensure the effective management of public finances by the state, citizens and the business community, embedded in the fundamental principles of the functionality of "Smart Government". The relevance of the problem under consideration is due to the need to assess the effectiveness of the implementation of national projects from the standpoint of using progressive tax tools that meet modern requirements for digitalization of the economy and embedding the mechanisms of functioning and results of the implementation of National projects into the "Smart Government". The purpose of this scientific research is to consider the directions for improving the systematic application of tools, mechanisms and elements of tax support in order to effectively implement national government ones. The methodological basis of the study is the principles of analytical justification, a systematic and process approach using general scientific methods—generalization, comparison, induction, deduction, statistical observations, system analysisThis work can serve as a theoretical basis for creating methods for the effective functioning of "Smart Government".

Keywords Digitalization · Federal Budget · National Government Projects · Performance Evaluation · Smart Government · Tax Tools

I. A. Zhuravleva (✉)
Financial University Under the Government of the Russian Federation, Moscow, Russia
e-mail: sia.mir67@mail.ru

1 Introduction

The concept of "Smart Government" characterizes the form of organization of the activities of public authorities, the hallmark of which is the use of electronic exchange and information processing technologies in the public administration sector. "Smart government" is an open government, the concept of which, in addition to the transparency of the work of government agencies and the use of new technologies, suggests that citizens and business communities can and should participate in the work of the state as external experts [7]. The idea of "smart" control originates from the well-known English term "smart" and was introduced into scientific circulation by Tikhonov and Bogdanov [1], which caused strong support from the scientific community, including:

- in the concepts of Vasilenko, Zotov, Zakharova of "smart" public administration and "smart" configuration of social networks in the context of digitalization [2];
- in the expression of Dudikhin and Shevtsova "the image of Big Data as a "fuel", and intelligent technologies as the "engines" of smart control" [3];
- Dutch scientists Jiang, Girtman, Witte consider smart management as a system of management structures that include an environment for citizen involvement within the management infrastructure using new technologies and communication channels for joint development and making political decisions on difficult situations and their effective and efficient implementation [4].

Bradul and Lebezova believe that "developed e-government based on open governance, integrating physical, digital, public and private environments for passive and active interaction and cooperation with citizens with the aim of sustainable and flexible development of the services and opportunities provided [5]. The development and implementation of "Smart Government" at all levels of the country's budget system would open up great financial, economic, social and managerial opportunities for effectively solving the problems of National Projects in action.

2 Methods

2.1 Literature and Research Review

National projects are integrated with the state programs of the Russian Federation. The national project consists of a set of federal projects that are included in the subprograms of the relevant state programs. At the same time, federal projects of the same national project, depending on its specifics, can be included in one or several state programs [6].

Kudelich in his scientific study notes that "the features are considered and the types of financial support for national projects at the expense of the federal budget are systematized, and the existing legislative requirements for assessing the effectiveness of national projects are analyzed, primarily in terms of the provision and use of budgetary funds aimed at their implementation" [7]. The author substantiates the conclusion that the legislation on national project activities in the Russian Federation pays insufficient attention to the requirements for evaluating the effectiveness of national projects/programs.

It should be noted that, in the context of the actual achievement of the goals set in the implementation of national projects and further from not only consolidation, but also development, it is important to analyze the problems of existing regulatory requirements for assessing the effectiveness of the implementation of national projects (programs) at the expense of the federal budget, the results of which can be used to improve the system of organization and effective functioning of national project activities in the Russian Federation as a whole [7].

Concerns about the prospects for the implementation of the "new wave" of national projects are also expressed in the scientific literature. So, earlier, Dynnik notes the presence of "uncertainty in understanding how the work on the formation of national projects, their coordination, and reporting will be built" [6]. The absence of a holistic mechanism for the effective management of national projects at this stage, despite "greater attention to the issues of resource provision of these projects and the effectiveness of the funds used for these purposes," is also emphasized in the works of other authors [8].

Taking into account that 1.7 trillion rubles were allocated for the implementation of new national projects in 2019 alone, which is 10% of the country's budget, it seems necessary to analyze the problems of existing regulatory requirements for assessing the effectiveness of national projects, primarily in terms of the conditions for providing and use of budget funds aimed at the implementation of the national project (program). At the same time, the requirements for evaluating the effectiveness of the provision of budgetary funds are considered in terms of the quality of the choice, the sufficiency of justification and the conditions for the implementation of the relevant types of federal budget expenditures, as well as the procedure for determining their volumes for the implementation of national projects (programs), and the requirements for the use of budgetary funds—in the context sufficiency of normative and methodological provisions regarding the intermediate and final assessment of the degree of achievement of goals, target indicators and additional indicators of national projects (programs) as a result of spending the allocated budget funds by managers and participants of national projects (programs) [7].

With regard to budgetary issues, the current legislation provides for separate requirements only for assessing the effectiveness of the use of budgetary system funds, but not for assessing the effectiveness of their provision. In particular, it provides for an assessment of the effectiveness of the use of funds from the budgets of the budget system aimed at the implementation of the national project by the head of the national project, and a similar assessment at the end of the federal project.

Also, Kudelich believes that "there is no methodological or regulatory framework for conducting this and other envisaged assessments (of the effectiveness of NP), as well as there are no normatively fixed consequences of recognizing the unsatisfactory efficiency of using budget funds of the budget system and (or) activities of participants in national projects or participants in federal projects" [7]. In essence, the only consequence of evaluating the effectiveness of the use of funds from the budget system of the Russian Federation at the completion of the federal project, which is part of the national project, is the indication of paragraph 86 of the Regulations on the organization of project activities that, following the results of its implementation, "if necessary, other federal projects aimed at achieving goals, target and additional indicators, fulfilling the tasks of the national project". In other words, the logic of the legislator proceeds from the fact that in case of inefficiency of the federal project, it is necessary to initiate other federal projects that will make it possible to achieve the necessary target and additional indicators. This approach seems to be flawed, since it does not involve drawing lessons from negative experience and the obligation to take them into account in the subsequent decision-making process [7].

Buchhwald notes that the key tool for achieving these goals is the use of accumulated big data that should be put at the state control service, including in relation to the implementation of national projects [9]. The authors propose to implement the "traceability" model for the implementation of NP on the principles of cascading information from general data on national projects to information about a specific result. The tool for the implementation of the "traceability" of information for citizens is the unified portal of the state integrated information system of public finances management "Electronic Budget" (hereinafter referred to as the GIS "Electronic Budget"), formed on the basis of the order of the Ministry of Finance of the Russian Federation of December 27, 2013 No. 141n "On the creation of the creation and maintaining a single portal of the budget system of the Russian Federation". Their idea is supported by Tsindeliani, Bi-Shabo and Gorbunova "All the results of the state's activities in federal and regional projects are presented in a convenient, understandable form, which is important, since the users of the Electronic Budget system are individuals and legal entities" [10].

Matveev notes that "for the successful implementation of national projects, it is very important to timely identify the risks of unattending their results in the process of implementing national projects [11].

Prokofiev He believes that "in the implementation of traceability of the results of national projects, it is very important to own information on the implementation of national projects at all levels of budgets of the budget system of the Russian Federation in online mode" [12].

The position of Zaporozhnaya in terms of the implementation of the NP at all levels of the country's budget system, it is that "budgetary financing in the areas of budget expenditures used to go through three, and now it should go through four channels. All this increases the budget document flow; complicates and increases the cost of the budget process; expands the scope, complicates and increases the cost of budgetary control" [13].

Another problem has arisen regarding the implementation of national project activities, how the costs of their implementation are calculated and how the effectiveness of these activities is assessed. How much, for example, should be allocated from the budget for the implementation of the programs "Strengthening Russian Civil Identity" or for the creation of a "Positive Image of an Entrepreneur" from the national project "Small and Medium Enterprises and Support for Individual Entrepreneurial Initiatives"? How to perceive the NP: is it an alternative to all other strategic planning documents of both the federal and regional levels; Is it some kind of target designation framework that should be further expanded and specified, including up to the level of the Strategy for the Social and Economic Development of the Russian Federation? Finally, national projects can be understood as a new form of structuring state programs in priority areas of economic and social development of the country [14]. However, it is noted that the transition to a project management system requires not only a serious methodological base, but also the training of appropriate personnel [14–17]. In addition, as studies show, for a clear positioning of national projects in the practice of strategic planning, including its spatial "cut", certain legal innovations are needed [14].

"The national project (each of them is a set of so-called "federal projects") contains a system of interconnected goal-orienting instructions and their corresponding by-laws, united by a single concept and ultimate goal. In a certain sense, national projects represent the testing of a new economic and budgetary strategy for the development of the country in the long term" [14]. Great importance is currently attached to the openness of national projects, the availability of the results of their implementation for the population of the country. To this end, 91 indicators for the target indicators of national projects have been included in the Federal Plan of Statistical Works.

2.2 Research Methods

We will consider the interdependence and interrelationship between the effective functioning of National Projects and the introduction of digitalization elements in the "Smart Government" using the example of tax tools to support national projects. Considering the functionality of the country's National Projects, we note that Decree of the President of the Russian Federation dated May 7, 2018 No. 204 "On the national goals and strategic objectives of the development of the Russian Federation for the period up to 2024" (hereinafter—Decree No. 204) defines 12 national projects that are, in fact, the key directions of the strategic development of the Russian Federation. National projects are tools for the implementation of national development goals that are of priority importance at a certain stage of the development of the state and require significant resources for their solution, assuming a clearly defined end result and ways to achieve them [8]. "The national project is understood as a systemic set of interrelated activities aimed at obtaining unique results in the conditions of time and resource constraints, ensuring the achievement of goals and targets, the fulfillment of the tasks defined by Decree No. 204, and, if necessary, the achievement of additional indicators and the implementation additional tasks" [18]. Table 1 shows examples of national projects with details by budget levels.

It should be noted that the general requirements for the system of financial support for national projects at the expense of the federal budget are established by clause 16 of Decree No. 204 and include, firstly, the requirement to annually provide for in

Table 1 Examples of national projects

National projects	Federal projects	Regional projects
Small and medium enterprises	Federal project "Acceleration of small and medium-sized businesses"	Regional project "Acceleration of small and medium-sized businesses in the Krasnoyarsk Territory"
	Federal project "Support for the self-employed"	Regional project "Creating favorable conditions for the implementation of activities by self-employed citizens"
	Federal project "Pre-acceleration"	Regional project "Creating conditions for an easy start and comfortable business"
International cooperation and export	Federal project "Systemic measures for the development of international cooperation and exports"	Regional project "Systemic measures for the development of international cooperation and exports"
Labor productivity	Federal project "Targeted support for increasing labor productivity at enterprises"	Regional project "Targeted support for increasing labor productivity at enterprises"
	Federal project "Systemic measures to increase labor productivity"	Regional project "Systemic measures to increase labor productivity"

Source http://krasmsp.krskstate.ru/nationalprojects

priority order, federal budget allocations for the implementation of national projects (programs);

Thus, the system of financial support for national projects at the expense of the federal budget is based on the principle of taking into account the results of the implementation of national projects and federal projects for the previous period for the purposes of subsequent budget planning, which should be recognized as a circumstance that can potentially have a positive impact on the effectiveness of implementation national projects (programs) [7]. It is not yet clear how this mechanism is regulated by Smart Government. Let us pay attention to the fact that the federal and consolidated budgets are formed from tax revenues (Table 2). And in financing for the implementation of national projects, the filling of budgets plays a key role, as well as the fact that the positive dynamics of tax revenues also provides additional financial resources in financing national projects.

No less important in the context of assessing the effectiveness of financial support for national projects (programs) will be the rules for granting subsidies to legal entities, individual entrepreneurs, individuals, including depending on the degree of competitiveness of their provision, the effectiveness of liability measures for failure to achieve results, and not the conditions for granting subsidies related to the timeliness and completeness of their use. In other words, when evaluating the effectiveness of providing such subsidies as part of the financial support of national projects

Table 2 Revenue part of the consolidated budget

No	Index	2014	2015	2016	2017	2018	2019	2020	2021	2022
SECTION I										
1	Income, total	26 766.1	26 922.0	28 181.5	31 046.7	37 320.3	39 497.6	38 205.7	48 118.4	53 073.8
1.1	Oil and gas revenues	7 433.8	5 862.7	4 844.0	5 971.9	9 017.8	7 924.3	5 235.2	9 056.5	11 586.2
1.2	Non-oil and gas revenues	19 332.3	21 059.4	23 337.5	25 074.8	28 302.5	31 573.3	32 970.5	39 061.9	41 487.6
1.2.1	VAT	3 931.7	4 234.0	4 571.3	5 137.6	6 017.0	7 095.4	7 202.3	9 212.6	9 553.0
1.2.2	Excises	1 072.2	1 068.4	1 356.0	1 521.3	1 589.5	1 792.3	1 800.3	2 095.5	2 367.9
1.2.3	Income tax	2 375.3	2 599.0	2 770.3	3 290.1	4 100.2	4 543.2	4 018.4	6 081.7	6 355.9
1.2.4	Personal Income Tax	2 702.6	2 807.8	3 018.5	3 252.3	3 654.2	3 956.4	4 253.1	4 883.9	5 729.1
1.2.5	Import duties	652.5	565.2	563.9	588.5	673.0	716.9	713.2	875.5	550.7
1.2.6	Insurance premiums for compulsory social insurance	5 035.7	5 636.3	6 326.0	6 784.0	7 476.9	8 167.2	8 286.1	9 018.3	9 397.2
1.2.7	Other	3 562.2	4 148.8	4 731.5	4 501.0	4 791.7	5 301.9	6 697.1	6 894.5	7 533.7
SECTION II										
2	Expenses, total	27 611.7	29 741.5	31 323.7	32 395.7	34 284.7	37 382.2	42 503.0	47 072.7	55 182.0
2.1	General government issues	1 640.4	1 848.2	1 849.9	1 952.6	2 131.6	2 334.8	2 551.7	2 852.0	
2.2	National defense	2 480.7	3 182.7	3 777.6	2 854.2	2 828.4	2 998.9	3 170.7	3 575.0	
2.3	National Security and Law Enforcement	2 192.9	2 072.2	2 011.4	2 034.1	2 110.5	2 233.6	2 392.4	2 504.4	
2.4	National economy	4 543.1	3 774.4	3 889.8	4 332.0	4 442.9	5 171.8	6 040.8	7 224.7	
2.5	Department of Housing and Utilities	1 004.7	979.9	992.6	1 209.9	1 324.1	1 574.9	1 590.5	2 172.0	
2.6	Environmental protection	70.2	71.7	84.0	116.3	148.3	250.3	303.9	438.4	

(continued)

Table 2 (continued)

No	Index	2014	2015	2016	2017	2018	2019	2020	2021	2022
2.7	Education	3 037.3	3 034.6	3 103.1	3 264.2	3 668.6	4 050.7	4 324.0	4 690.7	
2.8	Culture, cinematography	410.0	395.6	422.8	492.9	528.2	587.9	610.1	651.9	
2.9	Healthcare	2 532.7	2 861.0	3 124.4	2 820.9	3 315.9	3 789.7	4 939.3	5 167.3	
2.10	Social politics	8 803.3	10 479.7	10 914.2	12 022.5	12 402.2	13 022.8	15 121.7	16 002.3	
2.11	Physical Culture and sport	253.6	254.9	262.3	327.0	331.4	375.5	400.7	437.5	
2.12	Mass media	117.4	125.7	119.9	127.3	136.5	156.1	173.7	171.4	
2.13	Servicing state and municipal debt	525.4	661.0	771.8	841.8	916.1	835.4	883.5	1 185.1	
SECTION III										
3	Deficit (−) / Surplus (+)	−845.6	−2 819.5	−3 142.1	−1 349.1	3 035.6	2 115.3	−4 297.3	1 045.7	−2 108.1
3.1	Non-oil and gas deficit	−8 279.4	−8 682.1	−7 986.2	−7 321.0	−5 982.2	−5 808.9	−9 532.6	−8 010.8	−13 694.3

Source https://minfin.gov.ru/ru/document?id_4=93447-informatsiya_ob_ispolnenii_konsolidirovannogo_byudzheta_rossiiskoi_federatsii

(programs), importance should be given primarily to assessing the degree of achievement of planned results, and not assessing the completeness and timeliness of the development (use) of allocated funds [7]. It is this fact that indicates the targeted use of financial flows and allows you to increase the volume of tax revenues. It turns out, a kind of development spiral, but not closed, but increasing in the diameter of financial development and budget savings.

3   Results: Evaluating the Effectiveness of National Projects Based on the assessment of Budgetary Tax Expenditures as one of the Elements of the Effective Functioning of "Smart Government"

The currently generally accepted approach to assessing the effectiveness of budgetary and tax expenditures involves an assessment of three efficiency criteria: budgetary, economic and social.

When using the main criteria for assessing the effectiveness of the applied fiscal support measures, it is necessary to take into account the presence of a time lag, i.e. budget expenditures incurred over a number of years can only have an effect in the medium and long term. This is explained:

- firstly, the effect of the accumulation of financial resources by the enterprises of the industry for investing in modernization, technical re-equipment, expansion of production and other investments;
- secondly, the duration/duration of the production cycle (for example, in aircraft and mechanical engineering).
- thirdly, the implementation of fiscal support for one critically important non-primary industry can have a positive effect not only for this industry, but also for others (taking into account the larger time lag).

For example, stimulating the development of the radio-electronic industry will lead, among other things, to the development of CNC production, which is in dire need of the machine tool industry; the development of the radio-electronic industry will also have a certain positive effect on the aircraft industry.

The assessment of the budgetary effectiveness of state support measures is carried out on the basis of a calculation that determines the effect for the budget from providing support measures to enterprises in selected industries, which is expressed in an increase in tax revenues to the budget compared to the amount of shortfall in budget revenues/expenditures or tax expenditures. In addition, for a more accurate assessment, taking into account the impact of inflation, you can use the relative indicator of the difference between the growth rates of tax revenues to the budget and the inflation rate.

In contrast to the generally accepted approach, which involves a separate calculation of the effectiveness of tax incentives and the provision of direct budgetary support to business entities, it is advisable to consider the cumulative effect of the provision of fiscal measures to support critical non-primary industries. Moreover, the effectiveness of tax incentives is proposed to be assessed in aggregate by a range of taxes: VAT (since the increase in VAT amounts paid demonstrates an increase in value added), corporate income tax (since this indicator demonstrates an increase in the financial result of the activities of enterprises in the industries under consideration), tax on property of organizations paid on the basis of the average annual value of property (since this tax will reflect the increase in the value of real estate for industrial purposes from taxpayers or the acquisition of fixed assets: machinery, equipment). Therefore, the calculation of the budgetary efficiency of the provided fiscal support measures is carried out according to the formula (1):

$$CB_t^I = \frac{\sum_{i \in I} \left(T R_t^i - T R_{t-1}^i \right)}{\sum_{i \in I} \left(T E_{t-1}^i - B S_{t-1}^i \right)} \tag{1}$$

where:

I—a set of critically important non-primary industries;

I—industry from the list of critically important non-primary industries;

CB_t^I is the coefficient of budgetary efficiency of the measures of fiscal support provided by sectors I for period t;

TR_t^i—tax revenues from industry i for period t;

TR_{t-1}^i—tax revenues from industry i for period t–1;

TE_{t-1}^i—tax expenditures of the budget on benefits provided to enterprises of industry i for period t–1;

BS_{t-1}^i are direct budget subsidies provided to enterprises in industry i for period t–1.
    "Smart government" is not only the use of digital technologies in various government, social and administrative areas, which allows you to actively involve citizens and the business community in management, search for solutions, identify problems, but also, of course, find the source of filling the state treasury [19].
    We propose that Smart Government can also carry out an analysis of budgetary efficiency in the context of each selected industry separately.
    Analysis of budgetary efficiency can also be carried out in the context of each selected industry separately. But taking into account the effect highlighted above (providing support measures for one industry may well entail an increase in the performance of other industries), we consider it appropriate to conduct (among other things) an analysis of the budgetary efficiency of state support measures provided in aggregate for all selected industries.
    With the value $CB_t^I > 0$, the measures of fiscal support for critically important non-primary industries have a positive efficiency.

With a value of $CB_t^I \geq 1$, the payback of the state budgetary and tax expenditures carried out in order to support critically important non-primary industries has been achieved.

With the value $CB_t^I < 0$, the measures of fiscal support for critically important non-primary industries have a negative efficiency.

The calculation of economic efficiency is carried out through a study that determines the effect on economic growth in the industry from the provision of tax incentives to taxpayers, which is expressed in an increase in the aggregate economic indicators of enterprises in the industry. Achieving cost-effectiveness in providing fiscal support is a priority.

In contrast to the coefficient of budgetary efficiency, it is advisable to calculate the coefficient of economic efficiency in the context of each industry under consideration separately. The economic efficiency ratio (ER) is calculated as the ratio of the number of indicators for which growth occurred (or their level has not changed) to the total number of indicators according to formula (2):

$$ER = Np/Nc, \qquad (2)$$

where

- Np—the number of indicators for which there was an increase or their value remained the same;
- Nc is the total number of indicators.

As these indicators, such relative values were chosen as: the growth rate of proceeds from the main activity (in comparable prices), the growth rate of investments in fixed assets, the growth rate of the fixed production assets renewal coefficient. This list of indicators can be supplemented depending on the specifics of the individual sectors under consideration.

With the planned development of events ER $\geq$ 1. When the coefficient reaches the value of 1, we can assume that the measures taken to provide fiscal support to industries are effective. If ER $\geq$ 0, then the ineffectiveness of the provided support measures is stated.

We suggest that Smart Government supplement the calculation of the economic efficiency indicator with other indicators, the need for which predetermines the choice of criteria for an appropriate fiscal policy. In particular, criterion 7.4—control of targeted use of budgetary funds. Compliance with this criterion, as well as the principle of accountability, implies control over the targeted use of both budget funds directly received in the form of budget subsidies, savings from subsidizing interest rates on loans, etc., and tax savings resulting from the use of incentive tax benefits. In particular, additional funds received as a result of the implementation of fiscal support measures should be invested in production. Thus, the observance of the above principles and the application of the criteria of an appropriate budgetary policy presupposes the use of the following indicator of intended use [19] (3):

$$IU_{i \in I} = \frac{IS_t^i}{TE_{t-1}^i + BS_{t-1}^i} \qquad (3)$$

$IU_{i \in I}$ is an indicator of the intended use of funds received as part of the implementation of measures of state budgetary and tax support;

I—a set of critically important non-primary industries;

I—industry from the list of critically important non-primary industries;

IS_t^i—the volume of investments in fixed assets made in industry i for period t;

TE_{t-1}^i—tax expenditures of the budget on benefits provided to enterprises of industry i for period t–1;

BS_{t-1}^i are direct budget subsidies provided to enterprises in industry i for the period.

The closer the value of the CI indicator to 1, the higher the level of targeted use of the amounts of tax savings and direct budget subsidies and, accordingly, the higher the efficiency of the implemented state support measures for enterprises of critically important non-primary industries. This indicator can be calculated both for the totality of industries, and for analysis within one of the selected basic non-primary industries.

The social effectiveness of government support measures can be assessed using formula (2) based on the dynamics of such indicators as the number of employees in the industry, the average monthly wage of employees, and so on.

We will also analyze some of the existing tax instruments within the framework of the National Project (Table 3).

Analyzing the presented tax preferences, it should be noted that this is only part of the organizational measures in the field of application of the Special taxation regimes by small and medium-sized businesses. Figure 1 clearly shows the positive trend in the development of the SME in dynamics, as a factor in the implementation of this National Project. Speaking about "Smart Government", it should be noted that since January 2023, another taxation system for SMEs began to operate in an experimental mode: ASTS (Automated Simplified Taxation System), as another special taxation regime. The calculation of the tax base, the correctness and procedure for the application of rates and the calculation of the tax amount are carried out using digital technologies, according to the data of cash registers and banking operations, by the tax authorities.

Based on the data in Fig. 1, one can see a positive trend in the increase in tax revenue from year to year. In relation to 2021 to 2006, the relative increase in the tax payable when applying the simplified tax system is 1272%, which shows the effectiveness of the functioning of this tax regime and the relevance of its relevance. It is worth clarifying that this indicator was influenced by many factors, such as inflationary growth and the ongoing pricing policy of the state, changes in tax legislation in terms of the simplified tax system and an increase in the number of taxpayers in

Table 3 Passport of the national project "Small and medium-sized businesses and support for individual entrepreneurial initiatives"

No	Name of the task, result	Implementation period	Link
1	2	3	4
1.2	At least 1.2 million taxpayers—SMEs, applying the simplified taxation system with the object of taxation in the form of income and using cash registers in 2020–2024, were exempted from the obligation to submit a tax return, including (cumulatively): at least 0.8 million SMEs in 2020; at least 0.9 million SMEs in 2021; at least 1.0 million SMEs in 2022; at least 1.1 million SMEs in 2023; at least 1.2 million SMEs in 2024	July 1, 2020	4.1. Federal project (Improvement of business conditions)
1.4	The transitional tax regime for SMEs that have lost the right to use the simplified taxation system is legally fixed in case of exceeding the maximum level of revenue and / or the average number of employees	December 20, 2019	4.1. Federal project (Improvement of business conditions)
2.1	A pilot project was launched for self-employed citizens in 4 constituent entities of the Russian Federation on the basis of a mobile application in order to develop key parameters of a special tax regime, including: automatic transmission of sales information to the tax authorities, exemption from reporting obligations; payment of a single payment from the proceeds, which includes a contribution to the compulsory medical insurance fund; the possibility of forming tax capital for development to pay them a part of the amounts of the calculated tax	January 1, 2019	4.1. Federal project (Improvement of business conditions)

(continued)

Table 3 (continued)

No	Name of the task, result	Implementation period	Link
2.2	In order to improve the special tax regime for self-employed citizens, an analysis was made of the practice of implementing a pilot project	August 1, 2019	4.1. Federal project (Improvement of business conditions)
2.3	The introduction of a special tax regime for self-employed citizens throughout the Russian Federation is legislatively fixed based on an analysis of the practice of implementing a pilot project for the self-employed	December 20, 2019	4.1. Federal project (Improvement of business conditions)
2.6	The number of self-employed citizens who fixed their status, taking into account the introduction of a special tax regime for the self-employed, reached 2,400.0 thousand people in 2019–2024. (cumulative total), including: in 2019—200.0 thousand people; in 2020—800.0 thousand people; in 2021—1,400.0 thousand people; in 2022—1,800.0 thousand people; in 2023—2,100.0 thousand people; in 2024—2,400.0 thousand people	December 20, 2024	4.1. Federal project (Improvement of business conditions)
2.7	Launched a pilot project on the use of an automated simplified taxation system	January 01, 2023	

Source https://www.economy.gov.ru/material/file/65c7e743dffadf1f3f3a8207e31a0d99/Passport_NP_MSP.pdf

this regime, and others. Figure 2 shows the dynamics of shortfall in budget revenues/tax expenditures when applying the Simplified Taxation System.

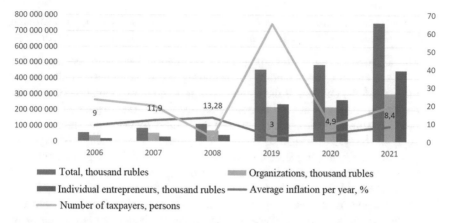

Fig. 1 Dynamics of the volume of tax revenues under the simplified tax system. *Source* compiled on the basis of data on statistical tax reporting forms—URL: https://www.nalog.gov.ru/rn77/related_activities/statistics_and_analytics/forms/ (accessed 01/19/2023)

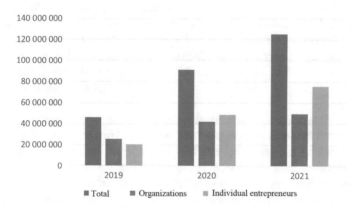

Fig. 2 Dynamics of the amount of single tax shortfalls due to the establishment of a reduced tax rate for 2019–2021. *Source* compiled on the basis of data on statistical tax reporting forms—URL: https://www.nalog.gov.ru/rn77/related_activities/statistics_and_analytics/forms/ (accessed 10.02.2023)

4 Conclusions

Thus, it should be recognized that the legislation on national project activities in the Russian Federation pays insufficient attention to the requirements for evaluating the effectiveness of national projects (programs), primarily in terms of the provision and use of federal budget funds allocated for their implementation. There is no analytical data and the authors propose a hypothetical model for evaluating the effectiveness of tax preferences offered by the state in order to assess their economic and financial efficiency and the effective use of "Smart Government". Based on the example of the National Project, according to statistical data, an analysis of tax preferences was carried out in order to implement and develop the SMEs. It should be noted that for the effective implementation of NP it is necessary to eliminate significant shortcomings:

1. The absence of any methodological or regulatory framework envisaged for the development and adoption for assessing the effectiveness of the use of funds from the budgets of the budget system in the implementation of national projects (programs), as well as other envisaged assessments (efficiency: shortfall in revenues of the country's budget system/tax expenditures, regardless of applying the system of taxation and tax preferences for the elements of the system, proposed activities, implementation mechanisms, activities of participants and managers). This urgent problem requires not only a solution, but also the need for a functional reflection in digital technologies of "Smart Government"

2. The absence of any requirements for assessing the effectiveness of the use of tax expenditures as an element of the country's budget system in terms of assessing the quality and validity of choosing the appropriate type of financial support for national projects/programs, regardless of the level of their implementation (federal and regional), the conditions for implementing the relevant types of expenditures of the federal budget, as well as the procedure for determining their volumes. In other words, the current legislation provides for separate requirements only for assessing the effectiveness of the use of budgetary system funds, but not for assessing the effectiveness of their provision and the result of the multiplier effect. The use of digital technologies will allow Smart Government to quickly solve this problem, understanding its financial and economic significance.

3. The principle of taking into account the results of the implementation of national and federal projects for the previous period for the purposes of subsequent budgetary and tax planning, which is incorporated into the system of financial support of national projects at the expense of the federal budget, does not find its practical implementation in the framework of regulating the procedure for assessing the effectiveness of the use of budget funds budgetary system in national projects/programs [19] and shortfalls in tax expenditures of the country's budgetary system. It is important for a "smart government" to be able to assess shortfalls in budget revenues, and for the business community to defend its tax preferences in order to further cooperation between business and the state.

4. The inability to change the types and (or) volumes of financial support for national projects (programs) in the course of their implementation, depending on the results of an intermediate (current) assessment of the effectiveness of shortfalls in expenditures of the country's budget system and (or) monitoring the implementation of national projects (programs). Digital technologies of "Smart Government" are designed to solve this urgent problem.

It seems that the elimination of these shortcomings will contribute to the improvement of the system of financial support for the implementation of national projects (programs) and will significantly increase the efficiency of the use of federal budget funds based on the proposed methodology [7], which will lead to an increase in economic and fiscal indicators in the country, the development of priority sectors the national economy, solving social issues and improving the welfare of citizens of the state, and of course to the positive dynamics of tax revenues and an increase in tax revenues to the budgets of all levels of the country.

This work was carried out as part of the applied research work "Transformation of tax policy and tools for its implementation in the conditions of the sanctions economy".

References

1. Tikhonov, A.V., Bogdanov, V.S.: From «smart regulation» to «smart management»: social issue of feedback digitalization. Sociol. Stud. **1**, 74–81 (2020)
2. Vasilenko, L.A., Zotov, V.V., Zakharova, S.A.: Use of the potential of social media in the development of participatory management. Bulletin of Peoples' Friendship University of Russia. Ser. Sociol. **20**(4), 864–876 (2020). https://doi.org/10.22363/2313-2272-2020-20-4-864-876
3. Dudikhin, V.V., Shevtsova, I.V.: Smart management—management using artificial intelligence. Public Administration. Electron. Bull. **81**, 49–65 (2020)
4. Jiang, H., Geertman, S., Witte, P.: Avoiding the planning support system pitfalls? What smart governance can learn from the planning support system implementation. Environ. Planning B: Urban Anal. City Sci. **47**(1), 1343–1360 (2020)
5. Bradul, N.V., Lebezova, E.M.: Conceptualization of the concept of "Smart Government": scientometric approach. Manager **3**(11), 33–45 (2020) (in Russian)
6. Dynnik, D.I.: National projects: main provisions and problems in management. Pravovestnik **6**(8), 11–13 (2018)
7. Kudelich, M.I.: The system of regulatory requirements for evaluating the effectiveness of national projects: current problems. Financ. Mag. **4**, 36–49 (2019)
8. Bukhvald, E.M.: National projects in the system of strategic planning in the Russian Federation. Theory Pract. Soc. Develop. **2**, 50–54 (2019)
9. Tsindeliani. I.A., Bit-Shabo, I.V., Gorbunova, O.N.: Financial Law in the Context of the Development of the Digital Economy, p. 320. Prospekt, Moscow (2019)
10. Prokofieva, A.I.: Information and analytical support for the management of national projects. Scie. Prism Time **1**(10), 97–99 (2018)
11. Matveeva, E.E.: Application of a risk-based approach in the field of public financial control. Econ. J. **1**, 41–56 (2018)
12. Prokofiev, E.S.: Modernization of the system of state financial control in the Treasury of Russia. Budget **2**, 38–43 (2017)
13. Zaporozhan, A.Y.: On the issue of the implementation of national projects. Manage. Consult. **5**, 18–23 (2019)
14. Ivanov, O.B., Bukhvald, E.M.: National projects of Russia: regional dimension. Stage: Econ. Theory, Anal. Pract. **1**, 37–52 (2019)
15. Ivanov, O.B., Bukhvald, E.M.: Investment priorities in the strategies of socio-economic development of Russian regions. ETAP: Econ. Theory, Anal. Pract. **2**, 31–47 (2018)
16. Ivanov, O.B., Bukhvald, E.M.: Strategic territorial planning in the regions of Russia. Stage: Econ. Theory, Anal. Pract. **3**, 7–21 (2018)
17. Kostareva, L.V.: Implementation of priority national projects in the Russian Federation: problems and prospects. Soc. Econ. Manage. **1**, 37–44 (2018)
18. Decree of the President of the Russian Federation dated May 7, 2018 No. 204 On the national goals and strategic objectives of the development of the Russian Federation for the period up to 2024. http://www.kremlin.ru/acts/bank/43027. Last accessed 30 July 2023

19. Kazantsev, N.M., Bukhvald, E.M., Bakhtizin, A.R.: Economic and legal institutions for regulating regional development of the Russian Federation: monograph. In: N.M. Kazantsev (ed) Institute of Legislation and Comparative Law under the Government of the Russian Federation, 468 p (2013)

Smart Governance: Digitalization of Tax Control in the Russian Federation to Increase VAT Collection

Irina A. Zhuravleva and Semyon A. Shamaev

Abstract The concept of "Smart Government" is increasingly reflected in the policies of various States. At the same time, the development of this concept is reflected in many areas, including in the field of tax relations. Thus, digitalization of the activities of tax authorities allows, on the one hand, to simplify the process of calculating taxes and fees payable to citizens and organizations, to provide them with up-to-date information about changes in legislation and current tax benefits, and, on the other hand, enables tax authorities to automate part of the work, increasing the efficiency of their activities. To date, a significant part of tax revenues in the Russian Federation are VAT receipts. At the same time, positive dynamics of the growth of these receipts has been observed for a long time. The positive trend was facilitated by the gradual introduction of the latest digital tools into the VAT calculation and payment control system. This chapter systematizes information and identifies the advantages and disadvantages of digitalization of this tax, as well as suggests possible measures to further increase the collection of VAT.

Keywords Digitalization · VAT · Tax control · Tax collection growth · Smart Governance

1 Introduction

Modern digital technologies are spreading to various branches of management, leading to serious transformations, the emergence of innovative developments, robotization and automation of processes. Digitalization has a beneficial effect on various structures, including state ones, among which tax authorities are one of the most important. The most widespread trend is the widespread introduction of digital technologies within the framework of the organization of effective interaction between

I. A. Zhuravleva (✉) · S. A. Shamaev
Financial University Under the Government of the Russian Federation, Leningradsky Prospekt, 49/2, 125167 Moscow, Russia
e-mail: sia.mir67@mail.ru

© The Author(s), under exclusive license to Springer Nature Switzerland AG 2024
M. Sari and A. Kulachinskaya (eds.), *Digital Transformation: What are the Smart Cities Today?*, Lecture Notes in Networks and Systems 846,
https://doi.org/10.1007/978-3-031-49390-4_9

taxpayers and tax authorities and the implementation of tax control, which is one of the most important elements of the tax system of the state. The level of budget revenues directly depends on the degree of effectiveness of the control activities of the tax control bodies.

Effective tax control can completely eliminate the existing shortcomings in the tax system and ensure that each subject of the tax system performs its duties properly. Tax structures are aimed at introducing digital technologies for a number of reasons. Firstly, they allow us to systematize and structure data on the status of the taxpayer, identify possible tax gaps in value added tax (hereinafter referred to as VAT) and reduce the risk of their occurrence.

Secondly, digital technologies contribute to reducing the administrative burden both on the tax authority and directly on tax payers. At the same time, innovative developments introduced in order to improve tax control lead to the transformation of the tax system as a whole, which determines the relevance of studying the peculiarities of their use at the present time.

The study of the mechanism of the impact of paying taxes on the economic condition of an economic entity, its solvency, and its financial results concerns all taxes. However, it must be stated that the study of the system mechanism of the procedure for calculating, paying and tax control of VAT is extremely relevant and at the same time quite complex.

The problems associated with the functioning of VAT and administration are caused mainly by the influence of subjective factors: corruption, lobbying of private interests, bureaucratization of management procedures, the variability of regulatory rules, the complexity and capacity of digital technologies, the diversity of interpretations of the norm of tax law, the multiplicity of amendments to existing rules.

In addition, digitalization of tax administration processes in general and VAT is an important task in the context of the development of the concept of smart governance. Increasing tax collection has a positive effect on the process of income redistribution and simplifying the processes of interaction between taxpayers and the state represented by the tax service reduces transaction, administrative and time costs of both parties.

At the same time, a whole range of problems remains unresolved the issues of improving individual elements of the functioning of the tax and the lack of systematic approaches to its administration, which suggests the need to develop areas of improvement, both tax elements and information technologies related to improving the efficiency of administration. All the above allows us to speak about the relevance of studying the trends in the development of tax control in the Russian Federation in order to increase VAT collection based on modern digital technologies.

Today, digitalization of taxation as a direction for further development of the concept of smart governance is a frequent subject for scientific research and discussion. Thus, I. Zhuravleva investigated the problem of countering aggressive tax planning [1], S. Shamaev considered ways to optimize the VAT tax control system in the context of digitalization [2], A. Sekushin analyzed the experience of various national tax administrations in the field of digitalization of tax control [3]. Foreign researchers are also talking about future VAT. Thus, the analysis of Peru's experience in the implementation of electronic invoices, carried out by a group of authors in 2022, is of interest [4]. Chartered Accountants Worldwide also write about the development and digitalization of tax administration [5], a group of PWC experts talked about the prospects for digital VAT administration [6]. In addition, various aspects of digital tax administration are also the main areas for international cooperation of national tax administrations. So, in December 2022, the European Commission published research data on the development of VAT in the European Union, analyzed changes in regulations, considered digital trends, etc. [7].

2 Theory and Current State of VAT Administration in Russia

2.1 Theoretical Foundations of VAT

VAT is an indirect tax with which the state can both stimulate consumer demand – during a crisis economy, and restrain – during an excessive economic recovery. VAT is charged at each stage of creating value added of economic activity of an economic entity and acts as an indirect tax on goods sold (works, services), the basis of which is the value added at each stage of production and sale of inventory [8].

The economic essence of VAT lies in the fact that the final consumption tax, which, subject to the application of economically weighted tax components, acts as a fairly effective element of the tax system from a regulatory point of view. Thus, VAT contains a built-in mechanism of fiscal devaluation, which allows achieving positive stimulating effects for the economy. This tax is fiscally significant for the public finance system, since it allows accumulating significant monetary resources in the form of tax revenues in the state.

A distinctive feature of VAT is that the tax enters the budget at the place of consumption [8]. In fact, it is a premium to the price established by the state, therefore, it belongs to the state, does not affect the formation of profits of economic entities, i.e., as a general rule, it is neutral to the financial indicators of the taxpayer. Along with the fiscal function of VAT, there is a regulatory one, which through incentive mechanisms (for example, the application of a reduced rate or exemption from taxation, other tax preferences) supports reduced rates on socially significant goods, which reduces the final cost of a good or service.

2.2 Features of Digitalization of the Sphere of Taxation

The introduction of digital technologies contributes to the transformation of the basic principles of calculating and paying taxes, which leads to the emergence of new fundamental research in the field of digitalization of taxation mechanisms. The totality of scientific research on this topic, from our point of view, can be divided into two approaches. Within the framework of the first of them, an attempt is made to determine the main trends and features of the impact of innovative technologies on tax control.

In this regard, it is impossible not to note the scrupulous review presented by the Institute of Chartered Accountants of England and Wales "Digitalization of Taxes: International Perspectives" [9], which quite fully presents the goals, directions of development and difficulties faced by tax authorities when introducing digital technologies. However, it is impossible not to note the rather poor coverage of the experience of Russia and other countries in this context. It should be noted that the most informative was the comparative analysis of the use of digital tools in the framework of tax control in Russia, the countries of the European Union, the USA, and the states of Southeast Asia.

On the contrary, a detailed study of the European experience of introducing innovative technologies in the framework of increasing VAT collection is presented by Professor of the Department of Tax Law at the University of Örebro (Sweden) Kristina Trenta. However, it lacks a comparative analysis with the experience of other States. In addition, the results of this study cannot be used in developing countries, as they were obtained as a result of the research of developed countries [10].

In relation to Russian-language sources, the work of O.M. Karova and I.A. Mayburov is significant, which allows you to get acquainted with the peculiarities of digitalization of the taxation sphere as a whole [11]. It is impossible not to agree with the position of the authors that the introduction of innovative technologies contributes to the emergence of specialized digital tools that have high potential due to its further development. The work of E.N. is no less informative. Golikova, in which she studies the specifics of the involvement of the state in the tax system through the introduction of digital technologies. However, the author considers this phenomenon rather one-sidedly, since digital technologies are not a strict state instrument [12].

The second approach involves identifying the areas that are most attractive for their implementation. First of all, in this regard, a study is interesting, according to the results of which an international group of researchers published a specialized guide on the use of electronic technologies for the organization of VAT payments and control. However, its main disadvantage is the study of only the experience of the European Union countries, which makes it inapplicable, for example, in Russia due to different levels of development, dissemination and implementation of information technologies. At the same time, a more comprehensive picture can be obtained by analyzing the reports provided by international consulting organizations [13] and OECD specialists [14] regarding digitalization and tax administration on the example of VAT.

Among domestic studies, the most informative in this area are reviews of analytical companies [15], as well as reports of tax services and authorized employees. It is impossible not to agree with the provisions of the work of E.E. Golova, I.V. Baranova, within the framework of which a full-fledged model of digital technologies in the organization of VAT payments is proposed [16]. However, the greatest disadvantage of modern research on this topic is the lack of structured concrete practical recommendations on the introduction of digital technologies to increase VAT collection, taking into account the introduction of anti-Russian sanctions and the current crisis situation. In this regard, it can be concluded that there is a lack of research that combines the two identified approaches and contains practical recommendations for increasing VAT collection using digital technologies in the current conditions.

It is necessary to note the fact of the influence of automated VAT control systems for business, both on the positive and negative sides. On the one hand, they increase the integrity of counterparties, simplify the submission of documents. On the other hand, they contribute to an increase in the burden of organizing tax audits due to the search for care schemes and the emergence of risks of hiding income, information, etc. Finally, VAT automation for the state also provides additional statistical information, which can help reduce the tax gap, improve the quality of budget planning and stabilize government revenues.

At the same time, a significant negative impact on the effectiveness of the implementation of digital technologies is caused by the refusal of organizations to implement automated control systems, poor communication quality. Despite the above-mentioned aspects, modern companies strive for transparency and are forced to adopt and implement automated VAT control systems. Thus, according to the forecasts of the Ministry of Finance of the Russian Federation, the percentage of tax collection, including through tax administration measures, should grow from 98.79% to 99% from 2021 to 2024. A direct confirmation of the appropriateness of the development of digital resources is the increase in the return on the funds spent on the digital transformation of tax administration. Over the past five years, the amount of taxes collected per ruble of costs invested in the digitalization of the activities of the Federal Tax Service of Russia has increased from 96.5 rubles to 123.4 rubles [17].

3 Results: Digitalization as an Element of Smart Governance

3.1 Features of VAT Digitalization: Causes, Functions, Advantages, Disadvantages

Currently, VAT is the tax that the most innovative methods and digital technologies are used to collect and control. This is due to the following aspects:

1. VAT is the most important and significant payment for the budget of Russia, which is proved by the relevant statistical data. An analysis of the total volume of the consolidated budget of the Russian Federation, including VAT, for the period from 2015 to 2022 is presented in Fig. 1.

Fig. 1 Dynamics of VAT receipts in the revenue structure of the consolidated budget of the Russian Federation in 2015–2022, trillion rubles [17]

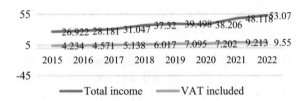

According to the data in Fig. 1, the tax revenue is growing every year by 0.34 trillion rubles. At the same time, VAT receipts from the sale of commodity-material values in the domestic market of the country, the second largest tax collection, amounted to 6.489 trillion rubles, which is 18.4% higher than at the end of 2021.

The observed tax increase directly correlates with the automation of the tax audit procedure. The relative value of VAT in the consolidated budget revenue structure is also increasing: from 15.7% at the beginning to 19.8% by the end of the period under review. Even despite the crisis in 2020, the amount of fees turned out to be more than in the previous year.

This fact confirms that VAT can bring stable income, despite unpredictable crisis phenomena. In recent years, the share of VAT on imported goods/customs VAT in the total revenue of the federal budget varied from 74 to 81%. Its revenues occupy a significant place in the revenue part of the budget of the Russian Federation. Moreover, the share of income from value added tax in the total tax revenues of the state is steadily increasing. Let's imagine the dynamics of VAT in Fig. 2.

In accordance with the data in Fig. 2, it can be noted that VAT receipts within the analyzed period increased in dynamics. Thus, the increase in internal VAT in 2022 compared to 2015 amounted to 4.01 trillion rubles, or the growth rate was 163%, respectively, and in comparison with 2021 by 0.98 trillion rubles, or the growth rate was 18%. VAT on imported goods showed a decrease of 18% in 2022 compared to 2021 and decreased to 3.063 trillion rubles, which is the result of the introduction of anti-Russian sanctions. The total amount of both taxes amounted to 9.552 trillion rubles. The share of VAT in the volume of federal budget revenues for the period from 2015 to 2022 is shown in Fig. 3.

Fig. 2 Dynamics of VAT receipts in 2015–2022, trillion rubles [17]

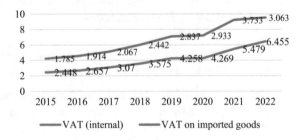

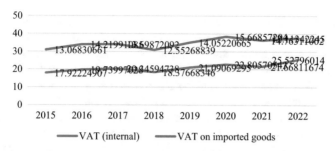

Fig. 3 The share of VAT in total federal budget revenues in 2015–2022, % [17]

Figure 3 shows that in the total structure of federal budget revenues, internal VAT increased from 17.9% in 2015 to 25.5% in 2022. It should be noted that the largest share of VAT on imported goods in the total volume of the federal budget was received in 2020 and amounted to 15.7%, while in 2022 it was the smallest. Thus, it can be concluded that during the period under review, the growth of the consolidated and federal budget was achieved due to an increase in VAT tax administration, namely, a combination of the following measures: extensive use of electronic administration systems and taxpayer services that allow reporting via virtual online communication channels instead of standard postal procedures for subsequent uploading to the databases of tax authorities, transform the format of determining the objects of taxation, i.e. to form it without the participation of the taxpayer himself in this process, to carry out many control measures remotely. All of the above allows us to conclude that the measures implemented by the state to introduce digital technologies into the VAT collection process show generally positive values.

As part of the further development of electronic services, the Federal Tax Service of Russia has gradually introduced and actively uses its own development - AIS "Tax-3" in its activities since 2019. This complex software product is a unified automated information system that allows you to significantly increase the level of automation of the functional activities of the Federal Tax Service of Russia within its competencies defined by the relevant Regulation "On the Federal Tax Service of Russia", and also makes it possible to receive, process, and provide data in contactless mode and analysis of information, formation of information resources of tax authorities, statistical data, information, necessary to provide support for managerial decision-making in the sphere of authority of the Federal Tax Service of Russia and the provision of information to external consumers.

The functioning of AIS "Tax-3" should ensure full automation of all spheres of activity of the tax service, allow to receive the necessary information from a single database of the system in an operational manner. As a result of the active use of new software products, according to the Federal Tax System of Russia, by the end of 2022, the VAT tax gap was only 0.43%, while a few years ago it was kept at the level of 8% [18]. Further introduction of the latest technologies will allow for more thorough implementation of control measures in relation to VAT;

2. Introduction of digital technologies, which include an automated VAT refund control system (VAT Control System), a System for marking and tracing goods (IP MPT)), a System for monitoring the use of cash register equipment (ASK KKT), an Information system for the Population Register and Civil Status Records (IS Registry Office) [18] in the collection process is spread over a significant number of participants in tax relations involved in this process, which include individuals and legal entities paying fiscal payments, government agencies, software co-creators;

3. The use of innovative technologies, which include electronic VAT declaration and electronic invoices, the goods tracking system, online cash registers, tax monitoring - integration into the taxpayers' accounting system, automatic exchange of financial information, the state accounting information resource (GIR BO), control of the issuance of EDS (item 1 of art. 15 of Federal Law No. 63-FZ dated 06.04.2011 "On Electronic Signature"), blockchain system of machine-readable powers of attorney, API integration of the functionality of service applications of the Federal Tax Service into software products and services of banks and operators of electronic platforms, online service "Transparent Business. Check yourself and the counterparty", the analytical complex "Tax potential"—to control the deviation of the level of profitability, tax burden, salaries in the organization from the average for the region, modeling the level of the tax base generated by the taxpayer based on its internal metrics, etc. [18], provides an opportunity to reduce the risk of VAT evasion by tax payers;

4. The use of various digital tools helps to increase the collection of VAT. For the purposes of this study, we will focus on the latter aspect in more detail. The use of digital technologies for calculating and systematizing receipts from individuals and legal entities poses the first task of forming a model in Russia of voluntary compliance with the norms of legislation for the growth of tax revenues. It implies increasing the transparency of the activities of all business entities, improving the effectiveness of budget planning and obtaining sustainable cash flows from fees and VAT in particular.

According to the Ministry of Finance of the Russian Federation, the level of tax revenues through the introduction of digital technologies, which include blockchain, Big Data, artificial intelligence, etc., will be increased from 98.79% to 99% for the period 2021–2024. In addition, an important criterion that testifies to the reasonableness of using innovations for tax control is the effectiveness of the funds spent on them. So, from 2018–2022. the volume of tax revenues per unit of expenses that were invested by the Federal Tax Service in digitalization increased in aggregate from 96.5 rubles to 123.4 rubles [18]. It should be noted that the introduction of digital technologies leads to an increase in the efficiency of tax control over the collection of VAT, on the one hand, but the transformation of related processes on the other, namely:

1. Transfer of the relationship between taxpayers and the tax authority to a remote basis, to the online sphere. For these purposes, electronic document management systems are being introduced, which allow minimizing the number of printed documents. It should be noted that a special Concept for the development of this area is currently being actively promoted in Russia, which sets the main goal of increasing the efficiency of tax control within the framework of increasing tax collection, in particular VAT. It provides for the digitization of the following documents:

   - at least 95% of invoices before the expiration of 2024;
   - 70% of transport and consignment notes by the end of 2024;
   - annual growth of digitized documentation, which is formed in the frames of sending organizations to each other by at least 20%;
   - reduction of the turnover of printed documentation by at least 10% annually, starting from 2023 [19].

   In addition to the above-mentioned indicators, the Concept of Tax Control Development implies the achievement of a predominant number of electronic documents over printed ones in order to reduce associated costs and improve safety, as well as the widespread introduction of an information system controlled by state bodies;

2. Expanding the coverage of digital technologies and their application both by the required authority and by tax payers. Within the framework of this direction, individuals and legal entities are being stimulated to use their personal account, providing the opportunity to receive electronic extracts and other information from online registries, the use of automated products in companies, etc. At the same time, the most important goal of introducing digital technologies to ensure tax control is to preserve the economic security of the country as a whole, the beginning of which can be considered 2015, when innovative software was first used - the automated control system (hereinafter - ASK) "VAT–3", which allows to identify cases of tax fraud [20]. This technology is based on automation of the process of reconciliation of data on counter-agent operations by studying incoming and outgoing invoice flows, which is carried out by using BigData and is a unique VAT accounting system that has no analogues in the world.;

3. The need to amend the current legislation. In relation to this aspect, we note that the current version of the Tax Code of the Russian Federation (hereinafter referred to as the Tax Code of the Russian Federation), which regulates the process of collecting and administering VAT up to the present time does not take into account the innovative technologies used in this area, which indicates the need to update it. For example, the Tax Code of the Russian Federation does not currently specify the tax nature of the electronic service and the specifics of VAT collection.

The changes outlined above, both in the framework of VAT collection and in relation to the ongoing transformations in the consumer, commercial, tax, and legislative spheres associated with it, lead to the emergence of a new phenomenon, which can be designated as digital VAT. It is impossible not to agree with the concept presented by O.M. Karpova, I.A. Mayburov, who for the first time cite the concept of smart VAT [11]. However, in our opinion, it would be more correct to name digital VAT, which will be understood as a special concept of indirect tax collection, which has a universal character and will be calculated and accrued fully with the help of digital technologies, in the online sphere, for example, blockchain [20].

Digital technologies, within the framework of increasing the collection rate and increasing the effectiveness of tax control, perform a number of specific functions. Among the most important of them, in our opinion, the following can be distinguished:

1. monitoring and supervision of the sphere of functioning of the results of the work of tax payers;
2. registration and registration of legal entities and individual entrepreneurs;
3. carrying out control measures in relation to the preparation and implementation of any type of gambling activity;
4. registration of cash register equipment;
5. providing tax payers with the necessary information;
6. transferred by tax payers;
7. creation and coordination of various forms of documentation.

The introduction of digital technologies contributes to the growth of VAT collection and increase the efficiency of tax control in general, this is due to the regulation of the fiscal function of the tax and the need for payers to increase the transparency of their activities.

3.2   Modern Digital Technologies for the Growth of VAT Collection and the Consequences of Their Application

Currently, various tools are used for the purposes of tax control and the growth of VAT collection. The main ones are:

- the methodology according to which possible risks are determined in order to identify the need for on-site tax audits, developed in accordance with the order of the Federal Tax Service of the Russian Federation "On approval of the Concept of the on-site tax Audit planning system") [18];
- Unified State Automated Information System (USAIS), which aims to control various transactions and systematize information about counterparties;
- ASK "VAT 3", which allow to study the possibility and amount of tax refunds and are based on a comparative analysis of the purchase books and sales books of VAT payers, which they need to transfer to the tax service through the electronic document management system along with the declaration. Its use makes it possible to increase the degree of control and determine the feasibility of conducting desk inspections;

- online cash registers, which are legally defined instruments and have become a consequence of the transformation in monetary transactions due to the impact of digitalization;
- enabling small and medium-sized businesses to use special tax regimes (automated simplified taxation system);
- development and further improvement of the functioning of digital technologies that were developed and implemented to increase the effectiveness of tax control during the coronavirus pandemic. An example is a special resource that is posted on the official website of the Federal Tax Service of the Russian Federation and contains information about methods and types of tax support [18];
- introduction of cash register equipment (hereinafter – KKT), which provides the opportunity to work with significant data sets, leveling the complex nature of the creation of tax reporting [19].

In addition to the digital technologies mentioned above, the model of remote control of taxes and fees through regular monitoring and control of tax obligations has now become widespread. Quite a significant number of companies (about 95 representatives of large businesses), which send about 30% of tax payments to the federal budget, apply the designated model on a regular basis.

A significant innovation was the concept developed by the Ministry of Finance, according to which VAT in the framework of trading on online platforms is collected by profiling countries according to the destination criterion, which allows hedging the risk of double taxation. Currently, special software equipment and financial technologies have not been developed in Russia that would allow them to be implemented and used in the case of tax control over online sales.

It should be noted that until the beginning of 2022, online trading platforms of the countries of the Eurasian Economic Union (hereinafter referred to as the EAEU) transferred VAT only in the state from where the products were shipped, without additional payments in the recipient country, that is, once. However, since 2022, Kazakhstan and Belarus have introduced this type of tax on the totality of imported goods, which leads to the appearance of double taxation, since online platforms must pay VAT both in the sending country and in the receiving country. According to the Association of Internet Trade Companies, at the rate of VAT in Kazakhstan (12%) and Belarus (20%), additional tax payments in the amount of 3.6 billion and 3.8 billion rubles are imposed on domestic online platforms.

In order to level out such a situation in Russia, amendments were adopted, according to which, when carrying out trade operations within the EAEU member states (Russia, Kazakhstan, Belarus, Armenia, Kyrgyzstan) in Russia, payers do not need to pay VAT. At the same time, if companies of the EAEU member states conduct transactions with Russian consumers using online platforms, VAT is assigned to be paid by the latter. From January 1, 2023, organizations and sole proprietors must transfer taxes and contributions in a single tax payment. The Federal Tax Service explained: VAT for goods imported to Russia from the EAEU countries is paid in a general manner. This means that it will be necessary to transfer it in the form of an

EPP no later than the 20th day of the month following the month of registration of goods or the payment deadline.

The developed draft amendments to the existing legislation are aimed not only at the abolition of double taxation, but also at leveling the advantages that such online platforms as AliExpress, eBay, Asos and others receive. When importing goods from the EAEU countries, instead of VAT in the amount of 10% or 20%, organizations will transfer 9.09% or 16.67%, depending on the type of product. It is obvious that the developed measures will help to increase the attractiveness of imports [17].

In addition to the above, the Federal Tax Service also assesses the possibility of introducing certain digital technologies into the process of tax control within the framework of VAT collection. In this regard, the following most promising areas are allocated to it:

– systematization of information from taxpayers' pages in social networks;
– use and further expansion of the functionality of a specialized application for mobile phones;
– implementation of information exchange between different devices;
– implementation of network analytics, which allows you to quickly filter and analyze information, as well as structure various types of data to detect hidden connections between them, identify suspicious patterns and transactions between individual elements, visualize data on supply chains or business processes, thoroughly study those areas that require increased attention. attention;
– receipt of all necessary documents in electronic form (primary, accounting, tax).

The designated set of measures will make it possible not only to form effective interaction between tax payers and all interested parties using Internet communication channels, but also will allow maximum automation of VAT calculation up to the cancellation of declarations and the transfer of all transactions to the online sphere with maximum transaction security and hedging threats of various types of fraud. Despite the active work of the state in improving the efficiency of tax control, there are negative consequences of this process. The most significant of them, in our opinion, are the following. Firstly, the rejection of digital communications by taxpayers. The indicated problem is a consequence of the presence of a stable habit of the target audience to the offline format of interaction with employees of the tax authority, the presence of an insufficient level of digital literacy, lack of necessary equipment or insufficient level of communication. To solve this problem, we offer the following recommendations:

1. the application of penalties to organizations that, for one reason or another, refuse to maintain the reporting format designated by the tax authorities, as well as the use of digital technologies, regardless of the reasons. Taxpayers, when deciding on the feasibility of tax evasion, compare their expected benefits and possible costs (fines). From the moment when the amounts of tax deductions become

greater than the amount of the expected benefits, tax evasion will become unprofitable, which indicates the need to increase the amount of penalties, simultaneously with which the state will provide all possible assistance for the introduction and use of digital technologies;

2. improving the quality of communication and expanding the coverage of the Internet in order to increase its accessibility to the population from different parts of the territories;

3. conducting special seminars by district public organizations to improve the digital literacy of the population, as well as conducting appropriate classes in schools, colleges, universities, institutes.

Secondly, the high cost and complexity of implementing the necessary software. The development of new online services, the use of digital technologies, the submission of new reporting forms, automation of business processes, etc. are high-budget activities for both the tax authority and the VAT payer himself. At the same time, there is currently a shortage of highly qualified specialists who had considerable experience in working and installing the necessary software equipment. However, in this regard, it should be noted that the expenses incurred by the tax authority are leveled with the help of a higher indicator of tax revenues. The costs of tax payers are compensated by the absence of penalties. To solve the identified personnel problem, it is important to conduct special advanced training courses for the involved personnel.

Third, the safety and protection of information. Digital technologies contribute to the emergence of a significant cyber risk, which, in turn, leads to an increase in fraud, the threat of data theft, etc. One of the explanations for this is also the lack of employees with experience in this field. To solve this problem, from our point of view, it is important to apply the following practical recommendations:

1. implementation of an internal control system that allows to supervise the activities of authorized employees who have access to confidential information;

2. regular updating of security systems regarding users' personal data.

Fourth, the translation of reporting and other necessary information into electronic form. In most organizations, reports and other quarterly, annual documents for the past years are filed in special folders and have no digital analogues. In this regard, in case of discrepancies and the need to search for relevant data, employees are forced to turn to printed media, while storing information in electronic form allows reducing time and labor costs and ensuring its high safety. Therefore, it is important to develop amendments to legislative acts, according to which documentation for past periods was necessarily digitized.

Fifth, the inconsistency of the norms of existing legislation with current trends. The Tax Code of the Russian Federation and other legal acts were developed before the digitalization of taxation and do not reflect the latest innovations, tools used, etc. Despite the amendments, the introduction of new software leads to the appearance of rights and obligations for payers and employees of tax services, which should be promptly reflected in the Tax Code of the Russian Federation and other legal documents. So, since 2014, tax payers are obliged to poison the VAT declaration at the end of the quarter using electronic document management [18]. To solve this problem, it is necessary to adapt the existing legislation, include in it the norms regulating the use of digital innovations in the framework of taxation and conduct regular monitoring for the appearance of new ones.

Sixth, the impact of anti-Russian sanctions and the need to use domestic software. The withdrawal of foreign IT companies from Russia and the ban on the use of their software leads to the need to create, test and implement domestic development and digital technologies that would be similar in efficiency and convenience. As part of solving this problem, it is necessary to form Russian software that was based on foreign algorithms and was low-cost, practical to use.

3.3   Practical Recommendations for Improving Tax Control in the Framework of VAT Collection Growth

The digital transformation of tax administration is a serious issue, the solution of which makes it possible to obtain significant positive shifts, using the potential of the latest digital technologies and relying on the experience of tax systems of foreign countries. The efficiency of automation of tax administration and the reorientation to the collection of significant amounts of data on the definition of tax obligations according to the principles of fairness and transparency are already being observed. This creates the basis for building a system of relations when performing procedures for collecting and controlling tax revenues, the importance and relevance is determined in accordance with the objectives of fiscal policy and the further development of economies.

It is proposed to include modern technologies such as blockchain and smart contracts based on blockchain, Data, artificial Intelligence (AI), machine learning (ML), cloud technologies, the Internet of things as technological tools of digital transformation of tax administration (IoT). The potential advantages of blockchain technology for tax administration, in our opinion, are as follows:

1. Increase of honesty and transparency in the fight against fraud and corruption by fixing the relations of taxpayers, ensuring the impossibility of changing and withdrawing data, which excludes the participation of third parties in making business agreements.
2. Reducing the cost of ensuring compliance with taxation rules, which can be recorded in the blockchain protocol using smart contracts, which allows transactions to be performed automatically, reduce the level of violations and reduce the need to submit tax documents and the need for tax audits.
3. Improving the efficiency and compliance with the deadlines for collecting taxes and fees due to automatic monitoring, high speed and accuracy of tax payments.
4. Stimulating compliance with tax legislation by increasing transparency and trust of taxpayers, providing cryptographic methods of protection and consensus mechanisms.

A practical application of blockchain technology for organizing business contractual relations and automating the implementation of business rules is a smart contract. For the purposes of tax administration, a smart contract ensures compliance with the conditions of economic relations between entities, simplifying control over the payment of taxes. The smart contract system fully automates the process of calculating the income tax of individuals. Consequently, blockchain technology and smart contracts make it possible to transform tax administration approaches in terms of the organization of tax relations, giving greater transparency and reliability, ensuring the exchange of information between various tax administrations in accordance with the harmonization of rules to the unified international requirements of the tax control system.

Artificial Intelligence Technologies (AI) and BigData allow forecasting and planning of tax revenues, tax control and compliance, administration of tax debt, tax consulting. Artificial intelligence – technologies based on the technology of expert systems, ML and natural language processing, virtualization and robotic automation (RPA), are used as a result in the tax administration system. At the same time, when implementing AI and Big Data, it is necessary to take into account the internal risks associated with the use of data, as well as unauthorized access and incorrect use of the technologies themselves.

Accumulation, constant support of the relevance and availability of information arrays of large amounts of data requires significant resource costs (technical, technological, financial) for its own IT infrastructure. The solution to the problem has become possible with the help of cloud technologies capable of ensuring continuity of security and data integrity through encryption and multi-factor authentication, flexibility and scalability at the level of resources and users, automatic backup and effective control, ubiquitous and convenient access to a common pool of computing and information resources. resources, using the power of the provider, the advantages of centralized storage, viewing and data transmission.

The Internet of Things (IoT) technology expands access to new sources of information through computer networks of physical objects. These may be cash registers that are disabled in the tax administration for the transfer and processing of cash receipts when paying for a purchase. The online data obtained allows for automatic risk assessment, as well as predictive analytics based on machine learning algorithms. Consequently, each of the considered digital technologies is implemented through a set of measures aimed at transforming the processes of tax administration and their transformation in accordance with the new digital model.

As a result of the above analysis, we noted that the effective functioning and use of innovative technologies is hindered by a low level of digital and tax literacy of the population. For this reason, we propose to create and implement a special online portal "VAT", which will provide all the necessary information, both on digital infrastructure, rules for filling in and providing documentation, and the specifics of the functioning of digital technologies, files with examples of documentation, etc. A link to it will be found on the main page of the website of the Federal Tax Service of the Russian Federation, the information is supposed to be posted both via the Internet and offline offices of the organization. Forums containing various topics related to

VAT taxation will be opened on the portal, as well as discussions will be opened both between participants and with representatives of the tax service. The opening of the online portal "VAT", from our point of view, will increase the level of public awareness about fees, tax calculation and involve in the use of digital technologies.

4 Conclusions

Thus, in this chapter, a voluminous study has been carried out, a scrupulous, generalizing, systematic analysis of the works of modern researchers devoted to the digitalization of taxation has been carried out. Thus, one of the shortcomings of foreign research is the lack of analysis of the experience of the Russian Federation, and the existing works of Russian authors—the lack of structured concrete practical recommendations for the introduction of digital technologies to increase VAT collection, taking into account the introduction of anti-Russian sanctions and the current crisis situation. In addition, the fact of the influence of automated VAT control systems for business, both on the positive and negative sides, is noted. On the one hand, they increase the integrity of counterparties, simplify the provision of documents. On the other hand, they contribute to an increase in the burden of organizing tax audits due to the search for care schemes and the emergence of risks of hiding income, information, etc. Finally, VAT automation also provides additional statistical information for the state, which can help reduce the tax gap, improve the quality of budget planning and stabilize state revenues.

The analysis of statistical data and their comparison with the chronology of the introduction of digital technologies allowed us to prove the positive impact of digitalization on the dynamics of tax revenues. In addition, several main problems arising in the process of VAT digitalization have been identified, such as the rejection of digital communications by taxpayers, the high cost and complexity of implementing the necessary software, the safety and protection of information, the translation of reporting and other necessary information into electronic form, the inconsistency of the norms of existing legislation with current trends, the impact of anti-Russian sanctions and the need to use domestic software, suggested possible measures to minimize their manifestations.

In addition, in order to further efficiency, we have presented the author's system for improving tax control and increasing VAT collection with a description of the necessary digital technologies, and also proposed the creation of an online portal "VAT" and justified the need for its functioning.

The results of this study can be used in business, as part of the implementation of government projects to increase the digital and tax literacy of individuals and legal entities, as well as by the tax service to form effective feedback between its employees, the public, and business.

References

1. Zhuravleva, I.: The value added tax at the present stage of economic development: the problems of aggressive planning. In: 2019 2nd International Workshop on Advances in Social Sciences (IWASS 2019). pp. 189–196 (2019)
2. Shamaev, S.: Optimization of the VAT tax control system in the context of the development of the digital economy in the Russian Federation. Financ. Bus. (3), 195–200 (2023)
3. Sekushin, A.: Digitalization and tax control: the experience of foreign administrations and the possibilities of its implementation in Russia. Taxes Tax. (3), 26–38 (2021)
4. Bellon, M. Dabla-Norris, M., Khalid, S., Lima, F.: Digitalization to improve tax compliance: evidence from VAT e-Invoicing in Peru. J. Public Econ. **210** (2022)
5. An official website of Chartered Accountants Worldwide https://charteredaccountantsworl dwide.com/digitalisation-tax-international-perspectives-2022-edition/, last accessed 2023/07/20
6. An official website of PWC https://www.pwc.com/mt/en/publications/vat/future-of-vat-digita lisation-is-here.html, last accessed 2023/07/20
7. An official website of the European Union https://taxation-customs.ec.europa.eu/taxation-1/ value-added-tax-vat/vat-digital-age_en, last accessed 2023/07/20
8. Zhukova, E.: Taxation of organizations: textbook, p. 66. KnoRus, Moscow (2021)
9. Aplin, P.: Digitalization of tax: international perspectives, https://www.icaew.com/-/media/cor porate/files/technical/digital-tax.ashx?la=en , last accessed 2023/07/06
10. Trenta, C.: European VAT and the digital economy, https://www.business.unsw.edu.au/About-Site/Schools-Site/Taxation-Business-Law-Site/Documents/imeTrenta-Re-thinking-EU-VAT-paper.pdf , last accessed 2023/07/06
11. Karpova, O., Maiburov, I.: Transformation of value added tax in the conditions of forced digitalization of the Russian economy. Bulletin of Tomsk State University. Economics **46**, 7–19 (2019)
12. Golik, E.: Digitalization of VAT administration in the system of ensuring economic security. Problems of the economy **2**, 103–109 (2021)
13. Geissbauer, R., Vedso J., Schrauf S. Industry 4.0: Building the digital enter-prise, https://www.pwc.com/gx/en/industries/industries-4.0/landing-page/industry-4.0-buildi ngyour-digital-enterprise-april-2016.pdf, last accessed 2023/07/06
14. OECD. Co-operative Tax Compliance: Building Better Tax Control Frame-works, https://doi. org/10.1787/9789264253384-en , last accessed 2023/07/06
15. Deeva, T.: Creation of digital platforms for the implementation of research and development in the field of tax security of the country: "end-to-end" digital technologies. Economy and society: modern development models **2**(11), 211–221 (2021)
16. Golova, E., Baranova, I.: Digitalization of tax processes as a way to increase the efficiency of the Russian tax system. Fundamental Research **4**, 30–34 (2021)
17. Official website of the Ministry of Finance of the Russian Federation, https://minfin.gov.ru/ru/ statistics/conbud/execute/?id_65=93449yezhegodnaya_informatsiya_ob_ispolnenii_konsoli dirovannogo_byudzheta_rossiiskoi_federatsiidannye_s_1_yanvarya , last accessed 2023/07/06
18. Official website of the Federal Tax Service of Russia, https://nalog.ru, last accessed 2023/07/06
19. Nikulina, E.: Analysis of problems and consequences of digital tax administra-tion, https://mcoip.ru/blog/2022/05/19/analiz-problem-i-posledstvij-czifrovogo-nalogovogo-administrirovaniya/, last accessed 2023/07/06
20. Maksimov, V., Oblatsova U.: The use of blockchain technology in the tax administration of the Russian Federation, http://edrj.ru/article/11-01-21, last accessed 2023/07/06

Methodology for Assessing the Effectiveness of Shortfalls in Revenues of the Country's Budget System on the Example of IT Companies: The Effectiveness of the Functioning of "Smart Government"

Irina A. Zhuravleva ⓘD and **Valentina V. Rzhevskaya** ⓘD

Abstract The dynamically developing tax legislation of the Russian Federation provides for a wide range of tax benefits aimed at achieving the goals of state programs of the Russian Federation or the goals of socio-economic development. Any tax break or relief, reduction of tax rates also affects tax revenues to the state budget system, forming tax expenses—shortfalls in state revenues. "Smart government" should fully evaluate both the effectiveness of the application of tax preferences by business and the loss of the budget, which the state as a whole receives as a result when granting tax breaks. In order to avoid unreasonable budget losses, it is necessary to timely and accurately assess the effectiveness of the tax preferences provided, however, at the moment, not all tax expenditures have been approved by Smart Government for appropriate methods. In this regard, the article proposes to consider a draft methodology for assessing the effectiveness of tax expenditures on new tax incentives provided for by tax law from 2021 for companies operating in the field of information technology. To identify the problems of effective assessment by "Smart Government" of this type of state expenditure. Currently, in Russia, special attention is paid to the digitalization of the economy, the development and application of the latest digital technologies by "Smart Government", and accordingly it is necessary to increase the competitiveness of the information technology industry, which is a supplier of digital products. The urgency of the problem confirms the need to assess the effectiveness of the provided preferences in order to obtain the multiplier effect of the tax benefit necessary for the Smart Government to use in its functioning.

Keywords Assessment of the effectiveness of tax expenditures · Digitalization · IT companies · Smart government · Tax expenditures

I. A. Zhuravleva (✉) · V. V. Rzhevskaya
Financial University Under the Government of the Russian Federation, Leningradsky Prosp., 49, 125167 Moscow, Russia
e-mail: sia.mir67@mail.ru

© The Author(s), under exclusive license to Springer Nature Switzerland AG 2024
M. Sari and A. Kulachinskaya (eds.), *Digital Transformation: What are the Smart Cities Today?*, Lecture Notes in Networks and Systems 846,
https://doi.org/10.1007/978-3-031-49390-4_10

145

1 Introduction

The concept of "Smart Government" can be characterized as a form of organizing the management activities of public authorities, the main feature of which is the use of the latest information technologies for electronic exchange, document management and information processing in the public administration sector. The country's tax service is a leading arsenal in terms of conducting Smart Government events, the adaptability of digital technologies and ways to improve emerging problems. The use of information and communication technologies is a necessary element of effective political management in the context of the rapid development of scientific developments in this area, which allow moving to a qualitatively new level of interaction between the government and society, optimizing the management of information flows within government departments [1]. The need for the effective functioning of "Smart Government" is also associated with the growing needs of users of public services, citizens and the business community in interactive services for interaction with the authorities, built on new generation digital technologies [1]. The Tax Service regularly replenishes and updates its Internet resources with new services that improve the exchange of information between taxpayers and thereby reduce the costs of tax control measures. The competence of the tax authorities also includes control over the legality of the application of tax preferences by various industries.

It is also relevant that the state is faced with the need to reduce costs when implementing digital tools of "Smart Government", trying to avoid the problems that other countries have in this area by conducting analysis and monitoring. The question is rhetorical: financing the development of "Smart Government" and reducing the cost of its implementation. In both cases, the source is taxes, the increase in their collection, savings on preferences/shortfalls in revenues of the country's budget system or tax expenditures.

To analyze current trends in the implementation of "Smart Government" tools, and the UN in its documents calls "electronic government", where a methodology was developed that is actively used by states to systematize digital, financial, economic and legal means in this area of development of government management. The E-Government Development Index, proposed by the UN Department of Economic and Social Affairs, has become one of the most authoritative indicators of the effectiveness of national policies in the field of digital and communication technologies. The 2014 UN Review Report defines e-government as "the use and application by public authorities of information technologies in public administration to optimize and integrate processes and procedures in order to effectively manage data and information, improve the quality of public service delivery and expand communication channels to involve people in political decision-making process."[2] According to the rating proposed in the study, Russia ranked 27th among 193 states in terms of e-government development [3].

In the tax policy of the Russian Federation for 2023–2025, the focus remains on improving the efficiency of the incentive function of the tax system with a concomitant increase in tax collection [4]. In this regard, the actual problem is the use by companies of the released cash flows in the company. On the one hand and on the other hand, what tools, mechanisms and methodology can be used by Smart Government.

It should be noted that in the scientific literature there are doubts about the appropriateness of the orientation of the UN member states to international meters to the detriment of the needs of society itself in public services [5]. This critical approach has its justified position: the presence of a direct relationship between the national specifics of political decision-making and the digital logic of the implementation of "Smart Government" in the public sector, the presence of vertical power leads to the mandatory and implementation of directives on the ground, in regions and municipalities, and only then is it taken into account opinion of users of citizens of the state and the business community. D. West, using the example of the United States as the parameters explaining the differences in the implementation of e-government by individual states, highlights their wealth and professionalism of legislators [6], in order to enrich the experience of other countries. The Spanish scientist Meyer A. considers in "Smart Government" the ability of management "to attract human capital in cooperation with the subjects of the object of smart management and the use of IT technologies" [7];

If we consider the idea of "Smart Government" from the standpoint of Russian scientists, then V.V. Lunev defines it as the concept of a "social multi-mind system" that combines a participatory organizational structure, a knowledge management system and a corporate culture based on trust and respect [8]. According to Kupryashin V. "Smart government" is considered as a system of cooperation between state, municipal, non-profit and business structures that ensure "the satisfaction of public interests and the solution of collective problems" [9].

According to the national development goals of the Russian Federation until 2030, the "Smart Government" provides for directions for stimulating sectors of the economy, which are reflected in the state programs of the Russian Federation. To achieve the relevant goals, it is envisaged, among other things, to provide tax benefits in the form of lower tax rates.

At the same time, lowering tax rates has a twofold impact on the financial and economic component of the state: the first is a decrease in income at all levels of the country's budget system, the second is the development of this industry. It should be noted that this influence gives its own positive, slightly protracted, effect of positive dynamics of tax revenues in the future. This is what Smart Government should count on when it comes to increasing efficiency. Thus, shortfalls in revenues of the budgets of the budget system of the Russian Federation, due to tax benefits, exemptions and other preferences, are defined by law as tax expenses [10].

Due to the growing number of tax benefits and preferences in the Russian tax legislation, the issues of assessing the effectiveness of tax expenditures are increasingly raised in the publications of Russian authors and need to be assessed by the Smart Government.

2 Methods

The topic of tax expenditures as a whole is quite common among scientific publications, and scientists present different opinions on what criteria should be used to determine tax expenditures. The current legislation applies the principle of compliance of tax expenditures with state programs of the Russian Federation, however, the scientific community does not allocate a sign of compliance of tax expenditures with state programs. And the "Smart Government" is faced with the task of attracting the scientific community, citizens and business to an adequate assessment of the use of tax preferences.

Russian researchers, in particular, T. Malinina, highlight the signs of tax expenditures, which include reducing government revenues, arising on the basis of preferential provisions of legislation regarding the basic structure of the tax, facilitating the implementation of non-fiscal policy goals, and the manifestation of an indirect measure of state support. At the same time I.A. Mayburov believes that the last two signs are not mandatory, and it is necessary to include in this list a sign of selectivity, meaning that tax expenditures are preferential provisions for certain groups of taxpayers [1].

The topic of tax expenditures as a whole is quite common among scientific publications, and scientists present different opinions on what criteria should be used to determine tax expenditures. The current legislation applies the principle of compliance of tax expenditures with state programs of the Russian Federation, however, the scientific community does not allocate a sign of compliance of tax expenditures with state programs.

Thus, it is noted that the assessment of tax expenditures in the context of compliance with state programs is an acceptable criterion, but not fundamental, since part of tax expenditures may not be taken into account in the assessment, or their effectiveness is not confirmed, due to non-compliance with state programs.

G.E. Karataeva and S.V. Chernova also proposed a number of methods for improving the system of assessing the effectiveness of tax expenditures, providing not only an assessment based on statistical data and mathematical calculations of the forecast, but also less formal methods, such as clarifying the opinion of recipients of tax benefits on the introduction, necessity and expediency of maintaining tax benefits [1].

At the same time, please note that the Decree of the Government of the Russian Federation No. 439 dated April 12, 2019 "On Approval of the Rules for the Formation of the List of Tax Expenditures of the Russian Federation and Assessment of Tax Expenditures of the Russian Federation" (hereinafter, respectively, Resolution No. 439, Rules No. 439) and in the Explanations on the Application of the Provisions of Rules No. 439 published on the official website the website of the Ministry of Finance of the Russian Federation provides for the consideration of these problems.

The List of tax expenditures of the Russian Federation (as of 04.07.2023) reflects not only tax expenditures corresponding to state programs, but also unallocated under state programs [3]. Such tax expenditures correspond to the goals of socio-economic policy reflected in strategic planning documents or decrees of the President of the Russian Federation defining the national development goals of the Russian Federation. Smart government needs to efficiently assess these financial flows in order to increase budget savings.

The assessment of the demand for benefits is not limited to certain methods and can be a calculated indicator, while the procedure for calculating it should be indicated in the methodology for assessing the effectiveness of tax expenditures. In addition, if the tax expense does not meet the criterion of demand, the curator of the tax expense should submit to the Ministry of Finance of the Russian Federation proposals on maintaining (clarifying, canceling) the benefits that cause the tax expense, which, in turn, can be stated taking into account the opinion of the direct recipients of tax benefits. The "Smart Government" is faced with the problem of systemic organization of the hierarchy of accounting for shortfalls in revenues of the country's budget system.

Although the legislation provides for a mechanism for the formation of methods for assessing the effectiveness of tax expenditures. However, a big problem is the lack of developed and approved methods for a number of tax expenditures due to the dynamic change in legislation in accordance with the economic and geopolitical situation in the country and in the world as a whole.

Currently, in the Russian Federation, special attention is paid to the digitalization of the economy, the development and increase in the competitiveness of the information technology industry by reducing the tax burden on those involved in the IT industry. Currently, in the Russian Federation, special attention is paid to the digitalization of the economy, the development and improvement of the competitiveness of the information technology industry by reducing the tax burden of companies involved in this area (hereinafter referred to as IT companies).

The provision of tax preferences for IT companies provides for the establishment of lower relatively generally applicable tax rates, which reduces the amount of taxes paid. IT companies can use the funds released in this way to invest in the development of new products, increase employee salaries, expand the scope of the organization's activities and strengthen its position in the market.

From 2021, the Russian Federation provides for a number of new tax preferences for companies in the IT industry, which is defined as tax expenses—shortfall in revenues of the budgets of the budget system of the Russian Federation—and is included in the List of tax expenses of the Russian Federation (as of 07/04/2023), corresponding excerpts from which are presented in Table 1 [3].

So, at present, there is no accepted methodology for estimating tax expenditures related to benefits for the IT industry, which is curated by the Ministry of Digital Development, Telecommunications and Mass Media. We assume that it is this methodology that will allow tracking important indicators of the introduction of tax incentives for the IT industry, as well as determining the subsequent need for their adjustment, expansion or addition.

The proposed draft methodology for assessing the effectiveness of tax incentives for IT companies presented in Table 1 will be described below, based on Resolution

Table 1 New tax preferences for IT companies

No.	Short name of the tax expense of the Russian Federation	Categories of payers for whom tax preferences are provided	Target category	Name of the state program of the Russian Federation	Curator of the tax expense of the Russian Federation
1.	Reduced income tax rate for Russian organizations operating in the field of information technology	Legal entities	Stimulating	Information Society	The Ministry of Digital Development, Communications and Mass Media of the Russian Federation
2.	Reduced insurance premium rates for IT companies				

No. 439, as well as Clarifications on the application of the provisions of Rules No. 439 and on the basis of the List of tax expenses of the Russian Federation.

Let's conduct a hypothetical assessment of the expediency of tax expenditures by determining the compliance of the above tax expenditures with the goals of the state programs of the Russian Federation and (or) the goals of the socio-economic policy of the Russian Federation, which can be used by "Smart Government". The result of the conformity assessment is shown in Table 2.

Next, we give a formula for assessing the demand for the benefits provided by payers (1).

$$W = \frac{\sum_{i=1}^{5} \frac{n_{used}}{N_{all}}}{5} \tag{1}$$

Table 2 Assessment of compliance of tax expenditure with the goals of state programs in the application of "Smart Government"

No.	Short name of the tax expense of the Russian Federation	Target category	Name of the state program of the Russian Federation	Name of the purpose of the state program of the Russian Federation
1.	Reduced income tax rate for Russian organizations operating in the field of information technology	Stimulating	Information Society	An increase in investments in domestic solutions in the field of information technology by four times compared to the indicator of 2019, (%)
2.	Reduced insurance premium rates for IT companies			

where:

W—the demand for the benefits provided by the payers, %;

i—the ordinal number of the year, which has a value from 1 to 5;

n_{used}—the number of payers (units) who have used the benefit; is determined based on the available data from the curator of tax expenditures or by requesting data to the relevant departments in case of insufficient data from the curator of tax expenditures.

N_{all}—the total number of payers eligible for the benefit (units); the source of information is the Register of accredited IT Companies, data from relevant departments on compliance by IT companies with the criterion of receiving at least 70% of income according to tax accounting from certain IT activities.

Demand is characterized by the ratio of the number of taxpayers who have used the right to benefits and the total number of taxpayers who are entitled to benefits over a 5-year period. If the tax benefit is valid for less than 5 years, then the assessment of its demand is carried out for the actual and forecast periods of the benefit, the amount of which is 5 years (for example, if the benefit is introduced in 2021, then the assessment of demand is carried out for 2021–2025). Formula (1) is applied separately for each of the tax expenses.

The threshold value of demand is proposed to be set as 70%. The value is taken in connection with the need to stimulate companies in the industry by using these tax benefits. A lower threshold value the threshold value will indicate that a smaller number of companies have taken advantage of these benefits, and their focus, economic feasibility for the entire industry will be under doubt. When assessing the effectiveness of tax expenditures, the threshold value at which the benefit is in demand should not be less than 70% of the total number of payers.

The next stage is to assess the contribution of the tax benefit to the change in the value of the indicator (indicator) of achieving the goals of the state program of the Russian Federation and (or) the goals of the socio-economic policy of the Russian Federation and the application of the results by Smart Government. This stage is the main one as an indicator of the essence of tax expenditure, since tax expenditures are aimed at achieving the goals of state programs. Table 3 presents indicators and sources of information about them on relevant tax expenditures necessary to assess the contribution to achieving the goals of state programs.

Calculation of values of indicators (indicators) without taking into account the benefits, it is proposed to be calculated by constructing a regression model based on data from indicators (indicators) for the years before the introduction of tax benefits—the period up to 2020 inclusive.

The contribution of the tax expense to achieving the actual value of the target indicator in year i is generally calculated as the difference between the value of the target indicator with benefits and the value of the target indicator without benefits, which is represented by the generalized formula (2). The formula is applied separately for each of the tax expenditures. Based on this formula, the formulas for assessing the contribution for each of the indicators are given below.

Table 3 List of indicators and sources of information

No.	Short name of the tax expense of the Russian Federation	Name of the indicator (indicator)	Name of the purpose of the state program of the Russian Federation
1.	Reduced income tax rate for Russian organizations operating in the field of information technology	Number of accredited IT companies	Register of accredited IT companies
2.	Reduced insurance premium rates for IT companies	Wages of employees in the relevant industry, thousand rubles	Federal State Statistics Service

$$Q_i = V_{iwith} - V_{iwithout} \tag{2}$$

where:

Q—contribution of the tax expense to the achievement of the actual value of the target indicator;

i—the ordinal number of the year, which has a value from 1 to 5;

V_{iwith}—the value of the indicator (indicator), taking into account the application of the benefit;

$V_{iwithout}$—the value of the indicator (indicator) excluding benefits.

3   Results

Table 4 provides a breakdown of the received value of the contribution of the tax expense to achieving the value of the target indicator for drawing conclusions about the contribution of the corresponding tax expense to achieving the goals of state programs.

Table 4 The value of the contribution of the tax expense to the achievement of the target indicator

No.	Name of the indicator (indicator)	The expected trend in the successful application of benefits	Contribution of tax expense (Q)		
			Q > 0	Q = 0	Q < 0
1.	Number of accredited IT companies	Increasing the number of accredited IT companies	Positive contribution of tax expense	There is no tax expense contribution	Negative contribution of tax expense
2.	Wages of employees in the relevant industry, thousand rubles	Wage growth of employees in the relevant industry	Positive contribution of tax expense	There is no tax expense contribution	Negative contribution of tax expense

The contribution level for indicator No. 1 "Number of accredited IT companies, units" is calculated according to the following formula (3):

$$Q_i = V_{iwith} - V_{iwithout} \tag{3}$$

where:

Q—contribution of the tax expense to the achievement of the actual value of the target indicator;

i—the ordinal number of the year, which has a value from 1 to 5;

V_{iwith}—actual accredited IT companies (taking into account the application of benefits), units.

$V_{iwithout}$—projected number of accredited IT companies, units.

The calculation of the contribution level for indicator No. 2 "Wages of employees in the relevant industry, thousand rubles" is made according to the formula (4):

$$Q_i = V_{iwith} - V_{iwithout} \tag{4}$$

where:

Q—contribution of the tax expense to the achievement of the actual value of the target indicator;

i—the ordinal number of the year, which has a value from 1 to 5;

V_{iwith}—the actual average salary of employees in the IT industry (taking into account the application of benefits);

$V_{iwithout}$—projected average salary of IT industry employees (excluding the application of benefits).

The grouping of methods and procedures for calculating the values of the contribution level of indicators (indicators) with and without benefits is presented in Table 5.

Next, an assessment of budget efficiency is needed. In order to assess the budgetary efficiency of tax expenditures, a comparative analysis of the effectiveness of the provision of benefits and the results of the use of alternative mechanisms to achieve the goals of the state program and (or) the goals of socio-economic policy is carried out. The comparative analysis includes a comparison of the volume of federal budget expenditures in the case of alternative mechanisms for achieving the goals of the state program and (or) the goals of socio-economic policy, and the volume of benefits provided.

To assess budget efficiency, it is necessary to compare the amount of tax expenditure and the amount of federal budget funds for the use of an alternative mechanism with identical values of the indicator (indicator) in the case of granting a benefit and in the case of using an alternative mechanism.

Table 5 Grouping of methods and procedures for calculating the values of the contribution level of indicators (indicators) with and without benefits

No.	Name of the indicator (indicator)	The formula for calculating the contribution level	The procedure for calculating the value of the indicator (indicator) without taking into account the benefit
1.	Number of accredited IT companies	$(3) Q_i = V_{iwith} - V_{iwithout}$	Regression model of forecasting the number of IT companies based on data on the number of accredited IT companies for the period before the introduction of tax benefits (up to 2020 inclusive)
2.	Wages of employees in the relevant industry, thousand rubles	$(4) Q_i = V_{iwith} - V_{iwithout}$	Regression model of the forecast based on data on the average salary of the IT industry for the period before the introduction of tax benefits (up to 2020 inclusive)

The calculation of the budget efficiency assessment is presented on the formula (5).

$$B_j = A_{jt} - C_{jt} \qquad (5)$$

where:

B_j—budget efficiency for the j-th tax expense;

A_{jt}—the amount of federal budget funds (taking into account the funds provided for the administration of state support measures) sent to recipients under the support mechanism alternative to the j-th benefit, which ensured the achievement of the t-th indicator in the reporting year;

C_{jt}—the total amount of tax benefits for the j-tax expense received by payers who ensured the achievement of the t-indicator (indicator) in the given year.

If $B_j > 0$, then it can be concluded that tax expenditures spend less federal budget funds than alternative mechanisms to achieve the same indicators of achieving the goals of state programs; and budget efficiency is positive for tax expenditures.

If $B_j < 0$, then it can be concluded that there are alternative mechanisms for achieving the goals of state programs that spend less federal budget funds for the same result, which is achieved by a more expensive method—tax benefits [12].

If $B_j = 0$, or its value is close to 0, then we can conclude that the budgetary efficiency of alternative mechanisms and tax expenditures is equal.

Thus, it is necessary to take into account that tax expenditures and alternative mechanisms are often used simultaneously to enhance overall budget efficiency and accelerate the achievement of the results of government programs and other favorable impacts on the industry.

Such alternative mechanisms include grant support for domestic IT companies, a moratorium on inspections, preferential lending, etc.

Grant support has been expanded by Government Resolutions No. 598, No. 599, and No. 601 of April 6, 2022. These resolutions regulate changes in the rules for granting subsidies to Russian foundations that provide grants to IT companies—grants can cover 80% of the cost of projects. Such measures are aimed at import substitution, reducing the negative consequences of sanctions and preventing the outflow of specialists abroad.

Preferential lending for IT businesses is also aimed at implementing a digital transformation project and involves limiting the size of the loan rate issued by banks to startups in the field of information technology. For example, Ak Bars Bank offers the following conditions: interest rate up to 5% per annum or preferential commission (remuneration) for discounting; currency—rubles of the Russian Federation; loan term—up to 31.12.2024; minimum amount—5 million rubles, maximum amount—5 billion rubles. VEB.RF offers similar conditions, but the minimum amount is 500 million rubles.

Table 6 shows the correlation of alternative mechanisms and tax benefits.

So, at present, there are enough alternative mechanisms that are used simultaneously with tax expenses for IT companies.

The assessment of the total budget effect (self-sufficiency) of tax expenditures is made according to the target expenditures of the incentive category, to which the considered tax expenditures belong separately for each tax expense. Self-sufficiency assessment is carried out in accordance with paragraphs 19–21 of Regulation No. 139. The assessment of the total budget effect is aimed at determining the payback of tax expenditures by increasing tax revenues to the budgets of the budgetary system of the Russian Federation from persons who have applied tax benefits that cause tax expenditures.

According to paragraphs 19–21 of Regulation No. 439, the assessment of the total budget effect (self—sufficiency) for the period from the beginning of the relevant benefits for payers or for 5 reporting years, and if these benefits are valid for more than 6 years,—on the date of the assessment of the effectiveness of tax expenditures of the Russian Federation (E) according to the following formula (6):

Table 6 Comparison of indicators (indicators) for achieving the goals of state programs and alternative mechanisms

No.	Short name of the tax expense of the Russian Federation	Name of the indicator (indicator)	Name of the alternative mechanism
1.	Reduced income tax rate for Russian organizations operating in the field of information technology	Number of accredited IT companies	Grant support for domestic IT companies
2.	Reduced insurance premium rates for IT companies	Wages of employees in the relevant industry, thousand rubles	

$$E = \sum_{i=1}^{5} \sum_{j=1}^{m_i} \frac{N_{ij} - B_{0j} \times (1 + g_i)}{(1 + r)^i} \qquad (6)$$

where:

i—the ordinal number of the year, which has a value from 1 to 5;

m_i—the number of payers who took advantage of the benefit in the i-th year;

j—the payer's serial number, which has a value from 1 to m;

N_{ij}—the volume of taxes, fees, customs payments and insurance contributions for compulsory social insurance declared for payment to the budgets of the budgetary system of the Russian Federation by the j-payer in the i-year.

If, as of the date of the assessment of the total budget effect (self-sufficiency) of the stimulating tax expenditures of the Russian Federation for payers entitled to benefits, the benefits are valid for less than 6 years, the amounts of taxes, fees, customs payments and insurance contributions for compulsory social insurance payable to the budgets of the budgetary system of the Russian Federation, are estimated (projected) according to the data of the curators of tax expenditures and the Ministry of Finance of the Russian Federation;

B_{0j}—the basic amount of taxes, fees, customs payments and insurance contributions for compulsory social insurance declared for payment to the budgets of the budgetary system of the Russian Federation by the j-payer in the base year;

g_i—nominal growth rate of budget revenues of the budgetary system of the Russian Federation in the first year in relation to the base year.

The nominal growth rate of revenues of the budgets of the budgetary system of the Russian Federation from taxes, fees, customs payments and insurance contributions for compulsory social insurance to the budgets of the budgetary system of the Russian Federation in the current year, the next year and the planning period is determined based on the real growth rate of gross domestic product according to the forecast of socio-economic development of the Russian Federation for the next financial year and the planned period, laid down in the basis of the federal law on the federal budget for the next financial year and the planned period, and also from the target inflation rate determined by the Central Bank of the Russian Federation for the medium term.

r—the estimated cost of medium-term market borrowings of the Russian Federation, assumed at 7.5 percent.

The basic amount of taxes and insurance contributions for compulsory social insurance declared for payment to the budgets of the budgetary system of the Russian Federation by the jth payer in the base year (B_{0j}) is calculated by the formula (7):

$$B_{0j} = N_{0j} + L_{0j} \qquad (7)$$

where:

N_{0j}—the amount of taxes and insurance contributions for compulsory social insurance declared for payment to the budgets of the budgetary system of the Russian Federation by the j-payer in the base year;

L_{0j}—the amount of benefits granted to the j-th payer in the base year.

The base year is the year preceding the year when the j-payer started receiving benefits.

4 Discussion

The issue of applying a hypothetically defined mechanism of the totality of the budget effect in order to increase the efficiency of the functioning of the "Smart Government" and the rational use of tax preferences in the industry directly related to the production of digital products is debatable.

Prospects for the further implementation of "Smart Government" in the public sector were outlined in the System Project for the Development of Electronic Government in the Russian Federation until 2020 [11], the development of which began in 2013. Although the document was not approved by 2017, its provisions allow assessing existing plans in the area under consideration for the effective application of tax preferences. The following can be identified as the problems of the effective functioning of "Smart Government" in Russia:

- the presence of organizational and technical obstacles and problems;
- imperfection of the legislative base and its complementarity between different authorities;
- lack of ease of understanding and functionality for the active participation of the business community in order to evaluate the results achieved, identify problems and proposals for their solution;
- the fundamental principles in taxation and synergy with "Smart Government: "zero paperwork", "zero officials", "zero interaction problems" have not been resolved;
- postulating the universal availability of electronic services as one of the key priorities, including the possibility of applying and justifying tax preferences for companies in the IT industry;
- the presence of a fragmented nature in relation to the methodology for assessing tax preferences for companies in the IT industry, which does not correspond to its designation as a systematic approach;
- despite the fact that one of the main tasks is to involve civil society and business in the processes of public administration, the System Project itself does not provide specific indicators by which it would be possible to assess the success of its solution, and even more so an assessment of shortfalls in state revenues in connection with the provision tax preferences, their effective evaluation.

5 Conclusions

In the final part of the methodology for assessing the effectiveness of tax expenditures, conclusions are needed about achieving the target characteristics, the contribution of the tax expenditure of the Russian Federation to achieving the goals of the state program of the Russian Federation, as well as about the presence or absence of more effective (less costly for the federal budget) alternative mechanisms for achieving the goals of the state program.

The format of performance evaluation results in the form of:

1. Conclusions (analytical note) containing conclusions based on the results of the assessment of the effectiveness of the tax expense;
2. Recommendations based on the results of the assessment made in case it is necessary to preserve, clarify or cancel the tax expense on certain grounds.

Thus, as a result of the conducted research, it can be concluded that at present there are many preferences for IT companies related to both taxation and other ways to develop this industry, affecting, among other things, the achievement of the goals of government programs. At the same time, it is necessary to carry out a timely assessment of tax expenditures in order to assess their contribution to achieving the relevant goals and the immediate need in comparison with the use of alternative methods in the form of grants, subsidies and other measures of structural elements of state programs aimed at supporting the IT industry.

Thus, we consider it expedient to timely introduce appropriate methods for assessing the effectiveness of tax expenditures by all supervisors of tax expenditures in order to obtain the most complete information about the effectiveness of tax expenditures of the federal budget and the subsequent possibility of improving tax legislation.

References

1. Vaslavsky, Y., Gabuev, S.: Options for the development of e-government. Experience of Russia, USA, China. Int. Process. **15**(1), 108–125 (2017)
2. United Nations E-Government Survey 2014: E-Government for the Future We Want. P. 2. URL: https://publicadministration.un.org/egovkb/Portals/egovkb/Documents/un/2014-Survey/E-Gov_Complete_Survey-2014.pdf
3. United Nations E-Government Survey 2014: E-Government for the Future We Want. P.2. URL: https://publicadministration.un.org/egovkb/Portals/egovkb/Documents/un/2014-Survey/E-Gov_Complete_Survey-2014.pdf
4. The main directions of budget, tax and customs tariff policy for 2023 and for the planning period of 2024 and 2025 (approved by the Ministry of Finance of the Russian Federation), https://minfin.gov.ru/common/upload/library/2022/11/main/2023-2025.pdf. Date of access 2023/07/08
5. Trutnev, D.R.: Influence of international ratings on the content of e-government development plans. Information systems for scientific research: Tr. XV All-Russian. merged. conf. Internet and modern society. SPb.: SPb NRU ITMO (2012)

6. West Darrell, M.: Digital Government: Technology and Public Sector Performance. Princeton University Press, Princeton, NJ (2005)
7. Meijer, A., Rodríguez, M. P.: Governing the smart city: a review of the literature on smart urban governance. Int. Rev. Adm. Sci. **82**(2), 392–408 (2016)
8. Lunev, V.V., Luneva, T.A., Modestov, F.A., Rakhinsky, D.V.: Smart management and organizational development. Mod. Sci.: Actual Probl. Theory Practice. Ser.: Econ. Law. **9**, 43–47 (2020)
9. Kupryashin, G.L.: Public Administration. Political Science. No. 2: Political science in modern Russia/ed.-comp, pp. 101–131. O.V. Gaman-Golutvina (2020)
10. Budget Code of the Russian Federation, article 6 https://www.consultant.ru/document/cons_doc_LAW_19702/055a71948dbf2a4fc2478437cd89cd864ee8e6e5/. Accessed 08.07.2023
11. System project of the electronic government of the Russian Federation. https://digital.gov.ru/uploaded/files/sistemnyii-proekt-elektronnogo-pravitelstva-rf.pdf. Date of access: 01/08/2023
12. Kudelich, M.: Problems of innovative science and technology center financing from budgetary funds. Financ. J. **13**, 76–87 (2021)

Development Trends of E-government and Public Finance Management in the Context of Digitalisation

Svetlana Demidova⑩, Anastasia Kulachinskaya⑩,
Viktoriya Razletovskaia⑩, and Stanislav Svetlichnyy

Abstract This chapter examines the key trends in the development of e-government based on international ratings. It analyses the state information systems of public finance management, which provide input to the development of the concept of "state as a platform". The information and technological pool of financial management and provision of public services is presented based on the Russian experience. The role of ICT in improving public finance management is described, the directions of technological modernisation are defined, risks and limitations are identified. Based on their analysis, a conclusion is made about trends in the development of government information systems and the need to move to a unified national platform for data interfacing. Real government platforms are still in the stage of formation. Transition to a single national platform will increase the availability of programme codes for all levels of public authorities, ensure transperancy and clarity of interaction between public authorities and citizens, ensure uniform standards of security of placement and reliability of published data, eliminate data incompatibility, reduce budget expenditures, in particular, for the integration of systems between each other and maintenance. It is proposed to create a unified budget platform, the approach to the formation of which can utilize the experience obtained in the development of a unified platform "Gostech".

Keywords E-government · State information systems · Digital transformation · Digital platform

S. Demidova
Financial University Under the Government of the Russian Federation, Moscow, Russia

A. Kulachinskaya (✉)
Peter the Great St. Petersburg Polytechnic University, St.Petersburg, Russia
e-mail: a.kulachinskaya@yandex.ru

V. Razletovskaia
MGIMO University, Moscow, Russia

S. Svetlichnyy
Moscow Technological Institute, Moscow, Russia

1   Introduction

The digital transformation of government activities is based on two fundamental concepts. The first concept or doctrine is New Public Management (NPM), within the framework of this concept, the objectives of methods and main directions of activity of public authorities are aimed at meeting the interests of society as a whole and the needs of citizens through optimisation and increasing the efficiency of public administration [1]. The second concept—the concept of e-government—characterises the form of organisation of public authorities' activity, the distinctive feature of which is the application of technologies of electronic exchange and processing of information in the public administration sector, flexibility and transparency of state processes [2, 3]. ICT development changes the consciousness of governments and citizens' expectations regarding the efficiency of public services [4, 5].

The global trend of changes generally corresponds to the model of "state for citizens", which is based on the principles of omnichannel delivery of public services, proactivity, and human-centredness. Technological contribution to the achievement of these principles is provided by digitalisation of public administration, the use of big data processing technologies, smart technologies, blockchain, artificial intelligence. At the same time, the creation of e-government requires a high level of trust on the part of citizens, which is ensured by transparency and accountability and guarantees of security of data and operations [6].

The theory of innovation distinguishes digital innovations associated with the ideology of "platforms". "Digital platform" as a conceptual category still needs to be clarified, but its main characteristics are already highlighted: the presence of digital infrastructure, functioning according to common rules, openness [7, 8]. In Russian practice, this approach formed the basis for the creation of the concepts of "e-state" and "state as a platform" [9]. According to experts' estimates, the creation of such a platform can reduce the level of expenditures on public expenditures by 0.3% of GDP by 2024. "Digital platforms" allow to implement such innovative approaches in the provision of state and municipal services and the performance of functions by the authorities as a "single window" and the principle of client-centredness.

E-government sets the outlines of the digital architecture, within the contour of which business processes related to public finance management are realised. The potential of digitalisation for the transformation of public finance management is very high and meets expectations [10, 11]. A new paradigm of "digital public financial management" is emerging, based on a rethinking of how digital technologies and public financial management intersect [12]. Digital public financial management is seen as a set of means to achieve the goal, and business processes (and the digital solutions that underpin them) remain flexible to respond to the needs of users [13, 14].

The purpose of this paper is to study the Russian experience of e-government development and public finance management in the context of technological transformation.

2 Materials and Methods

The main research method is a systematic review of the formation of information systems of public finance management. At the first stage, strategies, initiatives, problems of e-government, implementation of platform solutions were studied through the research of scientific publications, legislation and official websites. The selection of documents and articles was based on the relevance of the research topic. At the second stage, the digital system of public finance management is described using the example of the Russian experience. The matrix of evaluation of state information systems for the transition to a single centralised platform is proposed. The most urgent problems, risks and directions of development of digitalisation of the public sector are highlighted and discussed.

3 Results

Countries leading in e-government development can be examined through the EGDI ranking [15]. These countries are implementing a platform approach to public policy in the field of digital technologies (Table 1).

Denmark has been ranked higher than any other country in the world for three consecutive years and is leading top eight European countries as well as top six EU countries. In 2022, Iceland joined the top ten countries. Europe accounted for the largest share of the top-rated countries, totaling 53%. This impressive figure includes Denmark, Estonia, Finland, Iceland, Malta, the Netherlands, Sweden, and

Table 1 E-government rating EGDI

Country	EGDI			E-government ranking		
	2020	2022	2022/2020	2020	2022	2022/2020
Denmark	1	1	0	0.9758	0.9717	−0.0041
Finland	4	2	2	0.9452	0.9533	0.0081
Republic of Korea	2	3	−1	0.956	0.9529	−0.0031
New Zealand	8	4	4	0.9339	0.9339	0
Sweden	6	5	1	0.9365	0.941	0.0045
Iceland	12	6	6	0.9101	0.941	0.0309
Australia	5	7	−2	0.9432	0.9405	−0.0027
Estonia	3	8	−5	0.9473	0.9393	−0.008
Netherlands	10	9	1	0.9228	0.9384	0.0156
USA	9	10	−1	0.9297	0.9151	−0.0146
Great Britain	7	11	−4	0.9358	0.9138	−0.022
Russia	36	42	−6	0.8244	0.8162	−0.0082

the UK. Asia follows closely behind, contributing 27% with highly esteemed countries such as Japan, Korea, Singapore, and the UAE. Oceania makes up 13% of the top-rated countries, represented by Australia and New Zealand. Lastly, the America is responcible for 7%, with the USA leading the way.

Denmark, Finland, South Korea received the highest scores in terms of the volume and quality of online services, state of telecom infrastructure and human resource capacity. Due to notable improvements in telecom infrastructure and human potential development, the average EGDI in the world has increased—69% of UN states have a high to very high EGDI score.

Russia is not among the top 10 countries in terms of e-government development, although the country has implemented and operates similar platforms such as the Gosusgi services portal, the UIS public procurement portal, and the Unified Identification and Authentication System (UISA), which works as an "electronic passport" on various government resources. Denmark also has national portals: civic portal; business portal; health portal.

At the same time, according to the level of digitalisation of both government and public services GovTech Maturity Index (GTMI) [16], published annually by the World Bank, Russia ranked 10th out of 198. GTMI measures four areas: "Core Government Systems" (17 indicators), "Provision of Public Services" (9 indicators), "Digital Interaction of Citizens" (6 indicators) and "Institutional Support", which covers strategy, laws and regulations, the degree of IT implementation in public services, as well as programmes to develop new digital transformation projects (16 indicators).

The benchmarks of digital transformation in Russia were defined by the Strategy for the Development of the Information Society in the Russian Federation for 2017–2030, including goals, objectives, and measures in the field of information and communication technologies aimed at developing the information society and forming a national digital economy. To implement the Strategy, the following were adopted: the state programme "Digital Society", the national programme "Digital Economy of the Russian Federation", and the federal project "Digital Public Administration". An indicator of the national goal "Digital Transformation" is to increase the share of socially important services available in electronic form to 95 percent by 2030. State authorities are implementing departmental digital transformation projects.

To assess the openness and accessibility of information on public finances, the international practice uses the budget openness index; the data for Russia show significant progress in this area and substantial accessibility of information for external users (see Fig. 1).

The model of the public finance management system implemented in Russia belongs to the treasury-oriented group. Which makes treasury responsible primarily for processing budget payments. Since 2022, the Interregional Directorate for Centralised Data Processing has been functioning within the structure of the Russian Treasury, where a Data Laboratory and an Artificial Intelligence Laboratory have been established.

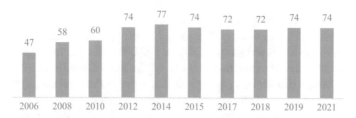

Fig. 1 Russia's openness index for the period 2006–2021 according to the International Budget Partnership (Open Budget Survey 2021) [17]

The volume of databases of all information systems of the Federal Treasury (9 GIS) in 2021 exceeded 5 PB (1 petabyte = 1015), and the volume of databases is constantly growing, on average by 0.5 PB per year [18].

The most important state information systems for public finance management in Russia include: State Integrated Information System for Public Finance Management "Electronic Budget" (GIIS "Electronic Budget")—budget.gov.ru; EIS—zakupki.gov.ru; State Information System on State and Municipal Payments (GIS GMP); State Information System of State and Municipal Institutions (GIS SMI)—bus.gov.ru; State Automated System (GAS "Management")—gasu.gov.ru.

GIIS "Electronic Budget" supports more than 500 thousand users and more than 700 end-to-end business processes of public finance management and project activities [19]. It is critical for the functioning of the entire budget system of the country. The operators of the GIIS "Electronic Budget" are the Ministry of Finance of the Russian Federation and the Federal Treasury. The following are required to register in the system and work within: state authorities, local self-government bodies, state extra-budgetary funds; participants of the budget process; legal entities receiving budget funds; organisations engaged in procurement activities.

Advantages of the GIIS "Electronic Budget": saving budgetary funds and increasing the efficiency of budget expenditures due to unification of procedures and reduction of their runtime, discontinueation of paper document flow (in favor of the digital one), reduction of expenditures on IT-specialists, ability to effectively control the allocation of budgetary funds. The principle of openness and accessibility has been implemented through an open part—the Unified Budget System Portal (UBSP). The portal functions as a single access point to the system for various categories of users to work in functional subsystems. For each category of users (the Ministry of Finance of the Russian Federation, the Federal Treasury, authorities, control and supervisory bodies), a separate personal account is provided, accessible via a mobile application. Information in analytical reports is presented in the form of tables, diagrams, cartograms, indicators, mind maps, Pareto diagrams, SWOT analysis and other representations.

Initially, Oracle solutions1 were widely used in the architecture of GIIS "Electronic Budget". And all state information systems worked with the Windows operating system. The implementation of import substitution plans is associated, among other things, with changing over to domestic software. Since 2021, the GIIS "Electronic Budget" started to work on the Astra Linux operating system, the developer of which is the Russian company OTR2. In 2022, the subsystem of state debt and financial assets management (maintenance of the State Debt Book, planning and accounting of operations of borrowing and placement of financial assets) was moved from Oracle to import-substituting technologies with open data.

GIIS "Electronic Budget" is integrated with the Unified Information System, which made it possible to effectively embed procurement processes into the public finance management system. Execution of customer functions, management and control of public procurement is carried out with the participation of the Unified Information System (UIS) (see Fig. 2).

Compiling documents for placement in the Unified Information System is carried out by federal customers in the procurement management subsystem of the Electronic Budget. Data from the library of standard contracts, as well as information on market prices for goods, works and services, are fed from the Unified Information System to

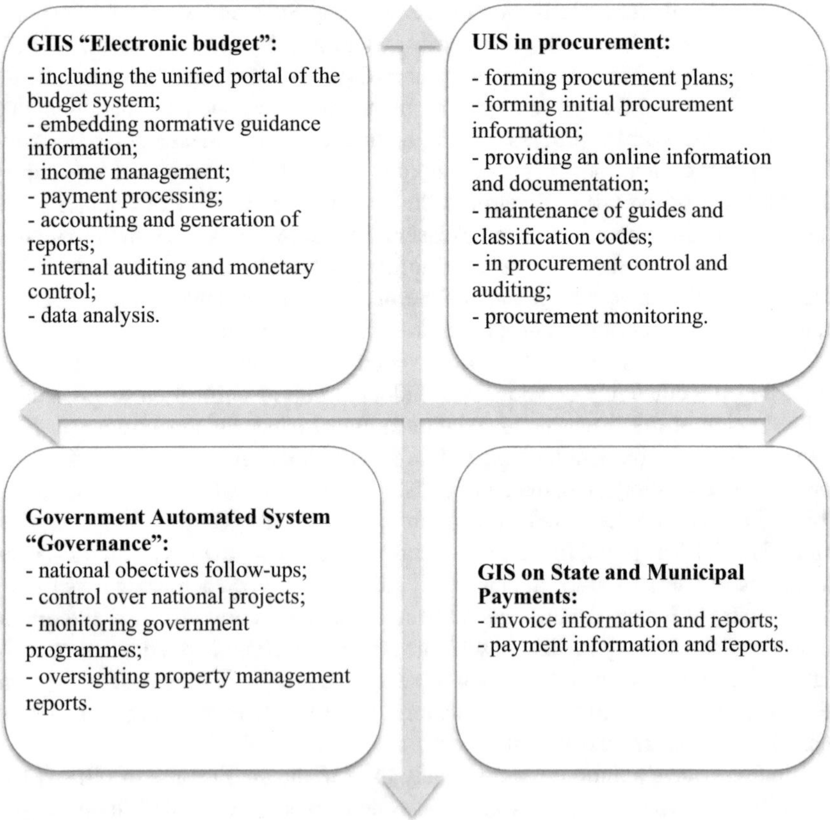

GIIS "Electronic budget":
- including the unified portal of the budget system;
- embedding normative guidance information;
- income management;
- payment processing;
- accounting and generation of reports;
- internal auditing and monetary control;
- data analysis.

UIS in procurement:
- forming procurement plans;
- forming initial procurement information;
- providing an online information and documentation;
- maintenance of guides and classification codes;
- in procurement control and auditing;
- procurement monitoring.

Government Automated System "Governance":
- national obectives follow-ups;
- control over national projects;
- monitoring government programmes;
- oversighting property management reports.

GIS on State and Municipal Payments:
- invoice information and reports;
- payment information and reports.

Fig. 2 State information systems in the field of financial management and their main functions

the E-Budget system. For the purposes of budgetary control, procedures are provided for checking the sufficiency of appropriations at the time of contract conclusion and the process of payment authorisation is automated. Payment for completed government contracts is made within 7–10 days (from the date of signing the acceptance act in electronic form), and in the short term it is expected to be reduced to 5 days. This is one of the best indicators in the world practice and became possible due to the introduction of electronic document flow at all stages of the procurement process. Statistical indicators on placed contracts show relative savings in contracting for the period 2019–2022. 6,5% [20].

The scheme of electronic interaction between GIS EIS and GIIS "Elektronny Budget", the Central Bank and commercial banks is presented in the figure below (see Fig. 3).

In order to shorten payment terms in respect of State contracts, orders to make treasury payments will be generated in the Unified Treasury Payment System (UTS).

The Russian Treasury is responsible for the policy of liquidity management on the Single Treasury Account (STA). Artificial intelligence algorithms are being developed to forecast account balances. This refers to treasury accounts and personal accounts opened in the GIIS "Electronic Budget". At the same time, the STA is opened in the Bank of Russia. The efficiency of the model depends on the number of criteria and attributes, for this purpose the Treasury began to use stock exchange indices, interdependencies between transactions within the budget process.

The GIS GMP is an example of technological innovation in the field of budget payments. Budget revenue administrators, organisations involved in the provision of state and municipal services must connect to the GIS GMP through the system of interdepartmental electronic interaction (SIEI). The participant then obtains a key certificate for an enhanced qualified electronic signature from an accredited certification centre. The number of credit organisations interacting with the GIS

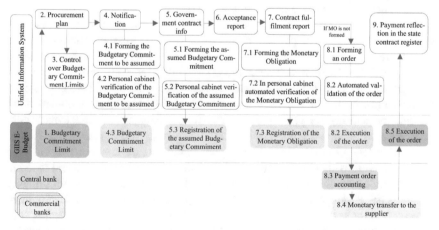

Fig. 3 Scheme of interaction and electronic document flow in GIS EIS and GIIS "Elektronny Budget"

GMP is almost 97.5% or 352 organisations. Credit organisations can accept payments with the indication of a single requisite—UAI (a unique accrual identifier, which is indicated in the payment document and replaces the other requisites).

A procedure has been established for making information from the GIS GMP available to credit, insurance and microfinance organisations and financial platform operators for the purpose of providing citizens and organisations with financial and other services at request.

Based on information from the GIS GMP, in 2022–2023, it is planned to ensure that citizens will be able to make payments in favour of government bodies in a fast payment system (banking service), including taxes, fines, levies and duties, and to implement a service for making payments from the budget.

According to information from the GIS GMP, the Federal Tax Service receives real-time information on the payment of taxes and levies throughout the country. When transaction is registered in the payer's personal cabinet, the accrual of penalties and fines is discontinued.

The process of coordinating informatisation requires maintaining a register of GIS with open data. For this purpose, the Federal Government Information System for Coordination of Informatisation (hereinafter—FGIS CI) with a portal solution was created, but currently the data is not relevant (some information has not been updated since 2015) and is not being updated.

Implementing GIS provides a number of automation advantages in public finance management: simplification of procedures in regard to collecting and analysing information, ensuring the appropriate use of budgetary funds; reducing the time required to carry out transactions; reducing the time required to analyse information and draw up proposals aimed at improving the efficiency of the budgetary funds usage and state property management; prompt access to data from previous periods; automation of control over the financial irregularities' elimination; recording and analysing data, providing high quality data to enhance the efficiency of budgetary funds, ensuring transparency of information on the authorities' performance.

One of the problems related to data storage and exchange is being resolved by using cloud technologies. The experiment of transferring Russian state information systems to a unified "Gosoblako" is underway and expected to be completed by the end of 2024. In Russia, it seems reasonable that the platform based solution for digital management of public finances should be a centralised budget platform, the development of which may be conditioned by the transition to a unified cloud technology platform for GIS—"Gostech". The development of digital markets in various countries shows a progressive trend towards the widespread application of platform-based solutions. Owners of digital platforms, government and business structures, work with a huge array of different data, both systematised and random. [21] The objective of the "Gostech" platform is to provide the capacity for bringing existing GIS to a single digital standard, for control over duplication of development projects and for scaling of the platform across all public authorities, including sub-federal systems. The advantages of the cloud platform are associated with accelerated launch of the system/subsystem/service, with new promt digital opportunities for interaction between the state and citizens, with multitenancy (capability of servicing users from different organisations on one platform), with increased cybersecurity and data security, and with the enhancement of system functionality through saving resources by means of replication.

In order to transition to a single platform, the existing GIS should be evaluated based on the following criteria: level of importance (shows what systemically important functions the GIS fulfils); technological maturity (assesses how secure, independent, reliable, infrastructural and up-to-date the GIS is); and cost-effectiveness (reflects how feasible the decision to transition to a single platform is). In the first phase, the first two criteria should be assessed taking into consideration the risks and priorities of moving to a single platform (see Fig. 4).

The risk level is linked to the possible non-harmful effects and their magnitude arising from the system disruption/malfunction. GIS with low and medium technological maturity level and medium importance level should be included in the group of high priority of transition to a platform solution. The prioritisation ranking of GIS transition should also take into account economical parameters, based on the estimated cost of GIS creation, development and operation. Among several scenarios of development, changes that maximise results in a short time and at a reasonable cost should be given the highest priority. GIS with low priority and low importance as well as with low cost-effectiveness are assessed as inappropriate for migration to a unified platform.

It is expected that in 2023 GIS of 24 federal level authorities will be transferred to "Gostech", and by 2026 the number of services on the platform should reach more than 100. In June 2023, an agreement was signed on the complete transition of all

Economic efficiency of the transition

GIS significance level		Technological maturity of GIS		Feasibility of transition
	low	**medium**	**high**	
high	Critical risk High priority	Minimal risk High priority	Zero risk Zero priority	
medium	Critical risk High priority	High risk High priority	Minimal risk Low priority	
low	Critical risk Zero priority	High risk Low priority	Minimal risk Low priority	
	low	**medium**	**high**	

Technological maturity of GIS

Fig. 4 Risk and priority assessment martrix for GIS transition to the unified platform "Gostech" by criteria

regional GIS of the Ulyanovsk region to the "Gostech" platform, other regional public authorities should transfer their information systems by 2025. The priority areas for migration to the platform solution are related to public services in healthcare, education, sports, transport, construction. Among federal agencies, subsystems of the Russian Treasury, Rosreestr, and the Social Fund of Russia (social treasury) are considered for migration; due to the complexity of the systems, the transition is planned to be phased.

The government's digitalisation trends are related to platform solutions and client-centric policies. Big data, artificial intelligence, blockchain, and neural network models remain the key technologies today. Centralisation on the basis of unified platforms is associated with more rigorous security requirements.

4   Discussion

A small number of OECD countries follow a decentralised approach in the organisation of the public financial management system, including the decision making in regard to approriate technology choices; such an approach leads to costly duplication and a coordination burden on the central financial authority.

Despite the fairly efficient functioning of state information systems in Russian practice, there are problems associated with the fact that each information system is initially formed as a stand-alone system and later integrated with other systems. And if at the federal level such tasks can be solved rather quickly and efficiently (given the availability of qualified IT-specialists and financial resources), at lower levels of government there are problems associated with the integration of systems. In case of changes in legislation and data requirements, information systems must be improved and fine-tuned, which requires time (search for performers, tender procedures) and additional budget expenditures.

The vector of development of e-government and digital public finance management is related to the transition to a single national platform solution.

The process of digitalisation affects the entire national financial system of Russia, primarily the banking sector and the institutional sphere. A basic law has been adopted that sets out the legal norms for the launch of the digital ruble as the third form of national currency in Russia, and the Bank of Russia's obligations as the operator of the digital ruble platform are legally enshrined.

In this regard, the issues of integration with external information systems in the field of public finance management, in particular with the Bank of Russia, are of particular relevance. This integration will affect the mechanism of e-government development, which includes the organisational structure of management, stages and participants, and will require improved coordination of participants' activities not only at the level of the budget sphere, but also within the national financial system as a whole [22].

The UK was one of the first countries to launch a strategy for shared public services in 2018. The goals were: to bring procedures and data to a standart, to improve efficiency, to balance price and quality, and to provide a better experience [23]. A similar platform oriented path is chosen by the USA. In Russian practice, in order to improve e-government performance providing public services, a unified platform "Gostech" is being created, to be integrated with regional systems in the future.

Finding a balance between the benefits of information systems standardisation and essential flexibility in public financial management is a complex task that reqires a thoughtful approach.

5 Conclusion

Platform based solutions are a necessary step in the direction of creating a digital institutional environment to improve the efficiency of management, transparency and control over the use of financial resources allocated to the development and refinement of information systems.

For state information systems in the sphere of financial management, due to the systematic nature of operations affecting all spheres of the economy and economic entities, it is advisable to form a unified budget platform, whose capabilities expantion is associated with the use of digital national currency.

In the circumstances of instability and external sanctions, the transition to import substituting technologies plays a significant role, and it is in this area that the risks of destabilisation of the digital circuit are being concentrated. In the Russian practice, the transition of the basic components of the GIS of the management of public finances to an independent technological stack is currently under way until 2025 and is still in the risk zone.

The creation, development and operation of GIS for federal systems related to providing public services should begin to be implemented as early as 2024 with the help of a unified platform "Gostech", subsequently integrating sub-federal level information systems. For the transition to the unified platform, the existing GIS should be evaluated based on the following criteria: level of importance; technological maturity; economic efficiency.

References

1. Osborne, D., Gaebler, T.: Reinventing government. In: How the Entrepreneurial Spirit Is Transforming the Public Sector. Ringwood: Penguin Books USA Inc., New York (1992)
2. Glyptis, L., Christofi, M., Vrontis, D., Del Giudice, M., Dimitriou, S., Michael, P.: EGovernment implementation challenges in small countries: the project manager's perspective. Technol. Forecast. Soc. Chang. **152**, 119880 (2020)
3. Aboulola, O.: Global E-government trends challenges and opportunities. SAR J. **4**(4), 175–180 (2021)

4. Bertot, J. C., Gorham, U., Jaeger, P. T., Sarin, L. C., & Choi, H. Big data, open government and egovernment: Issues, policies and recommendations. Inf. Polity **19**(1, 2), 5–16 (2014)
5. Sharma, S.K., Metri, B., Dwivedi, Y.K., Rana, N.P.: Challenges common service centers (CSCs) face in delivering e-government services in rural India. Gov. Inf. Q. **38**(2), 101573 (2021)
6. Alkhusaili, M. M. and Aljazzaf, Z. M. The Evolution of E-government Project in GCC Countries. In: Proceedings of the International Conference on Industrial Engineering & Operations Management, p. 13 (2020)
7. Pope, R. Playbook: government as a platform. Cambridge MA: Ash Center for Democratic Governance and Innovation. (2019) https://ash.harvard.edu/publications/playbook-government-platform
8. Ondrus J., Gannamaneni A., Lyytine K. The impact of openness on the market potential of multi-sided platforms. A case study of mobile payment platforms. J. Inf. Technol. **30**(3), 26 (2015)
9. Gosudarstvo kak platforma (kiber)gosudarstvo dlya tsifrovoy ekonomiki tsifrovaya transformatsiya, https://www.csr.ru/upload/iblock/313/3132b2de9ccef0db1eecd56071b98f5f.pdf, last accessed 2023/09/21
10. Gupta, S., Keen, M., Shah, A. and Verdier, G. Introduction: reshaping public finance. In: Gupta, S., Keen, M., Shah, A., Verdier, G. (eds.) Digital revolutions in public finance. IMF, Washington DC (2017)
11. Digital innovation in public financial management (PFM): opportunities and implications for low-income countries. Singapore and Sydney: Alphabeta (2018), www.alphabeta.com/wpcontent/uploads/2018/07/pfm-technology-paper-long-version.pdf, last accessed 2023/09/21
12. Long, C., Cangiano, M., Middleton, E., et al. Digital public financial management: an emerging paradigm. ODI Working Paper. London: ODI (2023), www.odi.org/en/publications/digitalpublic-financial-management-an-emerging-paradigm/, last accessed 2023/09/21
13. Cangiano, M., Curristine, T.R., Lazare, M.: Public financial management and its emerging architecture. International Monetary Fund (IMF), Washington DC (2013)
14. Manning, N., Hedger, E. and Schick, A. Advice, money, results: rethinking international support for managing public finance. New York NY: New York University (2020), https://wagner.nyu.edu/files/publications/NYU%20PFM%20Working%20Group%20WEB%206.6.20.pdf, last accessed 2023/09/21
15. E-Government Survey 2022, https://publicadministration.un.org/egovkb/en-us/Reports/UN-E-Government-Survey-2022, last accessed 2023/05/21
16. GovTech Maturity Index: Trends in Public Sector Digital Transformation. December 2022, https://www.worldbank.org/en/programs/govtech/gtmi. Accessed 21 May 2023.
17. Open Budget Survey 2021, https://internationalbudget.org/open-budget-survey-2021/, last accessed 2023/05/21
18. Informatsionnye tekhnologii v Federal'nom kaznacheystve RF, https://www.tadviser.ru/, last accessed 2023/06/02
19. Upravleniye obshchestvennymi finansami: novyye vyzovy i praktika, https://bujet.ru/article/433003.php#_ftn1, last accessed 2023/05/21
20. YEIS Zakupki, https://zakupki.gov.ru/epz/main/public/home.html, last accessed 2023/06/02
21. Yulia Kharitonova, Larisa Sannikova. Digital Platforms in China and Europe: Legal Challenges, 8(3) BRICS Law Journal 121–147 (2021).
22. Balynin I.V. The digital ruble as a tool to improve the operational efficiency of the use of budgetary funds in the Russian federation. Auditorskiye vedomosti. 3, 108–110 pp. (2021).
23. Cabinet Office (2018) A shared services strategy for government. London: UK Cabinet Office. www.gov.uk/government/publications/a-shared-services-strategy-for-government, last accessed 2023/05/21

Symbiotic Intelligence in Smart Configuration: Opportunities and Challenges

Liudmila A. Vasilenko and **Anastasiya D. Seliverstova**

Abstract Symbiotic intelligence is a form and result of the interaction of natural and hybrid intelligence, aimed at consolidating their potentials in order to solve the tasks. The digital twin of the network social community is presented in the form digital prototype built on an array of data about its activities. The simulation accumulates data about the interests and actions of the network community with a calculation of possible options for their behavior in real time. Smart configuration is considered by the authors as a way to implement positive feedback in decision support systems in public administration using symbiotic autonomous systems in the format of digital twins of collective subjects of social networks. Feedback is provided through their self-organized participation in a hybrid social environment in "self-service media" formats. The purpose of chapter is to use the digital twins in decision support systems in public administration. Symbiotic intelligence consolidates the potential of natural intelligence and hybrid intelligence of network society and decision making person (DMP) in the process of management decision. This process allows to aggregate the interests stakeholders in decision-making processes. The chapter presents an example of the possibility of using this approach in Russian national projects.

Keywords Digitalization · Digital society · Digital platforms · Media convergence · Public administration · Symbiotic intelligence · Smart configuration

L. A. Vasilenko (✉)
Russian Presidential Academy of National Economy and Public Administration (RANEPA), Moscow, Russia
e-mail: la.vasilenkola@igsu.ru

A. D. Seliverstova
Russian Presidential Academy of National Economy and Public Moscow, Moscow, Russia

© The Author(s), under exclusive license to Springer Nature Switzerland AG 2024
M. Sari and A. Kulachinskaya (eds.), *Digital Transformation: What are the Smart Cities Today?*, Lecture Notes in Networks and Systems 846,
https://doi.org/10.1007/978-3-031-49390-4_12

1 Introduction

The subject of symbiotic intelligence in management activities was actualized with reference to the rapidly developing global processes of digital transformation, the adoption of artificial intelligence systems, the advance of smart management ideas [1, p. 74], compelling in terms of Media Convergence and of the hybridity of the social control field [2, p. 123] to combine natural intelligence with hybrid systems based on artificial intelligence components.

The processes of Media Convergence and mediatization of society wiping out the boundaries of real and virtual reality. In agreement with number of studies, collective self-organizing subjects might be influential actors of forceful influence, able to compete for their audience.

Such self-organizing subjects influence their audience through special technologies that form meanings and fix them in the minds of their supporters. These meanings express interests of competing subjects [3, p. 16]. Monitoring changes in the interests and perceived meanings of each social group allows you to make more correct decisions.

In the work "Digital subjects as new power actors: a critical view on political, media- and digital spaces intersection", based on the methodology of a critical media theory, the authors focus on an intersection of various spaces (political, media- and digital ones), which gives rise to the complexity of social relations between power actors of each space and gives them additional dimensions of new opportunities. For the purposes of this article, the methodological tool concerning four situations in the digital society context may be useful [4, p. 233]. The authors propose the matrix conceptualizing new power balance between self-organizing power actors in the area of all three spaces intersection. Their different positions being formed as a consequence of this balance violations require different communicative strategies. For the purpose of this article, this methodological tool has a prognostic potential [4, p. 241]. There is a question about using Symbiotic Autonomous Systems (SASs) in the format of digital twins in processes management tasks of high complexity in a hybrid media environment.

The hybridity of the media communication environment is associated with the human interaction with many types of technical devices, the use of various software systems and huge interactive databases with information multiple forms of presentation.

The complexity increases due to the underdevelopment of the culture of geoinformation security the desire not so much to solve emerging problems, but to block access to information sources of potential dangers and a poor understanding of other management options in the network environment. It becomes necessary the application of the explanatory powers of complexity theory turns to necessity [5–7, pp. 15–16, 8, pp 300–309].

Studies have shown, that SAS was presented in a research article of Wang as autonomous collective intelligence, symbiosis of human interaction with intellectual and cognitive systems [9]. The authors of the model of symbiomemesis were perceiving it as a specific form of symbiosis machine learning and communication of human-autonomy symbiosis [10].

Pamela McCorduck is one of the very first researchers in the field of sociology of artificial intelligence, when this scientific direction was just emerging. She is the author of the work "Machines Who Think", which was written in the middle of the last century [11]. In connection with the topic of the application of symbiotic intelligence in a complex mendia-converged environment, стала востребованной the topic of the "smart configuration" in the context of democratization and network management [3. P. 29] as well as the issue of collision between artificial and natural sociality in management processes was updated. One of the famous works of the Nobel Prize winner Herbert A. Simon "The Sciences of the Artificial Science" considers the "artificial science" as general features of artificial systems, anticipating future artificial intelligence systems, which are the basis of their design. Subsequent achievements in the field of cognitive psychology and social informatics confirm and expand "the main thesis of the book: the system of physical symbols has the necessary and sufficient means for reasonable action" [12, p. 3]. Modern works deserve attention, in which we no longer meet slightly naive utopian formats for describing artificial intelligence systems of the first researchers in this field. Thus, in The series "Studies in Computational Intelligence" (SCI) we find new interdisciplinary advances, including both artificial intelligence and the theory and methods of in the fields of computer science, sociology and psychology, neural networks, genetic algorithms, self-organizing and hybrid intelligent systems and others [13: p. 3]. Of particular interest are the materials of the monographic collection "Artificial Intelligence in Control Systems and Decision Making" with the presentation of new theories and methods of artificial intelligence, including issues of managerial decision-making, applications and research materials [14].

Still, the question is by virtue of what to practically embed the symbiotic intelligence in managerial terms amid Governance and the theoretical basis for its application in public administration processes.

2 Materials and Methods

A conceptual approach to the definition of symbiotic intelligence based on the analysis of a number of theoretical sources on the following issues is disclosed in the article: the explanatory powers of complexity theory of Nobel Prize winner Ilya Prigozhin [6], Hermann Haken [7. pp 15–16] and Sergey Kurdyumov [8. pp 300–309]; sharing synergistic and fractal-evolutionary approaches in social management [15, pp. 62–63]; philosophical, cognitive and mathematical foundations of the symbiotic autonomous systems [9]; hybridity of the social field of management [2, p. 123], collisions of artificial and natural sociality in management processes (11); «smart management» in Governance of Tikhonov and Bogdanov [1, pp. 79–81]; «Smart Configuration» in Governance of Zotov [3, p. 16].

The investigation feasibilities of the processes of "smart management" by representatives of the authorities and citizens was carried out using materials from a sociological study. The questionnaire was conducted in September–November 2020 year, the respondents were the population of Moscow (450) and of the Kursk region (440) both through a field survey and using the Google service. The sample was quota by sex and age. During the Internet survey, the achievement of proportionality of quotas of the general population was ensured by sending out personal invitations to respondents satisfying the sampling parameters [2, pp 123–124].

3   Results

Symbiotic intelligence is a form and result of the interaction of natural and hybrid intelligences aimed at consolidating their potentials to solve the tasks assigned to symbiotic intelligence.

The reason for the use of symbiotic intelligence is the objective limitation of human capabilities. A. Fakhrutdinova draws attention to the revision of the postulates of the theory of rational decisions, since the decision-maker cannot be fully informed. Amina Zievna refers to the concept of asymmetric information by Akerlof, Spence and Stiglitz, who received the Nobel Prize in Economics in 2001 [16] The asymmetry of information is due to the fact that "all subjects of market relations have different amounts of information about the conditions of market transactions. So, the seller of the car has a much larger amount of information about the object of sale than the buyer, and the buyer of the insurance policy is much better informed about the state of his health than the seller [17]. Information asymmetry is a fundamental cause of uncertainty, which can lead to erroneous decisions. A decision-making model is needed that takes into account the subjective characteristics of the decision-maker, his selectivity of perception, limited memory, external social influences and much more.

As a result of the study, the authors identified possible approaches and methods for disclosing uncertainty in the processes of collective decision-making Decision Making Person (DMP):

- Accept the concept of "socionavigation", i.e. synergy of practices and technologies of orientation, self-determination and resonant control of the behavior of a person, group, team in the social world, organized on the basis of worldviews in the single creative field [18];
- Make a collective probabilistic forecast of changes in the internal and external environment;
- Develop a list of alternatives using collective peer review methods (brainstorming, Delphi method, decomposition method, game theory, Descartes square);
- Group the influencing factors into three groups-positive, negative or neutral, taking into account the degree of influence and select the criteria for evaluating

alternatives (criterion of "maximin", "maximax", "optimism–pessimism", "alpha criterion", criterion of losses from "minimax");
– Compile a decision matrix for a comparative probabilistic assessment of the merits and risks of each alternative.

It is proposed to develop hybrid intelligence as communication models of digital counterparts of collective subjects, and hybrid intelligence built on the principles of machine learning with the participation of highly professional experts in this subject area, systems based on artificial intelligence can be trained in the same way as people if their algorithms are properly developed and a training set is correctly formed.

The participation of highly professional experts in machine learning processes to assist the decision-maker will reduce the risk of errors caused by the imperfection of machine learning processes and the use of the principle of human independence from artificial intelligence algorithms.

The digital analogue of a collective subject is presented as an interactive self-organizing dynamic digital prototype of a network community. Digital modeling accumulates interactive 3D models of data about the views, interests and actions of the network community with the calculation of possible options in real time. The models are based on an array of data on the activities of an interactive self-organizing dynamic digital prototype since its creation.

Collective actors can accumulate the merits of social network leaders in combination with the interests of network communities that accumulate around them on issues related to decision makers, as well as those who have the ability to consciously choose their social roles, develop goals and ways to achieve them, participate in communication processes to develop alternative solutions. variants of management decisions.

It is important that all types of subjects fully participate in the periodic decision-making procedure by decision makers (Fig. 1):

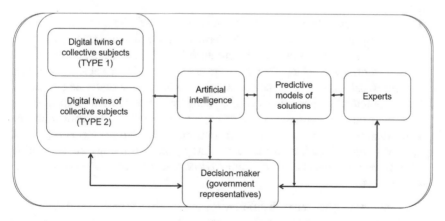

Fig. 1 The model of managerial decision-making with the use of symbiotic intelligence

- Representatives of the authorities responsible for the direction in which the decision is being discussed;
- Highly professional specialists in this field of activity;
- Several types of digital counterparts of collective subjects performing certain social roles of various groups of the population, supporting or opposing each other and authority on the issue under discussion.

Interestingly, highly professional experts can also unite to form a collective mind.

In general, the proposed approach can be applied within the framework of the implementation of national projects of the Russian Federation in terms of indicators and measures to involve citizens in interaction.

4 Discussion

Responding to the question is symbiotic intelligence in management processes in public administration processes in the format described in the results of the study, the authors discuss an option that is of great importance for the modern public administration system. A huge number of civil servants are engaged in the execution of the May decrees of the President of the Russian Federation.

The theoretical prerequisites of the observed results allow us to conclude that the intellectual potential of civil society is practically not included in these processes [2, pp. 123–124]. On the example of national projects in Russia, the authors suggested the option of forming an algorithm for a self-learning digital twin as a synergy of three sources of expert knowledge:

- Data of special studies on the development of changes in monitoring processes due to persistently low values of any indicator of the implementation of the national project;
- Data of interactive form filling by citizens, complaint/offer forms in a specially created Appendix on the Feedback Platform. Formalization of answers, accompanied by a conversation between a citizen and a digital twin to specify the expression of the interests and needs of a citizen. If the digital twin was able to determine that the situation is typical, then he issues a recommendation to the citizen and puts a complaint/proposal under control, and also supplements the corresponding section of the monitoring database. If the digital twin has not identified a standard solution path, then this situation is transmitted to a human specialist; data of specially organized actions using natural intelligence when a new situation arises with the involvement of civil servants and experts in the field of knowledge related to the issue.

Each option is associated with different organizational and meaningful algorithms and procedures for implementing symbiotic intelligence.

The answer to the question of by what means society's readiness for the use of symbiotic intelligence in the processes of smart government and configuration is seen was considered as part of a sociological study "Public administration as a configuration of relational networks in the public space of the digital society" [2, pp. 127–129]. Certain landmark cases of their application are clear and intelligible to citizens. However, the practical readiness of civil servants and civil society institutions to participate in interactions with such systems is assessed as insufficient.

5 Conclusion

The survey found that the study represents a previously not considered approach in public administration systems, since it is conceptual in nature, its practical and technological component does not have practical implementation.

Digital transformation processes also require significant regulatory attention in terms of ethics and rules for the application of symbiotic intelligence. Joint actions are needed to practically implement the presented concept with the participation of civil servants involved in the development and implementation of monitoring management processes, sociologists, experts in working with knowledge bases, IT specialists and representatives of various civil society institutions.

References

1. Tikhonov, A.V., Bogdanov, V.S.: From «smart regulation» to «smart management»: social issue of feedback digitalization. Sociol. Stud. **1**, 74–81 (2020). https://doi.org/10.31857/S01321625 0008325-0.
2. Vasilenko, L., Meshcheryakova, N., Zotov, V.: Digitalization of global society: from the emerging social reality to its sociological conceptualisation. Wisdom **21**(1), 123–129 (2022). https://doi.org/10.24234/wisdom.v21i1.720.
3. Zotov, V.V.: Configuration as a process of managing the capitalization of the social network space. Communicology **7**(4), 15–31 (2019). https://doi.org/10.21453/2311-3065-2019-7-4-15-31
4. Gavra, D., Dekalov, V., Naumenko, K.: Digital subjects as new power actors: a critical view on political, media-, and digital spaces intersection. In: Kotenko, I., Badica, C., Desnitsky, V., El Baz, D., Ivanovic, M. (eds.) Intelligent Distributed Computing XIII. Series Editor Janusz Kacprzyk, vol. 868. pp. 233–243. Polish Academy of Sciences, Warsaw, Poland.
5. Kashina, M., Vasilenko, L.: Fractality of gender relations and the use of gender resource of the public policy and administration in modern Russia. Woman Russ Soc **2**, 17–31 (2019). https://doi.org/10.21064/WinRS.2019.2.2
6. Prigozhin, I.: From Being to Becoming. W. H. Freeman and Company, New York (1980)
7. Haken, H.: Synergetics, p. 325. Springer-Verlag, Berlin-Heidelberg-New York (1977). https://doi.org/10.1002/bimj.4710200213.
8. Kurdyumov, S.: Evolution and self-organization laws in complex systems. Int. J. Mod. Phys. **1**(4), 299–327 (1990)
9. Wang, Y., Karray, F., Kwong, S., Plataniotis, K.N., Leung, H., Hou, M., Tunstel, E., Rudas, I.J., Trajkovic, L., Kaynak, O., Kacprzyk, J., Zhou, M., Smith, M.H., Chen, P., Patel, S.: On the philosophical, cognitive and mathematical foundations of symbiotic autonomous systems. Philos. Trans. R. Soc. A: Math., Phys. Eng. Sci. **379**, 1–16 (2021) https://doi.org/10.1098/rsta.2020.0362. Phil. Trans. R. Soc. A.3792020036220200362.

10. Abbass, H., Petraki, E., Hussein, A., McCall, F., Elsawah, S.: A model of symbiomemesis: machine education and communication as pillars for human-autonomy symbiosis. Philos. Trans. R. Soc. A: Math., Phys. Eng. Sci. **379**, 1–9 (2021). https://doi.org/10.1098/rsta.220.0364.
11. McCorduck, P.: Machines Who Think: a Personal Inquiry into the History and Prospects of Artificial Intelligence. Includes Bibliographical References and Index/Pamela McCorduck, 2nd edn, p. 554. A K Peters, Ltd. Natick, Massachusetts (2004)
12. Simon, H.A.: The Sciences of the Artificial. Reissue of the Third Edition. With a New Introduction by John Laird, p. 256. Publisher, The MIT Press, Cambridge (2019).
13. Kacprzyk, J.: Students in Computational Intelligence (SCI), vol. 868. pp. 233–243. Polish Academy of Sciences, Warsaw, Poland (2020). Springer, Heidelberg. http://www.springer.com/series/7092. Accessed 9 February 2023
14. Kondratenko, Y.P., Kreinovich, V., Pedrycz, W., Chikrii, A., Gil Lafuente, A.M.: Artificial Intelligence in Control and Decision-making Systems, vol. 1087 (2023). https://doi.org/10.1007/978-94-011-0305-3.
15. Vasilenko, L.: Fractal-synergetic approach to the research of entrepreneurship in the non-profit organizations. Wisdom **12**(1), 62–72 (2019)
16. Amina, Z.: Fakhrutdinova: models of rationality in the bases of the theory of decision-making. Philos. Sci. Technol. **24**(1), 131–144 (2019)
17. Akerlof, G., Shiller, R.: Phishing for Phools, p. 280. Economics of Manipulation and Deception. Princeton University Press. Princeton, New Jersey, USA (2015)
18. Fedotova, M.A.: System management of team work: evolution of representations and development prospects. Res. Result. Sociol. Management. **4**, 137–151 (2018)

Identification of Trends in the Digital Methodological Toolkit for the Analysis of Public Expenditure Financing to Improve the "Electronic Budget" System

Nadezhda Yashina⊙**, Natalia Pronchatova-Rubtsova**⊙**, Oksana Kashina**⊙**, and Sergey Yashin**⊙

Abstract The research is devoted to improving the methodological aspects of the analysis of the execution of budget expenditures, taking into account the assessment of budgetary stresses associated with the execution of the expenditure part of the consolidated budgets, in order to form financial mechanisms for preventing threats for social stability in the face of internal and external challenges that can be implemented using digital tools. The digitalization of the state budget system allows you to increase the effectiveness of managerial decisions in unstable economic conditions. For this purpose, many countries of the world are developing state integrated information systems for managing public finances. In particular, in Russia, a similar information system was developed for the digitalization of the budget system, called the "Electronic Budget", which contributes to the transparency of information and openness of planning and monitoring of state expenses. This technology implies the transfer of all key stages of the budget process to an electronic format, in addition to servicing budget accounts. The aim of the research is to develop a theoretical approach and digital methodological tools for the practical implementation of an automated mechanism for assessing the sustainability of regional and municipal finances. The methodological basis of the analysis was economic methods and methods of mathematical statistics, including methods of optimal solutions. The proposed methodological tools were tested on the example of the consolidated budget of the Russian region, its susceptibility to budgetary stress in the execution of budget expenditures by sections of the budget classification was assessed, taking into account the proposed levels of stability in financing expenditures. The directions of "anti-stress" budget management are formulated. Methodological tools can be integrated into the information system "Electronic budget" to improve the algorithm for analyzing budget expenditures.

N. Yashina (✉) · N. Pronchatova-Rubtsova · O. Kashina · S. Yashin
Lobachevsky State University of Nizhny Novgorod, Gagarin Avenue 23, 603022 Nizhny Novgorod, Russia
e-mail: sitnicof@mail.ru

Keywords Public financing of expenditure · Budget risks · E-budget (Electronic budget) · Budgetary stress · Indicators' Standardization

1   Introduction

The important socio-economic changes are taking place very quickly in Russia in the context of the global crisis. The proposed theoretical approach makes it possible to solve the debatable issues of the execution of public financing, taking into account budgetary stresses.

The pandemic has turned out to be a new challenge for the global economy and the scientific community; therefore, the number of scientific works in the field of assessing the consequences of such large-scale crises is somewhat limited. Nevertheless, many researchers have already noted the global impact of the Coronavirus pandemic (comparable to the economic situation after World War II) not only on the global health system, but on all spheres of society [1]. The need for an urgent assessment of the impact of COVID-19 on society and the global environment was emphasized in their works by researchers Chakraborty, Maity [2], Husain [3], Lucchese, Pianta [4]. In addition to the importance of assessing the economic consequences of the pandemic, the problem of developing possible actions that can lead society to a more stable, healthy, equitable and sustainable development trajectory is relevant [4].

In addition to the importance of assessing the still deeply unexplored economic consequences of the pandemic, the urgent problem of developing possible actions that can lead to a more stable and sustainable trajectory for the development of territories in modern conditions remains relevant, as noted by Gamukin [5], Zonova and Kislitsyn [6], Kolesov [7], Kireeva [8]. In the context of crisis phenomena, the key in the field of budget policy is the transformation of the budget mechanism into an effective tool for macroeconomic stabilization and the use of all reserves of budget expenditures to finance the development of the economy and human capital [9].

In Russian economic science, the definition of the concept of "budget risk" is debatable [10]. Scientist and practitioner A. Kudrin characterize the risk by the uncertainty of obtaining budget revenues. Former Minister of Economic Development A. Ulyukaev, on the contrary, defines budget risk as the probability of non-implementation of planned budget expenditures [11]. M. Chichelev gives a comprehensive definition of budget risk as non-receipt of income and underfunding of expenses [9, 12].

Currently, in many countries of the world, digital methods of public administration using various integrated information systems have become widespread [13]. For example, in the Russian Federation, a digital information system "Electronic budget" is being developed, in which only certain functions of budget administration are subject to automation. However, in some countries of the world, public authorities are gradually moving towards a complete or almost complete digital transformation of a number of their control functions [14–16].

One of the successful examples of successful digitalization of public adminis-tration is the transition of public services from an analog to a fully digital model of service delivery in Denmark [13]. As a result, Denmark ranked first among United Nations countries in 2022 in providing online government services and policy development.

At the same time, a number of countries face many problems with the digitalization process of managing budget funds, primarily related to the lack of a unified strategy for digitalizing the economy and insufficient funding [17]. There are practically no methodological tools for automated monitoring of budget indicators that allow to make prompt solutions and to develop anti-crisis measures to stabilize the economic situation in the country.

The purpose of the research is to develop a theoretical approach and methodolog-ical tools for the practical implementation of the mechanism for assessing the sustain-ability of regional and municipal finances in comparison with the "pre-pandemic" period and during a pandemic, taking into account budgetary stresses based on deter-mining the risk of financing expenditures to improve the algorithm for analyzing budget expenditures in an integrated information system "Electronic budget".

Approbation of the proposed methodological tools for analyzing budget expendi-tures is carried out on the example of the consolidated budget of one of the regions of the Russian Federation, in particular, the Nizhny Novgorod region [18]. The subject of the study is a set of theoretical and methodological issues in the analysis of budget expenditures based on the identification of budgetary stresses associated with the execution of the expenditure part of the budget, and the development of financial mechanisms to counteract social instability in the regions of Russia. It is advisable to implement the proposed methodological tools for analyzing costs on a monthly basis through the digital system "Electronic budget". This approach will help increase the transparency of information and the efficiency of the activities of state bodies.

2   Materials and Methods

The analysis of state financing of expenses was carried out on the basis of information from the Ministry of Finance of the Nizhny Novgorod Region [18]. The database included the estate dynamics of budget expenditures. The data were standardized and transformed taking into account the requirements for indicators (minimizing or maximizing them).

The risk of budget expenditures is calculated on the basis of the average deviation of the execution of budget expenditures by formulas (1) and (2).

The average deviation as a risk of financing budget expenses (σ) determines by formula (1):

$$\sigma = \sqrt{\frac{\sum_{i=1}^{N}\left(\Pi_j - \overline{\Pi}\right)^2}{N-1}} \tag{1}$$

where $\prod_i$–is the indicator of execution of expenditures by subsections of the corresponding section of the budget classification; $\prod$–is the average indicator of the execution of expenditures by subsections of the corresponding section of the budget classification; N–is the number of sections, subsections, in the context of which the costs are analyzed.

The relative risk is calculated using the coefficient of the variation of budget execution according to the relevant subsections of budget expenditures (CV) according to the formula (2):

$$CV = \frac{\sigma}{\overline{\prod}} * 100\% \tag{2}$$

The normative value of the indicator is an expert assessment. The higher this coefficient, the higher the variability, the discrepancy between the planned and actual values of the performance indicators of budget expenditures. A qualitative assessment of the values of the coefficient of variation is possible.

Formulas (3) and (4) are used to standardize the indicators characterizing the aggregate budgetary stress: execution of expenditures, risks of execution and growth rates of expenditures of the consolidated budget. Based on the interpretation of the criterion: when minimizing (decrease is treated as a positive factor) and maximizing (growth is treated as a positive factor), formulas (3) and (4) are used to standardize indicators.

$$K_{ij}^* = \frac{K_{ij} - K_{imin}}{K_{imax} - K_{imin}} \tag{3}$$

$$K_{ij}^* = \frac{K_{imax} - K_{ij}}{K_{imax} - K_{imin}} \tag{4}$$

where K_{ij}^*–is the standardized indicator of the i-th proposed indicator of the budgetary stress assessment system in the j-th month, K_{ij}–is the estimated value of the i-th proposed indicator of the budgetary stress assessment system in the j-th month, K_{imax}–is the highest calculated value of the i-th indicator in the analyzed period, K_{imin}–is the lowest calculated value of the i-th indicator in the analyzed period [19, 20].

The chosen standardization method results in a change in the values of the indicators in the range from 0 to 1. The standardized indicators of the budgetary stress assessment system are summed up in each month in order to determine the cumulative standardized coefficient. The lower the value of the aggregate standardized coefficient for assessing budgetary stresses by period, the higher the ability of the budget to maintain the stability of its current execution [19].

The monthly dynamics of the formed system of indicators of the local and regional budget of the Nizhny Novgorod region is analyzed on the basis of monthly reports on the execution of budgets for 2019–2021 [18].

The following indicators are used to assess budgetary stresses:

- Monthly spending of budget funds in 2019–2021, their horizontal, structural, comparative, factor analysis;
- Coefficients of the budget execution and the revised plan;
- Coefficients of variation in budget execution and the revised plan, their comparison in 2019–2021;
- The rate of growth (decrease) in budget revenues and expenditures compared to the same period last year;
- Number of amendments to the budget law in 2019–2021 etc.

The set of indicators can be expanded. The current values of the assessment indicators are determined on the basis of data on the execution of budgets. Formulas (1)–(4) are used to assess the stress levels of financing expenditures by sections of the budget classification based on the results of the budget reports.

The level assessment of current and potential stress in financing expenditures by sections of the budget classification was carried out based on the results of the 2021 budget execution report. The final standardized indicator for the level of financing is determined by summing up the standardized indicators of the percentage of execution to the plan for 2021, execution to the plan of 2021, percentage of execution for 2020–2021, risks (actual and potential) budget execution.

The current risk is based on the discrepancy between the actual and planned budget indicators. Potential risk is calculated with the extension of the percentage of 2021 by 2020. Ranking on the final standardized indicator by level of funding indicates the level of exposure to stress. The level of exposure to budgetary stress is determined based on the indicator of exposure to budgetary stress (Table 1).

The indicator is calculated step by step as follows.

(1) The rank is determined by the value of the final standardized indicator by the level of financing (ascending values).
(2) The average value of the rank is determined as the arithmetic mean of the ranks of all sections.
(3) The budget stress exposure indicator is defined as the ratio of the rank of a particular section to the average value obtained for all ranks, then the optimal value of the budget stress exposure indicator is 1.
(4) Value limits expert interpretation of the indicator of exposure to budget stress (Table 1) makes it possible to form the levels of stability of expenditure financing.

3 Results and Discussion

The test of the proposed methodological tools for analyzing budget expenditures, as mentioned earlier, was carried out using the example of the consolidated budget of one of the regions of the Russian Federation, in particular, the Nizhny Novgorod region. This choice was due to a number of factors. The Nizhny Novgorod region is one of the central regions of the European part of the Russian Federation that does

Table 1 Determination of the expenditure financing stability level and exposure to budgetary stress

Level of stability of expenditure financing item, characteristics	Lower bound	Upper bound	Deviation of the indicator's average value	Range	Expert interpretation of budgetary stress exposure
First level–high level of expenditure financing stability	0.75	1.25	|0.25|	[-0.25, 0.25]	Low level
Second level–average level of expenditure financing stability	0.5	1.5	|0.5|	[|0.25|, |0.5|]	Allowable (reasonable) level
Third level–problems with financing budget expenditures	<0.5	>1.5	>|0.5|	>0.5; <−0.5	High level

not have raw materials. At the same time, the region is an industrial center with a high share of industry in the economy, and has a unique scientific and technical potential.

The expenses of the consolidated budget of the Nizhny Novgorod Region, taking into account the budget of the Territorial Fund for 2021, were executed in the amount of 326,940.1 million rubles or 96.3% of the updated plan for the year, excluding the budget of the Territorial Fund-283,670.3 million rubles or 95.9%.

In general, the non-execution of the expenditure part of the budget, taking into account the budget of the Territorial Fund, amounted to 12,496.4 million rubles.

Regional budget expenditures taking into account the budget of the Territorial Fund, increased by 23,181.6 million rubles (growth by 7.6%), excluding the budget of the Territorial Fund for 22,077.4 million rubles (growth by 8.4%) in comparison with 2020.

Let's calculate the level of budget stress in the Nizhny Novgorod region for 2019–2021. Based on budget indicators in accordance with the proposed methodology of a comprehensive assessment of budget stress. Table 2 provides a comparative analysis of monthly execution, execution risk and growth rates of the consolidated budget expenditures in 2019–2021.

The results of a comprehensive assessment of budget stress by means of a cumulative standardized coefficient in 2019–2021 are presented in Table 3.

An analysis of the indicators in Table 3 makes it possible to understand that, according to the results of the budget stress assessment, the least stressful budget execution in terms of expenditures in 2019 was recorded from September to December. Based on the results of the assessment of budget stress in 2020, based on

Table 2 Comparative analysis of monthly execution, execution risk and growth rates of expenses of the consolidated budget of the Nizhny Novgorod region, 2019–2021

Name of the month, indicators	2019			2020			2021		
	Execution, %	CV, %	Growth rate versus previous year, %	Execution, %	CV, %	Growth rate versus previous year, %	Execution, %	CV, %	Growth rate versus previous year, %
1	6.2	12	108.4	6.2	11	114.6	5.3	40.88	94.5
2	11.7	10	104.8	12.9	16	113.2	12.1	19.89	102.1
3	19.1	17	102.4	19	19	115.5	12.1	16.33	110.2
4	28	20	107.3	27.7	22	116.3	29.4	14.37	113.6
5	33.5	25	105	33.8	25	117.4	33.9	12.4	108.7
6	41.5	24	103.6	40.4	26	116.7	41.6	10.12	110.1
7	48.5	23	105.6	49.3	25	119.2	50.6	9.63	110.4
8	55.1	23	107.5	56	28	119.7	57.4	9.27	111.2
9	61.9	24	109.9	61.5	27	118.6	63.4	9.23	111
10	72	8	110.9	69.5	8	116.4	72.8	8.72	112.5
11	78.9	7	111.3	76.4	8	117	78.9	7.3	110.4
12	95.3	2	115.6	96.2	3	121.5	78.9	2.76	88.4
Ratio	46	16	107.7	45.7	18	117.2	44.7	13.4	106.9
Max value	95.3	25	115.6	96.2	28	121.5	78.9	40.9	113.6
Min value	6.2	2	102.4	6.2	3	113.2	5.3	2.8	88.4
Range of variation	89.1	23	13.2	90	25	8.3	73.6	38.1	25.2

the values of the aggregate standardized coefficient, budget execution became the least stressful in the second half of the year. An assessment of budget stress in 2021 based on the values of the aggregate standardized coefficient showed that the budget execution in terms of the expenditure side of the budget was the least stressful in April, June, July, August, September, October and November. Sufficiently high stress was observed in January, February, March, May 2021. December is characterized by a satisfactory level of stress in terms of financing the expenditure part of the budget. Table 4 demonstrates the main results of budget execution.

The execution of expenditures by 2021 has increased, expenditures in the social sphere and the payroll expenditures to employees have increased. Expenditures under the regional budget of the Nizhny Novgorod Region for 2021 are provided in the amount of 250,290.6 million rubles, executed in the amount of 241,315.9 million rubles, which is 96.4% of the revised plan of the year. The deviation of cash expenditures from planned annual appointments amounted to 8,974,661.1 thousand rubles. Cash execution by sections of expenditure classification was in the range from 77.7%

Table 3 Standardized indicators for assessing the budgetary stress of financing expenditures (*) based on monthly execution, execution risk and growth rates of expenditures of the consolidated budget, 2019–2021

Month	2019				2020				2021			
	Performance*	CV*	Growth rate versus previous year*	Aggregate index*	Performance*	CV*	Growth rate versus previous year*	Aggregate index*	Performance*	CV*	Growth rate versus previous year*	Aggregate index*
1	1.000	0.423	0.547	1.970	1.000	0.325	0.841	2.166	1.000	1.000	0.757	2.757
2	0.938	0.335	0.820	2.092	0.926	0.524	1.000	2.449	0.908	0.449	0.457	1.814
3	0.855	0.645	1.000	2.500	0.858	0.643	0.729	2.229	0.908	0.356	0.134	1.398
4	0.755	0.778	0.629	2.162	0.761	0.762	0.630	2.153	0.673	0.305	0.000	0.977
5	0.694	1.000	0.802	2.495	0.693	0.881	0.497	2.072	0.611	0.253	0.197	1.061
6	0.604	0.956	0.909	2.469	0.620	0.921	0.587	2.128	0.507	0.193	0.138	0.838
7	0.525	0.911	0.760	2.197	0.521	0.881	0.287	1.689	0.385	0.180	0.127	0.691
8	0.451	0.911	0.611	1.973	0.447	1.000	0.218	1.664	0.292	0.171	0.098	0.561
9	0.375	0.956	0.432	1.762	0.386	0.960	0.349	1.695	0.211	0.170	0.103	0.483
10	0.261	0.260	0.354	0.875	0.296	0.225	0.622	1.143	0.083	0.156	0.043	0.282
11	0.184	0.197	0.328	0.710	0.221	0.203	0.544	0.967	0.000	0.119	0.128	0.247
12	0.00	0.00	0.00	0.00	0.00	0.00	0.00	0.00	0.00	0.00	1.00	1.00
Ratio	0.554	0.614	0.599	1.767	0.561	0.610	0.525	1.696	0.465	0.279	0.265	1.009

Table 4 Execution of the consolidated budget in terms of expenses in billion rubles, 2020–2021

Period	Expenses, total	Expenditures on social sectors	Including payroll costs	Share of payroll expenditures (%)
2020	261.6	177.2	88.7	50
2021	283.7	183.5	90.2	49
Growth rate % to 2020	108.4	103.6	101,7	

to 99.7% of annual appointments. For individual subsections and target items of expenditure, there was a deviation of cash expenditures from planned appointments. Let's assess exposure to stress (Tables 5 and 6), taking into account the proposed levels of stability in financing costs. The highest proportion of expenditures are expenditures on the national economy, education, health care, and social policy.

Determining the level of stress exposure based on the stress exposure indicator (Table 6).

The development of a single methodological and information space using information technology will improve the quality of state financial management.

Expenses by sections of the budget classification, related to insignificant and acceptable levels of exposure to budgetary stress, are highly likely to finance top-priority and socially significant expenses in a timely manner and in full. The results of the calculations demonstrate that the sections responsible for financing human capital (for example, "education" and "healthcare") are characterized by high exposure to budget stress. On the contrary, financing of the sections "National Economics", "National Security and Law enforcement" is slightly subject to budget stress.

The financing of welfare is characterized by an acceptable exposure to stress, which allows the state to control and prevent social instability in the regions.

Table 5 Ranks formation of execution of budget expenditures by sections of the budget classification based on the final standardized indicator by the level of financing

Expenditures	Standardized indicators				Final standardized indicator by funding level	Rank by the final standardized indicator on the level of funding
	% of performance to plan	% performance 2021 to 2020	Current risk 2021 (%)	Potential risk 2021		
Mass media	0	0.615	0	0	0.615	1
МБТ бюджетам муниципальных образований общего характера	0.014	0.553	0.122	0.234	0.923	2
Education	0.136	0.495	0.292	0.155	1.078	3
Housing and public utilities	0.236	0	0.538	0.498	1.272	4
Welfare	0.077	0.535	0.664	0.092	1.369	5
Environmental protection	0.014	1	0.008	0.385	1.407	6
National defence	0.005	0.601	0.84	0.032	1.478	7
Physical education and sport	0.345	0.502	0.671	0.076	1.595	8
National Security and Law enforcement	0.064	0.514	0.164	1	1.742	9
National economics	0.15	0.311	1	0.317	1.778	10
Ratio	0.15	0.499	0.989	0.208	1.846	11
Culture and cinematography	0.214	0.855	0.687	0.181	1.937	12
General government issues	0.141	0.348	0.644	1	2.134	13
Public health service	0.218	0.586	0.789	0.55	2.143	14
Public debt management	1	0.716	0.84	0.032	2.589	15

Table 6 Levels of exposure to budget stress related to the execution of budget expenditures

Expenditure	Rank by the final standardized indicator on the level of funding	Budgetary Stress Exposure Indicator	Deviation	Stability level of expenditure financing	Level of exposure to budgetary stress
Mass media	1	0.125	0.875	Third	High level
Interbudgetary transfers to the budgets of municipalities of a general nature	2	0.25	0.75	Third	High level
Education	3	0.375	0.625	Third	High level
Housing and public utilities	4	0.5	0.5	Second	Allowable (reasonable) level
Welfare	5	0.625	0.375	Second	Allowable (reasonable) level
Environmental protection	6	0.75	0.25	First	Low level
National defence	7	0.875	0.125	First	Low level
Physical education and sport	8	1	0	First	Low level
National security and law enforcement	9	1.125	−0.125	First	Low level
National economics	10	1.25	−0.25	First	Low level
Ratio	11	1.375	−0.375	Second	Allowable (reasonable) level
Culture and cinematography	12	1.5	−0.5	Second	Allowable (reasonable) level
General government issues	13	1.625	−0.625	Third	High level
Public health service	14	1.75	−0.75	Third	High level
Public debt management	15	1.875	−0.875	Third	High level

4   Conclusion

The proposed methodological toolkit for automated analysis of public expenditure financing allows to determine the stability of the state financial system during crises and also promptly developing and making appropriate financial decisions. The research made it possible to formulate the directions of "anti-stress" budget management:

- To increase the execution of budget assignments due to force majeure circumstances, such as the spread of the coronavirus infection COVID-19, a difficult epidemiological situation, and a mobilization situation;
- To improve the disbursement of funds in connection with savings in appropriations as a result of competitive procedures, non-fulfillment of contractual obligations by contractors;
- To review investment plans and work out borrowing opportunities.

We believe that the integration of this methodological toolkit into the "Electronic Budget" information system will help to improve the algorithm of analyzing the financing of public expenditures in conditions of financial instability and budgetary stress.

Acknowledgements The study was carried out within the framework of the realization of the Strategic Academic Leadership Program "Priority 2030", project H-426-99_2022-2023 "Socio-economic models and technologies for the creative human capital development in the innovative society".

References

1. Nicola, M., Alsafi, Z., Sohrabi, C., Kerwan, A., Al-Jabir, A., Iosifidis, C., Agha, M., Agha, R.: The socio-economic implications of the coronavirus pandemic (COVID-19): a review. Int. J. Surg. **78**, 185–193 (2020)
2. Chakraborty, I., Maity, P.: COVID-19 outbreak: migration, effects on society, global environment and prevention. Sci. Total. Environ. **728**, 138882 (2020)
3. Husain, A.: Coronavirus Pandemic: Effects, Prevention and Management. The Readers Paradise, New Delhi (2020)
4. Lucchese, V.M., Pianta, M.: The coming coronavirus crisis: what can we learn? Intereconomics **55**(2), 98–104 (2020)
5. Gamukin, V.V.: Budgetary risks: options for trajectories. Quest. Econ. Regul. **6**(4), 120–130 (2015)
6. Zonova, A.V., Kislicyna, V.V.: Methodology for conducting a SWOT analysis in the process of considering the issue of the development of the Russian federation. Reg. Nal Econ.: Theory Pract. Practice **4**(19), 27–33 (2005)
7. Kolesov, A.S.: On the integrated assessment of the financial position of budget financing objects. Finance **6**, 9–11 (2000)
8. Kireeva, E.V.: Improving the efficiency of the regional finance management system. Reg. Econ. Manag.: Electron. Sci. J. **2–1**(50), 30–34 (2017)

9. Bogolib, T.: Fiscal policy as an instrument of macroeconomic stability. Econ. Ann.-XXI **3–4**(1), 84–87 (2015)
10. Gamukin, V.V.: Budgetary risks: introduction in the general axiomatics. Terra Economicus **11**(3), 52–61 (2013)
11. Nikulina, E.V., Fedyushina, I.G.: Characteristics of budget risks: economic essence and measures to minimize them. Young Sci. **1**, 412–416 (2014)
12. Yanov, V.V.: Substance of budget risk: theoretical and methodological aspects. Sib. Financ. Sch. **6**, 96–101 (2012)
13. Scupola, A., Mergel, I.: Co-production in digital transformation of public administration and public value creation: the case of Denmark. Gov. Inf. Q. **39**(1), 101650 (2022)
14. Mergel, I., Edelmann, N., Haug, N.: Defining digital transformation: results from expert interviews. Gov. Inf. Q. **36**(4), 101385 (2019)
15. Panagiotopoulos, P., Klievink, B., Cordella, A.: Public value creation in digital government. Gov. Inf. Q. **36**(4), 101421 (2019)
16. Tangi, L., Janssen, M., Benedetti, M., Noci, G.: Barriers and drivers of digital transformation in public organizations: results from a survey in the Netherlands. In: Pereira, G.V., et al. (eds.) Electronic Government, EGOV 2020, LNCS, vol. 12219, pp. 42–56. Springer, Cham (2020)
17. Dunleavy, P., Margetts, H., Bastow, S., Tinkler, J.: New public management is dead–long live digital-era governance. J. Public Adm. Res. Theory. **16**(3), 467–494 (2006)
18. Ministry of Finance of the Nizhny Novgorod region, http://mf.nnov.ru/. Accessed 5 May 2023
19. Yashina, N.I., Kashina, O.I., Pronchatova-Rubtsova, N.N., Yashin, S.N., Kuznetsov, V.P.: Assessment of budgetary stresses for socio-economic development of regions. In: Popkova, E.G. (ed.) Imitation Market Modeling in Digital Economy: Game Theoretic Approaches, ISC 2020, LNCS, vol. 368, pp. 620–631. Springer, Cham (2022)
20. Yashina, N.I., Kashina, O.I., Pronchatova-Rubtsova, N.N., Yashin, S.N., Kuznetsov, V.P.: Financial monitoring of financial stability and digitalization in federal districts. In: Popkova, E.G., Sergi, B.S. (eds.) Smart Technologies" for Society, State and Economy, ISC 2020, LNCS, vol. 155, pp. 1045–1051. Springer, Cham (2021)

Triple and Quadruple Helix Econometric Models for Solving Applied Problems of Innovative Economies of Countries

Nikolay E. Egorov⊙ and **Tatiyana V. Pospelova**⊙

Abstract The Chapter presents the methodology and tools of econometric analysis and assessment of innovative development of territories within the framework of the concept of the Triple Helix theory. Based on the proposed econometric model of the Triple Helix and a system of 16 main statistical indicators in the field of innovation, numerical calculations were performed to assess the level of innovative development on the example of Russia for 2021. The results of the calculations showed that in Russia, compared with other indicators under consideration, there is a relatively large number of advanced production technologies used, a high proportion of the costs of implementing and using digital technologies to the gross regional product, as well as a significant ratio of exports and imports of technologies and services of a techno-logical nature. The results of the rating express assessment of the level of innovative development of European countries are presented. On the basis of correlation and regression analysis within the framework of the quadruple helix model, the possibility of predictive assessment of the level and quality of life of the population of the territory depending on the level of innovative development of European countries is shown. In general, the presence of a certain relationship between key indicators in the innovation sphere and civil society allows us to use a regression formula to model predictive predictive estimates of the impact of innovation on improving the quality of life of the population.

Keywords European countries · Innovative development · Triple and Guadruple Helix · Model · Quality of life of the population

N. E. Egorov (✉)
The Research Institute of Regional Economics of the North, North-Eastern Federal University, Yakutsk, Russian Federation
e-mail: ene01@yandex.ru

T. V. Pospelova
The National Research University Higher School of Economics, Moscow, Russian Federation

1 Introduction

It is common knowledge that innovation is the main factor determining which countries are the most technologically advanced ones in the world. The successful development of a country's innovative economy depends mainly on the activities of its government, fiscal policy, education policy and the environment of innovation (contribution of innovation), as well as on intellectual deliverables and their introduction into the manufacturing business sector (results of innovation).

At present, very relevant is the issue of assessing the impact of the innovation component of the economy on the level and quality of life of the population. The need for an in-depth and systematic study of the relationship between the prospects of innovative development and quality of life is primarily due to the development in modern conditions of digital transformation of the economy and society [1]. For example, in the context of the relationship analysis "innovative development—quality of life", the territorial heterogeneity in the quality of life of the population and indicators of innovative activity occurs in the territorial subjects of the Russian Federation [2]. At the same time, the trend of lagging behind in the development of the knowledge-based economy, the annual decrease in domestic spending on research and development in the Russian economy may contribute to increasing regional differentiation and further increasing social stratification of the population due to falling real income [3].

A close relationship between the development of innovative activity and the quality of life of the population is confirmed by the works of a number of researchers [4–10]. The analysis of indices assessing the level and quality of life and innovative development of countries shows that a high level of innovative development does contribute to maintaining a high level and quality of life for some countries (e.g., Sweden, Switzerland) [6]. According to the correlation analysis results, there is an absolute positive relationship between the indicators of scientific and technological potential and positive human potential factors in most regions of Russia. Moreover, such factors of scientific and technological potential as research and development costs, technological innovation costs, and the number of advanced production technologies used showed the greatest correlation dependence on the quality of life [7, 8].

Modern scientific literature describes various methods for ranking the level of innovative development of regions (IDR) [11–14]. According to the Global Innovation Index (GII) rankings, Switzerland is in the lead for the twelfth time in 2022, followed by the United States of America, Sweden, the United Kingdom and the Netherlands. At the same time, there is almost no information in the economic scientific literature about methods for quantifying the level of contribution of the scientific and educational complex (SEC), industry and government to the general innovative development of an economic entity based on the concept of the well-known Triple Helix model [15–17].

Currently, a number of research papers address issues of *quantitative measurements* of the level of synergy of TH partners, for example, based on publication activity data [18–20] and on information from high-tech industries [21]. In [22], an econometric analysis of the quantitative relationship between innovation activity indicators based on statistical data from Federal State Statistics Service [23] was performed using the TH model. As the authors L. Leydesdorff et al. note, while scientometric indicators are seen as a means to 'objectify' the quality debate, the quality of the indicators themselves is also an issue in the results debate that needs attention [24]. At the same time, policies can be justified, not just legitimized by evidence-based triple, quadruple, etc. models [25].

Nevertheless, given the complexity of the analyzed processes no unambiguous approach to measuring the processes occurring in the Triple Helix model has been formed at present [26]. This may be due to the fact that the specific application of the TH model in quantitative assessments was not quite obvious, primarily because of the complexity of modeled relationships. If measurements of physical quantities in physical environments do not cause fundamental difficulties, then practical measurements in complex socio-economic environments are characterized by significant difficulties [27].

Thus, as a literature review of researchers' works shows, at present there are actually no practical tools to *quantify* the IDR level based on the theoretical TH model, except for the simulation model of relations between TH actors [28, 29]. In this regard, the development of approaches to understanding the patterns of development of spiral harmonics of the TH model and their interaction, and then applying this material to quantitative assessments of innovation processes is a highly relevant task [30].

2 Triple Helix Econometric Model

As far as is known, the TH model of innovation is based on the interaction between the three main actors of the innovation economy (*actors are significant subjects which play a prominent role in a particular process*): universities engaged in basic research, industries producing commercial goods, and governments regulating markets. Researchers and practitioners from various fields and interdisciplinary research areas such as artificial intelligence, political theory, sociology, professional ethics, higher education, regional geography, and organizational behavior are working with the three-spiral model worldwide, finding opportunities for integration and new directions in TH theory research [31].

Graphically, the relationship between actors can be represented as a three-dimensional geometric representation of the components of a rectangular parallelepiped (Fig. 1).

The presented TH econometric model makes it possible to quantify the integral level of IDR and the contribution of each of the actors to the innovative development of a country on the basis of known trigonometric expressions [32–34]. At that, the quantitative assessment of the IDR level is based on three aggregated blocks of quantitative indicators that characterize the innovative activity of actors.

Fig. 1 Econometric model
for quantitative assessment
of the level of innovative
development of a country
based on the TH model

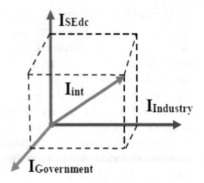

In accordance with Fig. 1, the total integral value of the IDR index can be calculated using the well-known mathematical formula for determining the radius-vector of the three components of a rectangular parallelepiped (the diagonal of a rectangular parallelepiped equals the square root of the sum of the squares of its dimensions):

$$\text{I int} = \sqrt{(\text{Ij SEdr})^2 + (\text{Ij Ind})^2 + (\text{Ij Gov})^2} \tag{1}$$

where

$I_j{}^{SEdr}$—assessment of the impact of state projects on the innovative development of the j-th region;

$I_j{}^{Ind}$—assessment of the impact of industries on the innovative development of the j-th region;

$I_j{}^{Gov}$—assessment of government support of the innovative development of the j-th region.

The value of I_j for each actor is calculated using the arithmetic mean formula in the form:

$$I_j = \sum_{i=1}^{3} K_{in}/n_i \tag{2}$$

where i—group number of actors' indicators; n—number of internal indicators of the i-group; K_{in}—internal indicator calculated by the standard normalization formula for bringing into a unified measurement scale:

$$K_{in} = \frac{x_{in} - x_{in}^{min}}{x_{in}^{max} - x_{in}^{min}} \tag{3}$$

Based on formula (1), the share of each actor's contribution (as %) to the general innovative development of the j-th region is estimated by the ratio:

$$C_j = \left(I_{Int}/I_j\right)^2 * 100 \tag{4}$$

Thus, the presented econometric model makes it possible to implement numerical calculations and quantify the shares of each TH actor's contribution to the general IDR. Essentially, this methodology can also be used to assess the role of R&D, industry and government in the innovative development of the whole country, region, as well as in the context of a separate municipality, industry of the real economy, etc. In this case, the calculation results will depend on the choice of the system of economic indicators of the economy subject under study.

The Russian Federation is considered as an example for performing an integral assessment of the IDR level index using the TH econometric model. Numerical calculations according to formula (1) were performed for the following key (significant) statistical indicators characterizing the innovative activity of the territory (Table 1).

Trends in the Innovation Development Index (IDI) and actors of the TH model for 2015–2021 shows the positive changes of Block C actor (Government, 8.5%), although in 2018 there was a significant decrease in its value by 19.2% (Fig. 2). The rest of the actors and the IDI as a whole saw a decrease in their values, with Block A actor (SEdC) by 20.3%.

Figure 3a illustration of the innovation profile for 2021 shows that Russia, in comparison with other indicators under consideration, has a relatively large number of advanced production technologies in use (B4, 0.98), a high share of costs for introducing and using digital technologies to the gross regional product (GRP) (C3,

Table 1 System of key indicators for assessing the IDR index based on the TH econometric model

Block A. Science and Education (SEdC)	
A1	Number of personnel engaged in research and development, to the average annual number of employed in the economy, %
A2	Number of issued patents for inventions and utility models per 10,000 of the labor force (LF), units
A3	Coefficient of inventive activity, %
A4	Number of students enrolled in bachelor's, specialist's and master's degree programs per 10,000 people
Block B. Industry	
B1	Level of innovative activity of organizations, %
B2	Share of innovative goods, works and services in the total volume of shipped goods, works and services, million rubles
B3	Share of organizations' costs for innovative activity in the total volume of shipped goods, performed works, services, %
B4	Number of advanced production technologies used per 100 units of enterprises and organizations
Block C. Government	
C1	Share of federal budget expenditures on civil science to GRP, %
C2	Share of domestic spending on research and development to GRP, %
C3	Share of costs for introducing and using digital technologies to GRP, %
C4	Ratio of exports to imports of technologies and technology services

Fig. 2 Innovation development index and TH model actors' trends

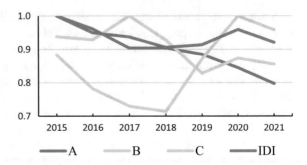

Fig. 3 Innovation profile (**a**) and share of actors' contribution (**b**) in the development of innovation activity

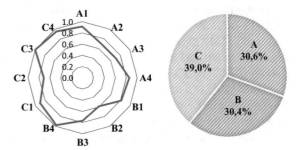

1.0) and the ratio of exports to imports of technologies and technology services (C4, 0.95). At the same time, the contribution of government support for innovation (39.0%) exceeds the shares of the other two actors of the TH model (Fig. 3b).

In general, the research results confirm the opinion of the authors [35] that the Russian Triple Helix concept is characterized by too extensive legal influence of the state on the innovation environment (C indicator), which has a detrimental effect on the development of network interactions, which is why many components of this system remain extremely underdeveloped at the moment.

Essentially, using this methodology it is also possible to perform numerical calculations for rapid assessment of the IDR level index on the basis of the following minimum key indicators of innovative activity:

X—share of R&D personnel and researchers in total active, %;

Y—share of innovative goods, works and services in the total volume of shipped goods, works, and services, %.

Z—share of government budget appropriations or outlays on research and development, %.

Table 2 includes only those countries whose indicators are available in the Eurostat database [36] and the statistical collection of HSE University [37] for 2021. The results of the rapid ranking of European countries by the innovation potential index, calculated using the TH econometric model based on three indicators (Table 2), revealed the leading positions of Iceland, Ireland and Finland (Fig. 4).

Table 2 Indicators of innovative activity of European countries (as %), 2021

Country	X	Y	Z
Austria	2.06	13.0	1.44
Belgium	2.49	15.1	1.31
Denmark	2.23	15.0	1.81
Finland	2.29	19.3	1.60
France	1.84	6.2	1.20
Germany	1.87	14.0	2.19
Greece	1.58	20.3	1.49
Iceland	2.19	5.8	2.43
Ireland	1.52	42.4	0.90
Italy	1.64	13.5	1.18
Luxembourg	1.01	3.8	1.37
Netherlands	1.92	8.9	1.72
Norway	1.94	6.0	1.89
Portugal	1.51	14.5	0.76
Spain	1.28	21.7	1.23
Sweden	2.39	12.7	1.58

Fig. 4 Distribution histogram of European countries by innovation potential index, 2021

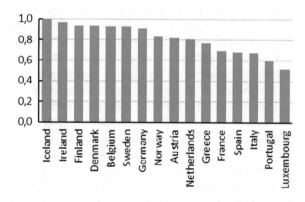

Thus, the above Triple Helix econometric model will allow performing quantitative assessments on monitoring and evaluating the level of innovative development of countries on the basis of statistical data, which is of applied significance for practical activities in science, education, industry, and governance of socio-economic development of a country.

3 Quadruple Helix

The innovative and sustainable development of a country's economy depends not only on having a strong government, scientific community, universities, and industries, but also on how they interact to improve the level of life. In this regard, the evolution of the interactions of innovative models increasingly puts forward the need to consider civil society in the Triple Helix model (TH model).

In recent years, researches has looked at more complex configurations of the innovation process—the "Quadruple Helix", which includes the relationship with civil society [38], and the quintuple, which considers the environmental component in innovative development [39]. This development has now turned the Triple Helix into a Quadruple Helix [40]. Thus, the Quadruple Helix (QH) model, based on the TH model, adds a fourth component to the structure of interaction between university, industry, and government: the public consisting of civil society and the media. Indeed, the Triple Helix model and the Quadruple Helix model are competing concepts, the transition between them is an evolutionary process, and they can complement each other [41].

The QH model concept was first proposed by Elias G. Carayannis and David F. J. Campbell [42] for the purpose of engaging society in the innovation process. The need to add a fourth component is due to the fact that society, specific people play today too large role to be excluded from this model. It is believed that the QH better characterizes the modern post-industrial economy than the TH, since in the twenty-first century, due to globalization, civil society acquires a critical role in the creation and dissemination of new goods and values. Accordingly, the four-link model extends the TH paradigm by adding the function of society that helps to understand the mechanism of knowledge and technology dissemination for diffusion and innovation [43]. The QH model conceptualizes government, science, industry, and civil society as key actors promoting a democratic approach to innovation through which strategy development and decision-making are subject to feedback from key stakeholders, leading to socially responsible policies and practices [44]. In practice, the modern "quadruple helix" concept has been implemented not so long ago, mainly by the countries of Northern Europe and some US states, but governments and the scientific community in other states also became interested in it.

To date, the world practice also lacks a methodology for quantitative assessment of the relationship between the QH actors. Therefore, it seems relevant to develop the Quadruple Helix econometric model adding the fourth link to the Triple Helix model, i.e. civil society with the appropriate definition of its system of indicators in the social sphere, and conduct numerical test calculations to identify correlation links between the partners of these models with a subsequent assessment of the levels of their interaction (Fig. 5).

In the figure above, the intersection area of the four actors forms the QH model core, which represents users of innovation: this model stimulates the creation of innovation important for users (civil society). Users (i.e. citizens) define and drive the innovation process [45]. Citizens not only participate in the actual development

Fig. 5 Quadruple helix structural model of actors' interaction in innovation process

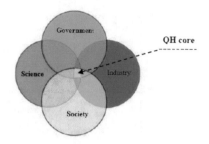

process, but also can propose new types of innovation, as a result of which users are connected with other participants of the quadruple helix in the scientific and educational sphere, business structure and regional executive power. In turn, representatives of these other three helix links support the innovative activity of citizens (provide them with tools, information, platforms for development, and skills necessary to create innovation). Industry and the public sector will be able to further benefit from innovation created by citizens for the further development of the innovation economy of a region.

It is common knowledge that improving the quality of life of the population is the main priority of any state, since the socio-economic well-being of a country directly depends on the national well-being of citizens living in this territory. At the same time, regions with an increased level of innovativeness have a higher competitive attractiveness for the consumers of the territorial space in terms of economy and quality of life and are developing more steadily [46]. It should be noted that in 2022 the list of Top 5 countries with the highest quality of life (Quality of Life Index (QLI) by Country) is occupied by Switzerland, Denmark, the Netherlands, Finland and Australia [47]. The analysis of indices assessing the innovative development and quality of life in some European countries also shows the positive effect of a country's high level of innovative development on improving the quality of life of the population [6]. In this regard, for a certain scientific and practical interest, a quantitative assessment of the relationship between the level of innovative development of countries and the quality of life of the population of a particular region is carried out.

As far as is known, to measure the strength of the relationship between two variables, the Pearson correlation coefficient (R) is used, usually determined by the Chaddock scale [48]. The results of the correlation analysis between the Innovative Development Index (IDI), calculated by the authors according to the TH econometric model using formula 1 and the data in Table 2, and the Quality of Life Index (QLI) for European countries in 2021 are shown in Fig. 6.

The presented illustration shows that there is a moderate degree of regression dependence between IDI and QLI according to the Chaddock scale (range 0.3–0.5) with a correlation coefficient of R = 0.43. This fact is quite consistent with the results of a similar analysis between the Russian Innovation Development Index and the Quality of Life Index of the regions of Russia [10, 49].

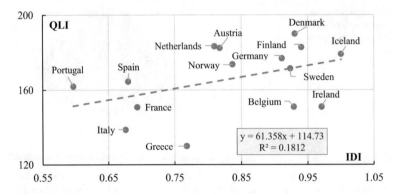

Fig. 6 Relationship between the innovation development index (IDI) and the quality of life index (QLI), 2021

In general, a certain relationship between key indicators in the innovation sphere and civil society will make it possible to use the standard regression formula of the y = ax + b type to model forecasts of the effect of the level of innovative development on improving the quality of life of the world population.

4   Discussion and Conclusion

The Chapter presents the methodology and tools of econometric analysis and assessment of a country's innovative development within the Triple Helix theory. Econometric calculations according to this methodology allow carrying out rapid assessment of the level of innovative development and the contribution of scientific and educational complex, industry, and government to the consolidated integral index of innovative development of Russia and European countries.

Currently, the issues of studying the relationship between levels of innovation activity and quality of life of the population are highly relevant tasks, especially in the modern conditions of digital transformation of the economy and social sphere. For example, in the subjects of the Russian Federation, there is territorial heterogeneity in the quality of life of the population and indicators of innovation activity.

As the results of the correlation analysis of the regions in the Arctic Zone of the Russian Federation (AZRF) show, there is a moderate degree of correlation according to the Chaddock scale between the level of innovation development and the quality of life of the population observed only in the Republic of Karelia (0.45), while in the Komi Republic, Yamalo-Nenets Autonomous Okrug, and Krasnoyarsk Krai, the correlation is weak (0.1–0.3). In the rest of the regions in AZRF, there is a negative correlation between these indicators. Therefore, this article demonstrates the possibility of predictive assessment of the quality of life of the population in a territory based on the level of innovation development in the country within the framework of the quadruple helix model. Thus, the research results demonstrates a certain relationship between the indices under consideration, which makes it possible to use the regression dependence formula of the $y = ax + b$ type to model forecasts of the effect of innovative development on improving the level and quality of life of the population.

The research results obtained in this paper are quite reliable, as it uses statistical data from official sources for publication in the public domain. The process of quantitative assessment is based on a system of indicators in the innovative and social spheres, which can be adjusted depending on the goals and objectives of the research.

The outlined methodology of the research will improve the level and quality of strategic planning and management of the development of innovative economy and civil society of macro-regions. The research results can be useful to executive authorities, businesses and scientific and educational institutions of regions to analyze, forecast the formation and development of the socio-economic system of territories.

Further research on this topic will be aimed at studying the effect of the innovative activity results on improving the quality of life of the population of territories in the conditions of digital transformation of industry, social sphere and government based on the presented Quadruple Helix model.

Acknowledgements The article is prepared within the framework of the state assignment of the Ministry of Education and Science of Russia under the project No FSRG-2023-0025 "Modern Methods of Mathematical Modeling and Their Applications".

References

1. Gagulina, N., Samoylov, A., Novikov, A., Yanova, E.: Innovation-driven development and quality of living under conditions of digital economy. E3S Web Conf. **157** (2020). https://doi.org/10.1051/e3sconf/202015704037
2. Okrepilov, V., Gagulina, N., Getmanova, G.: Factors of innovative development of regions in the concept of quality economics. In: The European Proceedings of Social and Behavioural Sciences, pp. 409–418 (2020). https://doi.org/10.15405/epsbs.2020.10.03.46
3. Okrepilov, V., Gagulina, N.: Structural transformations of innovative development and quality of life in modern environment. E3S Web Conf. **258** (2021). https://doi.org/10.1051/e3sconf/202125806008
4. Arkhipova, M., Bazhenova, T.: Study of the effect of innovative development of the region on the main characteristics of the quality of life of the population. Drukerovskij vestnik **4**(18), 176–191 (2017)
5. Dubrova, T., Klochko, Y.: Analysis of the quality of life of the population, taking into account the development of innovative activity in the regions of Russia. J. Econ. Entrep. **9**(74), 908–914 (2016)
6. Nikulina, N., Bagaryakov, A.: Impact of innovative development on the quality of life of the population of the region. Proc. High. Educ. Inst. Sociol. Econ. Polit. **1**, 70–72 (2015)
7. Fedotov, A.: Human and scientific and technical potential: correlation regional analysis. Population **23**(4), 61–70 (2020). https://doi.org/10.19181/population.2020.23.4.6
8. Fedotov, A.: Scientific and technological potential and the quality of life of the population: search for interrelationships. Int. J. Hum.Ities Nat. Sci. **11–3**(62), 251–257 (2021). https://doi.org/10.24412/2500-1000-2021-11-3-251-257

9. Fedotov, A.: Quality of life and human potential in the concepts of sustainable and human development (Part Two). Population **24**(3), 42–50 (2021). https://doi.org/10.19181/population. 2021.24.3.4

10. Egorov, N.: Relationship between innovative development and quality of life in the Asia-Pacific region and the Far East. Far East. Stud. **5**, 47–61 (2022). https://doi.org/10.31857/S01312812 0022348-3

11. The Global Innovation Index. https://www.globalinnovationindex.org/gii-2022-report. Accessed 28 June 2023

12. European Innovation Scoreboard. http://www.proinno-europe.eu. Accessed 26 June 2023

13. Regional Innovation Scoreboard. http://ec.europa.eu/growth/industry/innovation/facts-figures/ regional_en. Accessed 28 June 2023

14. Kudryavtseva, S., et al.: The methods of national innovation systems assessing. Int. Rev. Manag. Mark. **6**(2), 225–230 (2016)

15. Etzkowitz, H., Leydesdorff, L.: The Triple Helix—university-industry-government relations: a laboratory for knowledge based economic development. EASST Rev. **14**, 14–19 (1995)

16. Etzkowitz, H., Leydesdorff, L.: The dynamics of innovation: from national systems and "'Mode 2'" to a Triple Helix of university–industry–government relations. Res. Policy **29**(2), 109–123 (2000). https://doi.org/10.1016/S0048-7333(99)00055-4

17. Etzkowitz, H.: Innovation in innovation: The Triple Helix of university–industry–government relations. Soc. Sci. Inf. **42**(3), 293–337 (2003)

18. Mêgnigbêto, E.: Measuring synergy within a Triple Helix innovation system using game theory: cases of some developed and emerging countries. Triple Helix **5**(6), (2018). https://doi.org/10. 1186/s40604-018-0054-8

19. Leydesdorff, L., Park, H.: Can synergy in Triple Helix relations be quantified? A review of the development of the Triple Helix indicator. Triple Helix **1**(4) (2014). https://doi.org/10.1186/ s40604-014-0004-z

20. Nurutdinova, A., Dmitrieva, E.: Experience of the TH model: comparative analysis (in the context of Russia and China). Mod. Probl. Sci. Educ. **6**, 201 (2018). https://doi.org/10.17513/ spno.28296

21. Leydesdorff, L., Perevodchikov, E., Uvarov, A.: Measuring Triple-Helix synergy in the Russian innovation systems at regional, provincial, and national levels. J. Am. Soc. Inf. Sci. **66**(6), 1229–1238 (2015). https://doi.org/10.1002/asi.23258

22. Istomina, S., Lychagina, T., Pakhomova, E.: Econometric analysis of innovative development drivers in the Russian economy. Natl. Interest: Prior. Secur. **14**(10), 1943–1960 (2018). https:// doi.org/10.24891/ni.14.10.1943. Accessed 28 June 2023

23. Regions of Russia. Socio-economic indicators. Statistical Collection (2022). http://www.gks.ru. Accessed 28 June 2023

24. Leydesdorff, L., Wagner, C., Zhang, L.: Are university rankings statistically significant? A comparison among Chinese universities and with the USA. J. Digit. Inf. Sci. **6**(2), 67–95 (2021). https://doi.org/10.2478/jdis-2021-0014

25. Leydesdorff, L., Smith, H.: Triple, quadruple, and higher-order helices: historical phenomena and (Neo-)evolutionary models. Triple Helix **9**, 6–31 (2022). https://doi.org/10.2139/ssrn.381 7410

26. Popodko, G., Nagaeva, O.: Conditions for the implementation of the Triple Helix model in resource-type regions. Issues Innov. Econ. **9**(1), 77–96 (2019). https://doi.org/10.18334/vinec. 9.1.40494

27. Drobot, D., Drobot, P., Uvarov, A.: Prevailing role of universities in the Triple Helix model. Innovations **4**(15), 93–96 (2011)

28. Ivanova, I., Leydesdorff, L.: A simulation model of the Triple Helix of university-industry-government relations and the decomposition of the redundancy. Scientometrics **99**(3), 927–948 (2014). https://doi.org/10.1007/s11192-014-1241-7

29. Ivanova, I., Leydesdorff, L.: Rotational symmetry and the transformation of innovation systems in a Triple Helix of university-industry-government relations. Technol. Forecast. Soc. Chang. **86**, 143–156 (2014)

30. Drobot, P., Drobot, D.: Analysis of the main U-harmonic (university) and B-harmonic (business) in the model of the "Triple Helix." Innovations **11**(229), 101–105 (2017)
31. Cai, Y., Etzkowitz, H.: Theorizing the Triple Helix model: past, present, and future. Triple Helix J. **6**(1), 1–38 (2020). https://doi.org/10.1163/21971927-bja10003
32. Egorov, N.: Method of assessment of level of contribution of Triple Helix participants in innovative development of economy subject. Helice **4**(2), 27–31 (2015). https://doi.org/10. 17686/sced_rusnauka_2015-1323
33. Egorov, N., Pospelova, T., Yarygina, A., Klochkova, E.: The assessment of innovation development in the arctic regions of Russia based on the Triple Helix model. Resources **8**(2), 72 (2019). https://doi.org/10.3390/resources8020072
34. Egorov, N., Babkin, A., Babkin, I., Yarygina, A.: Innovative development in Northern Russia assessed by Triple Helix model. Int. J. Technol. **12**(7), 1387–1396 (2021). https://doi.org/10. 14716/ijtech.v12i7.5355
35. Udaltsova, N., Krutskikh, D.: Features of the formation and development of the innovation system of Russia in the context of the "Triple Helix". Issues Innov. Econ. **11**(1), 33–46 (2021). https://doi.org/10.18334/vinec.11.1.111894
36. Database Eurostat. https://ec.europa.eu/eurostat/web/main/data/database. Accessed 28 June 2023
37. Indicators of Innovative Activity 2023: Statistical Collection. HSE University (2023). https:// doi.org/10.17323/978-5-7598-2749-8
38. Leydesdorff, L., Etzkowitz, H.: Can "the public" be considered as a fourth helix in university–industry–government relations? Report of the fourth Triple Helix conference. Sci. Public Policy **30**(1), 55–61 (2003)
39. Campbell, D., Carayannis, E., Rehman, S.: Quadruple helix structures of quality of democracy in innovation systems: the USA, OECD countries, and EU member countries in global comparison. J. Knowl. Econ. **6**(3), 467–493 (2015)
40. Kimatu, J.: Evolution of strategic interactions from the triple to quad helix innovation models for sustainable development in the era of globalization. J. Innov. Entrep. **5**(16) (2016). https:// doi.org/10.1186/s13731-016-0044-x
41. Cai, Y., Lattu, A.: Triple Helix or Quadruple Helix: which model of innovation to choose for empirical studies? Minerva **60**, 257–280 (2022). https://doi.org/10.1007/s11024-021-09453-6
42. Carayannis, E., Campbell, D.: 'Mode 3' and 'Quadruple Helix': toward a 21st century fractal innovation ecosystem. Int. J. Technol. Manag. 46, **3**(4), 201–234 (2009)
43. Shestak, V., Tyutyunnik, I.: Financial and legal support for innovative activity. Financ. Theory Pract. **21**(6), 118–127 (2017). https://doi.org/10.26794/2587-5671-2017-21-6-118-127
44. Carayannis, E., Campbell, D.: Mode 3 knowledge production in quadruple helix innovation systems. In: Mode 3 Knowledge Production in Quadruple Helix Innovation Systems, pp. 1–63. Springer, New York (2012)
45. Carayannis, E., Grigoroudis, E.: Quadruple innovation helix and smart specialization: knowledge production and national competitiveness. Foresight STI Gover. **10**(1), 31–42 (2016). https://doi.org/10.17323/1995-459x.2016.1.31.42
46. Grinchel, B., Nazarova, E.: The influence of regions' innovativeness on competitive attractiveness and stability of economy and quality of life. Innovations **8**(226), 105–113 (2017)
47. Quality of Life Index by Country. https://www.numbeo.com/quality-of-life/rankings_by_cou ntry.jsp?title=2022. Accessed 28 June 2023
48. Methods for Calculating the Correlation Coefficient. https://ekonometrik.ru/cheddock-scale. Accessed 28 June 2023
49. Egorov, N., Pavlova, S.: Joint research of the innovative development and quality of life of region on the example of the far eastern federal district. Population **25**(4), 92–103 (2022). https://doi.org/10.19181/population.2022.25.4.8

Digital Transformation Tools for Sustainable Supply Chains in the Life Cycle Processes

Anna Sinitsyna◉ and Alexey Nekrasov◉

Abstract In this chapter reveals innovative methods and tools for the digital transformation of supply chain processes. A comprehensive methodological approach to the sustainability of complex organizational and technical objects, in particular, transport and logistics terminals (TLT), is considered, based on system engineering and an architectural approach to the process of cost incurrence in conditions of emergency situations. A dynamically changing technological environment involves the development of a program for analyzing risks and critical points of cargo delivery management, which ensures highly efficient integration of TLT processes, a control mechanism based on digital technologies. A systemic basis is created for fast reconfiguration of processes in conditions of failure situations. Life Cycle Models (LC) form the backbone of next-generation integrated supply chains powered by smart technologies such as Big Data, Block Chain, Internet of Things, Cloud Services, Mobileapps, Artificial Intelligence, RFID, 3D Printing, Industry 4.0 (Robotics), Omni-Channel Logistics and process principles. The creation of digital platforms and ecosystems is considered as a traditional tool of digital transformation, which allows to systematically solve a number of tasks related to the functions of transportation and delivery of goods for a unified risk management strategy, as well as for the formation of supply chains of smart cities. With the help of digital life cycle modules, linked to the identification, assessment and monitoring of risks, interaction between all technological and logistic processes is ensured. Everything is ready for smarter, faster, safer supply chains.

Keywords Digital transformation · System engineering · Supply chain · Risk management · Homeokinetic plateau · Key control points

A. Sinitsyna (✉) · A. Nekrasov
Russian University of Transport, 9b9 Obrazcova Ulitsa, Moscow 127994, Russia
e-mail: acc-lgkr@mail.ru

A. Nekrasov
e-mail: tehnologistic@mail.ru

© The Author(s), under exclusive license to Springer Nature Switzerland AG 2024
M. Sari and A. Kulachinskaya (eds.), *Digital Transformation: What are the Smart Cities Today?*, Lecture Notes in Networks and Systems 846,
https://doi.org/10.1007/978-3-031-49390-4_15

1 Introduction

The main trend of the modern economy and business is digitalization and the use of innovative methods of systems and software engineering. Distributed databases, classifiers, Standards and integration of processes using information and communication technologies, including in supply chains, form the basis for the digital transformation of transport infrastructure, the introduction of the concept of smart cities and logistics processes, which ensure sustainability and intelligent mobility. A smart city is not only about improving infrastructure, transport and utilities. It is a cyber-physical engine for optimization/innovation of any human activity.

The digital supply chain management system in the context of the application of innovative solutions related to the concept of Industry 4.0 should acquire the properties of a «living system», transforming not individual functions and processes, but the state of the entire system [1, 2].

A «Living system» is a hierarchically organized complex system that has its own development program, the implementation of which ensures the preservation of the system on the basis of maintaining a certain level of entropy that is different from the maximum. Any «living system» has a complex hierarchical network structural organization, which is the highest form of system's ordering and is the source of its anti-entropy [3].

A new approach is required, based on the systemic mechanism of integrated management not only of the supply chain, but also on the basis of the interaction of intelligent and Internet technologies (such as «Internet of Things»), using the capabilities of combining processes and resources for the delivery of goods in the context of global risks.

2 Methods

As a scientific hypothesis, the authors consider the integrated application of digital technologies and business models, which implies a significant revision of traditional approaches. The mechanism for managing complex objects in a digital transformation based on an integrated architecture of the supply chain provides for the coverage of all stages of the life cycle, including conception, design, creation, application (operation), support and writing off. The actualization of digital transformation, mechanisms and smart digital technologies (artificial intelligence, blockchain, big data analysis and the internet of things), revolutionary changes in the creation of platform-type models and development strategies within the framework of Industry 4.0 allow creating new potential for the development of enterprises in the transport industry [1, 2]. In this regard, a number of significant problems can be identified:

- Different modes of transport are not sufficiently integrated with each other
- Access to high-quality and reliable transport services is not provided

- Such tasks as reducing the harmful impact on the environment, reducing accidents, ensuring the safety of transport facilities have not been resolved
- The share of transport costs in the price of products is large, the specific level of which should decrease by 2030 from 20 to 13%.

In 2018–2019, the situation on the world commodity market worsened due to a significant increase in tensions in trade relations between the largest countries of the world. The total cost of transportation and logistics in the world is estimated at 9.25 trillion US dollars (about 10.8% of global GDP). In 2020, the drop in cargo turnover in all segments of the transport services market was from 6 to 16%, depending on the type of transport, as well as due to a reduction in cargo turnover, production and cargo base, disruption of international supply chains, reduced import demand and deterioration of export conditions on the world market. The COVID-19 pandemic has disrupted global supply chains, as well as reduced freight traffic and industrial production. To develop the analysis of the problem, the chapter considered a pronounced crisis tendency to reduce the efficiency or disintegration of traditional supply chains and transport processes. For example, in the first two months of 2020, the transshipment of foreign trade cargo in containers in Russian ports decreased by 2.7%. The global container transportation market in the 1st quarter of 2020 decreased by 4.7%; freight transportation by rail decreased by more than 3%. The main reason for the reduction is a reduction in imports due to quarantine measures in China and some European countries, as well as a decrease in consumer demand. The overall decline in GDP in 2020 ranged from 3.8 to 10.2% [4, 5].

As digital transformation tools, the authors of the article consider the integration of cargo delivery processes in the operation of supply chains (SC) and transport and logistics terminals (TLT) with an automated risk monitoring system based on a technological map of key control points (KCP).The automated control mechanism covers the entire cycle of cargo movement from initialization to the assessment of the service provided. The technological environment involves the development of a program for analyzing risks and critical points of cargo delivery management, which ensures highly efficient integration of TLT processes and a control mechanism based on digital technologies.

The control mechanism covers the entire cycle of cargo movement: from initialization to assessing the level of services provided. The risk map for various processes covers not only the level of an individual participant (event), but also involves the development of a comprehensive program for analyzing risks and critical points for managing the delivery and handling of goods, which ensures highly efficient integration of processes of transport and logistics terminals based on various digital tools, thereby an innovative systemic basis is created for the rapid reconfiguration of processes in the face of rapidly changing events, including failures.

The general trend is a high risk of not only failure of local networks for transportation of goods, but also the destruction of global supply chains. For this reason, the crisis situation of instability in the functioning of supply chains should imply a revision of approaches, including the development of digital transformation strategies and scientific and methodological principles for the organization and management of existing supply chains. In the very near future, bimodal supply chains will be formed: «first mode» (traditional)—lean efficiency, low risks, high predictability; «second mode»—the need for agility, speed and exploring new possibilities.

At the same time, the main task is seen as the digitalization of the entire supply chain and the widespread introduction of Industry 4.0 technologies that do not depend on the human factor [4–6]. This will require innovative methods to reformat approaches to digital transformation, which are taking into account an integrated nature and a risk-based approach to supply chains in relation to the transport services and freight market [7–10].

3   Theory and Practical Application

Due to the constant increase in the complexity of systems created by man, a number of scientific and methodological problems arise at different stages of the life cycle of complex organizational and technical systems (COTS) and at different levels of architectural detail. A system-wide approach is needed that ensures both the effective interaction of processes throughout the entire life cycle (LC) and the search for mechanisms for the sustainable functioning of all TLT objects and the supply chain as a whole. Thus, the problem of researching new tools for digital transformation of supply chains is largely related to the integration of information and logistics technologies (ILT) and the life cycle processes of delivery and handling of goods. When developing an ILT strategy, attention is focused on three main approaches:

- Increased cargo security in the supply chain
- Facilitate the distribution network within the extended security structure
- Provide for rapid recovery of the supply after an incident that disrupts the supply integrity system.

All these approaches can be implemented on the basis of the formation of innovative tools within the framework of integrated transport and logistics systems (ITLS) that function steadily at various stages of the life cycle. ITLS provide a more efficient and safer level of operation [10, 11]. In the integrated model, further changes are taking place for the development of transportation (including multimodal) based on large TLT or with mixed types of transportation models. Figure 1 shows the main elements and relationships of such a cargo terminal focused on air and road transport. The creation of digital platforms and ecosystems is considered as a traditional digital transformation tool associated with transportation functions and a unified risk management strategy [12].

The key requirements are minimizing the cost of transportation, ensuring the safety and timeliness of cargo delivery, and ensuring the sustainability of the processes themselves, including the forecast of possible negative events. This approach to security is natural within the framework of modern globalization and multimodal approach in transport logistics [13, 14]. For the optimal functioning of the supply chain of goods, an approach is required that provides for the assessment

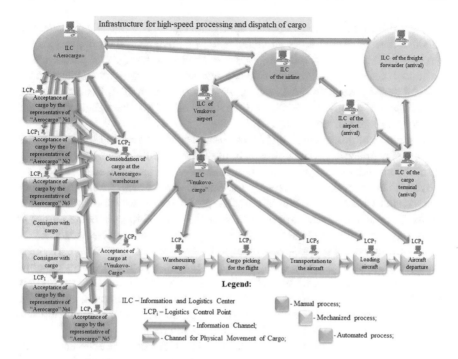

Fig. 1 A complex system of TLT with a mixed model of handling and delivery of goods

and optimization of potential hazards and risks, i.e. it is necessary to move from the principles of «zero risk» to the principles of «acceptable risk» [15–18].

Sustainability analysis allows selecting a plan with the required performance guarantee, identifying bottlenecks in the plan and measures to strengthen them, and develop scenarios to support operational decisions on reconfiguring supply chains based on an analysis of key performance indicators and tolerances of plan parameters.

The use of this tool in the model of supply chain management and TLT, in addition to the development of theoretical foundations, has practical significance, such as improving the quality and accuracy of planning and management and decision support by management at the levels of goal setting, planning, monitoring and regulation of supply chains.

To ensure the competitiveness of TLT, the focus of management of purely technical processes is changed to the creation of sustainable organizational and technical structures for managing life cycle models, including production assets, business processes, and after-sales service of products. Supply chain risk management integrates the risk management program with digital life cycle processes. At the same time, when considering the overall reliability of the supply chain, it is necessary to consider the combination of all elements interacting within the functional logistics cycle in the form of a sequential chain of events.

Event-driven interaction between all technological and logistic processes is provided through digital lifecycle modules (design, planning, delivery, service and support) related to risk definition, assessment and monitoring. The interaction of the modules is ensured by means of electronic «dossiers» with an architecture for combining risk resources and assessing the performance of TLT. For example, the management of a deliver «dossier» will be linked directly to the logistic import and export operations carried out by TLT.

In addition to processing individual «dossiers» that correspond to consignment notes, the integrated digital modules can be used to manage the grouping of «dossiers» or their consolidation based on the combination of various resources. If there are delivery and transportation schemes in the database, the choice of a specific scheme is associated with the formation of a tracing profile for the execution of sequential events of an administrative nature. The «Dossier Profitability» module allows to prepare a profitability forecast for individual operations and expected costs. At the same time, after the completion of each logistics cycle and the operations performed, the expected costs and receipts that have not yet been registered in the accounting department are established. For each delivery scheme, it is necessary to obtain not only financial results, reflected through the operating income margin, but also the possibility of comparing the obtained data on productivity and risks, taking into account the dynamics of their development in the future.

The indicators of the reliability of production and logistics processes are the data of probabilistic values in the interval $0 \leq P \leq 1$. In this case, «0» is an indicator of complete breakdown of functioning (failure), and «1» is an indicator of complete interaction. The reliability of processes in a transport and logistics terminal refers to the probability that the normative («effective») values of the parameters will be achieved within a certain period of time and within the specified tolerances. An order for the delivery of goods is considered completed if it is implemented within the area of permissible deviations, or the area of positive events. Probability values are determined based on statistical data and design values of system parameters using the formula:

$$P_i = A/B, \tag{1}$$

where: A is the amount of absenteeism outside the tolerance; B—the total amount of available data.

4   Results and Discussion

The basic result of the research is to obtain new characteristics of the sustainability of the life cycle processes of TLT handling and delivery of cargo in a «mixed model», to establish a range of key risk points for the entire logistics system. Risks are various probabilistic events that can positively and/or negatively impact goals. The State Standard «Risk Management. Terms and definitions» [19] risk refers to the combination of the probability of an event and its consequences. The risk arises from the possibility of deviation from the expected result or event. Risk elements include sources of hazards, events, consequences and likelihood.

Given the multifaceted nature of security and sustainability issues, it is recommended to identify those risks that have the most significant impact on the management of cargo delivery processes. Consequently, they must be controlled and / or mitigated through the implementation of programs to improve and increase the sustainability of processes. An analytical tool for assessing the risks of TLT sustainability is a methodological approach based on the identification and recognition of risk levels. In particular, 5 levels of risk were identified, which were assessed in terms of the probability of events and the rating of danger (Table 1).

As a SC risk, a set of criteria such as the efficiency of order fulfillment in terms of keeping of delivery times, quality of services provided, assortment of supplies and costs throughout the life cycle will be considered. To ensure the sustainability of the SC relative to control points, it is necessary to create stability mechanisms, which will allow linking the problems of the effective functioning of processes within the functional and logistics cycle with the terminal's technological environment.

One of the difficult problems is the establishment of critical points (boundaries) of sustainability in relation to various stages of the life cycle of TLT. The problem of technological transformation of all systems will come to the fore on the basis of the system sustainability criterion as a complex indicator of the quality of TLT functioning, on the basis of the risk management mechanism [20–24].

As transformation tools, this research examines the integration of delivery and handling processes in TLT with a risk monitoring system based on a flow chart of key critical points (KCP). The control mechanism covers the entire cycle of cargo movement: from initialization to assessing the level of services provided. The risk map for various processes covers not only the level of an individual participant (event), but also involves the development of a comprehensive program for analyzing risks and critical points for managing the delivery and handling of goods, which ensures highly efficient integration of processes of transport and logistics terminals based on various digital tools, thereby an innovative systemic basis is created for the rapid reconfiguration of processes in the face of rapidly changing events, including failures. Technological methods of their description with the help of special organizational-digital forms are of particular importance, which provide the following main advantages:

Table 1 Correlation of risks and hazard ratings of SC/TLT

Risk level	Event probability	Seriousness of the consequences	Hazard rating	Color indicator
1	0.8–1.0	Very high	Critical	Red
2	0.6–0.8	High	Considerable	Red
3	0.4–0.6	Average	Small	Yellow
4	0.2–0.4	Low	Acceptable	Yellow
5	0.1–0.2	Negligible	Minimal	Green

- It becomes possible to electronically check the processing of service results and analysis of data reflected in the forms
- The initial forms can be used to generate input and output data, arrays and procedures of the system being implemented
- Convenience for the subsequent use of electronic data that are collected in the form of predetermined forms, are of particular importance.

Consider the KCP used forms. For example, the «Pre-project stage» begins with the stage of diagnostics of the survey of existing business processes and information system (IS). The result of this work is an analytical note and the structure of existing business processes. The collected information is used by the designers of IS to get a complete picture of the activities of TLT. The main directions of service activities, corporate (cross-functional) problems associated with data processing should be assessed. In the second phase of the pre-project stage, more detailed data on the logistics infrastructure and its facilities are collected. The structural and cost form of risk analysis is shown in Table 2.

To ensure successful interaction in the life cycle processes, both strategic and operational information about the state and changes of all components of the system is required—stage «Measuring and evaluating performance».

It's about the opportunity the possibility of obtaining quantitative assessments (indicators) for all factors (models) and parameters based on the accepted assessment criteria and converting them into calculated indicators. When managing the operational and coordinating circuits of SC, the information flow is able to perceive internal and external signals (influences), providing through system factors the adaptation (adaptation) of the entire after-sales service system. The information system acts as a feedback system based on an automatic control mechanism. A prerequisite is the inclusion in the regulation of parameters («critical points») only those

Table 2 Structural form of risk analysis for the sustainability of SC

The structure of key critical points	Key competencies scheme			Total logistics costs
	Transport «Dossier»	Transport and logistics processes	Result A	
			Result B	
Personnel				Σ
Cargo				
Hierarchical levels				
Processes				
Results/failures				
Task (business process): preparation of proposals to reduce risks				Σ
Total				

The amount of total costs for tasks (business processes) to identify and assess the contribution of each process to the overall result created. After that, the level of SC stability is assessed that determines the final results, productive interaction of processes and hierarchical levels. The basis for the systemic processing and integration of digital data is information modeling based on the analysis of the behavior of system factors, processes and results within the framework of the SC process engineering methodology. In the process of converting information from the comparison of the planned and reporting (threshold) data with the final data, new (expanded, aggregated) information appears, which should ensure the interaction of the inputs-outputs of information systems.

A feature of the received tools for digital transformation of sustainable supply chains is their multi-component nature, coverage of all stages of the life cycle, the presence of a customizable adaptive management system focused on the risk management program. In turn, the function of sustainability management, based on the methodology of systems engineering with a constantly decreasing period of the need for reconfiguring processes, will allow linking all parties and resources involved in the transport and logistics process. The coordinated interaction of processes, for example, considered in a transport and logistics terminal with a mixed model, involves the integration of life cycle events with proactive actions throughout the entire risk management cycle based on the flow chart of key critical points.

Technological maps form the basis for highly efficient sustainable functioning of SC processes, which is especially important in the context of digital transformation. In this context, the existing approaches to logistics and its traditional infrastructure within the CPU are considered as the basis for the integration of life cycle processes, which is focused on engineering risk-oriented solutions. When forming a transportation scheme, data on the order for delivery and handling of goods, data on logistic labels of units of transported goods are grouped, which serve as a reliable source of changes in events throughout the use of the entire flow chart, and key critical points are the basis for monitoring event risks. Life cycle models form the basis for the creation of a new generation of integrated supply chains, including SC with a mixed model of transportation and handling of goods, based on digital technologies and principles of self-regulation.

The use of flow charts with key critical points creates conditions for an engineering and design approach to risk management of production and digital assets of the terminal, and technical processes through key critical points integrate organizational and technical measures for handling cargo with the equipment used throughout the entire life cycle.

5 Conclusions

As a result, the formation of a new image of design and mechanisms for the sustainability of the logistics system and the sphere of supply chain management is an integral part of the digital transformation process of any complex object. As a key factor in ensuring the competitiveness and sustainability of SC functioning, it is necessary to consider the concept and mechanism of the homeokinetic plateau [23], which makes it possible to implement the stated scientific and methodological principles and approaches to risk management based on systems engineering. The formation of an integrated model of digital transformation of SC should be carried out not within the framework of a single enterprise, but in the interaction of all participants in the supply chain of products, delivery and handling of goods, who ensure a comprehensive and rapid transformation of decision-making processes in conditions of high dynamics of change.

The proposed interdisciplinary approach to digital transformation of theoretically and practically justified stages of building models and risk management tools for the SC model allows monitoring the multistructural state of a transport and logistics facility in a turbulent environment.

Process systems engineering methods and tools, combined with digital technologies; provide holistic and sustainable management of physical and digital objects. The basis for modeling complex organizational and technical systems, supply chains are the principles of adaptation and integration of system elements to overcome the problems of uncertainty and functioning in conditions of failure situations.

Technological systems design methods and tools, combined with digital technologies, ensure the holistic and sustainable management of physical and digital assets. Modeling of complex organizational and technical systems, supply chains is based on the principles of adaptation and integration of system elements to overcome the problems of uncertainty and functioning in emergency situations.

Adaptability, stability in development, and the ability to recover quickly are emergent properties of complex adaptive systems [25, 26]. Smart systems allow to make informed decisions on the go to adapt and optimize production and adapt to individual customer needs without interrupting processes. Finally, smart technologies open up new opportunities for democratic governance and decision-making based on multi-stakeholder participation [27]. These approaches can be considered as an example of the most complete realization of the potential of the Fourth Industrial Revolution, including the development of the Smart City concept, which is the locomotive of the digital economy and the most dynamic segment of the global digital market.

References

1. Aleinik, N.: What is digital transformation and how does it differ from digitalization and Industry 4.0. https://rb.ru/story/what-is-digital-transformation/. Accessed 09 February 2020
2. Tang, C.S., Veelenturf, L.P.: The strategic role of logistics in the industry 4.0 era. Transp. Res. Part E: Logist. Transp. Rev. **129**, 1–11 (2019). https://doi.org/10.1016/j.tre.2019.06.004
3. Hintjens, P.: Social Architecture: Building On-line Communities. Chapter 6. Living systems. https://habr.com/ru/company/philtech/blog/342036/. Accessed 09 February 2020
4. The impact of COVID-19 on the Russian economy. McKinsy & Company (2020). http://www.kovernino.ru/?id=24309. Accessed 6 July 2020
5. Wilding, R.: The coronavirus will reshape logistics and global supply chains. GMK Center (2020). https://gmk.center/opinion/koronavirus-izmenit-logistiku-i-globalnye-cepochki-postavok/. Accessed 25 January 2020

6. Samarin, A.: Digital transformation of organized systems: the main thing for ministers, programmers and critics. http://digital-economy.ru. Accessed 26 January 2020
7. Drobot, V.S.: Prospects for the development of cyber-physical production systems. https://controlengrussia.com/magazine/control-engineering-rossiya-october-2018/. Accessed 25 January 2020
8. Andreeva, L.A.: Innovative processes of logistics management in intelligent transport systems. In: Levin, B.A., Mirotin, L.B. (eds.) Educational and Methodical Center for Education on Railway Transport, Moscow (2015)
9. Nekrasov, A.G., Melnikov, D.A.: Supply Chain Security in the Aviation Industry. Publisher, PrintUp, Moscow (2006)
10. Nekrasov, A.G., Ataev, K.I., Nekrasova, M.A.: Management of Security and Risk Processes in Supply Chains: A Training Manual. Technical Polygraph Center, Moscow (2011)
11. Nekrasov, A.G.: A paradigm shift: transition to the PBL methodology. Logistics **1**, 54–57 (2015)
12. Sokolov, B.V., Yusupov, R.M., Ivanov, D.A.: Conceptual description of integrated risk modelling problems for managerial decisions in complex organisational and technical systems. Int. J. Risk Assess. Manag. **3–4**, 288–306 (2015)
13. Tyurin, V.: Nine challenges faced by the digital platform ecosystem. https://www.itweek.ru/idea/article/detail.php?ID=196238. Accessed 10 November 2020
14. Mirotin, L.B., Gudkov, V.A., Zyryanov, V.V.: Management of freight flows in transport and logistics systems. In Mirotin, L.B. (ed.) Hotline Telecom, Moscow (2010)
15. Rezer, S.M.: Ensuring the safety and insurance of risks of cargo transportation. Transp. Sci. eq-nt, ma-nt (5), 3–12 (2016)
16. Kravchenko, A.: Supply chain management in the era of digital transformation. https://vc.ru/transport/78912-upravlenie-cepyami-postavok-v-epohu-cifrovoy-transformacii. Accessed 15 April 2020
17. Laaper, S., Yauch, G., Wellener, P., Robinson, R.: Embracing digital future. How manufacturers can unlock the transformative benefits of digital supply networks. https://www2.deloitte.com/us/en/insights/focus/industry-4-0/digital-supply-network-transformation-study.html. Accessed 16 April 2020
18. Nekrasov, A.G.: Fundamentals of supply chain security management. MADI, Moscow (2011)
19. State Standard 51897-2011. Risk management. Terms and definitions. Standartinform, Moscow (2011)
20. van Gig, J.: Applied general theory of systems. In 2 books. Per. from English. Publishing House «Mir», Moscow (1961)
21. Sundeev, P.V.: Functional stability of critical information systems: the basics of analysis. http://ej.kubagro.ru/2004/05/pdf/03.pdf. Accessed 25 January 2020
22. Rodionova, L.N., Abdullina, L.R.: Sustainable development of industrial enterprises: terms and definitions. http://ogbus.ru/files/ogbus/authors/Rodionova/Rodionova_5.pdf. Accessed 5 July 2020
23. Dedkov, S.M., Turko, V.A.: Balancing the economy as a condition for enhancing innovation: a system analysis and experience of the Republic of Belarus. J. Probl. Territ. Dev. **5**(91), 44–57 (2017)
24. Supply Chain Sustainability. A Practical Guide for Continuous Improvement. Business for Social Responsibility, June 2010. http://www.bsr.org/reports/BSR_UNGC_SupplyChainReport.pdf. Accessed 17 April 2020
25. Rzevski, G., Skobelev, P.: Managing Complexity. WIT Press, Southampton, Boston (2014)
26. Rzevski G.: Intelligent multi-agent platform for designing digital ecosystems. In: V. Marik, P. Kadera, G. Rzevski, A. Zolti, G. Anderst-Kotsis, A. Min Yjoa, I. Khalil (eds.) Proceedings of the 9th International Conference, HoloMAS 2019, Linz, Austria, August 26–29, pp. 29–41 (2019)
27. Smart cities: how rapid advances in technology are reshaping our economy and society. https://www2.deloitte.com/tr/en/pages/public-sector/articles/smart-cities.html. Accessed 13 July 2023

Assessment of the Effect of Road Infrastructure on Traffic Accidents Based on the Fuzzy Linear Regression Method

Marta A. Muratova⊙**, Dmitry G. Plotnikov**⊙**, Lilia V. Talipova**⊙**, Angi E. Skhvediani**⊙**, and Aushra V. Banite**⊙

Abstract This study deals with the impact of transport infrastructure facilities on traffic accidents. The article presents an algorithm for assessing the impact of the location of transport infrastructure on road accidents occurrence and a universal method based on the algorithm. The method includes a detailed description of the steps made in the Python programming language code. The method was tested in the Vasileostrovsky district of St. Petersburg. As a result, the authors determined the probabilities of road accidents and developed the map with marks of all accidents for the given period, the map with marks of cluster centers with the largest number of road accidents, the map with marks of the influence of the location of certain transport infrastructure objects on road traffic accidents and the histogram for general visualization of the results obtained. The table with traffic density on each street in the area for additional assessment of the impact of transport infrastructure on road accidents occurrence at different traffic densities was compiled.

Keywords Road infrastructure · Traffic accidents · Fuzzy linear regression method · Road safety indicators · Mathematical modeling

1 Methods of Mathematical Modeling for the Development of a Software Package for Assessing Factors and Managing Road Safety Indicators

The road network is a complex system that faces various problems at all stages from design to operation [1, 2]. The main problem is road traffic accidents (RTAs). The consequences of low level road safety go beyond direct health risks, as they impede the socio-economic development of regions and countries [3, 4].

M. A. Muratova · D. G. Plotnikov · L. V. Talipova · A. E. Skhvediani (✉) · A. V. Banite
Peter the Great St. Petersburg Polytechnic University, Saint Petersburg 195251, Russia
e-mail: shvediani_ae@spbstu.com; aeskhvediani@mail.ru

© The Author(s), under exclusive license to Springer Nature Switzerland AG 2024
M. Sari and A. Kulachinskaya (eds.), *Digital Transformation: What are the Smart Cities Today?*, Lecture Notes in Networks and Systems 846,
https://doi.org/10.1007/978-3-031-49390-4_16

One of the approaches to reducing road traffic accidents is to predict the conditions for the occurrence of accidents to prevent them [5, 6]. This approach is prospective, but the lack of existing methods becomes a limiting factor for improving road safety.

Currently, statistical methods using information about the location and severity of the accident as initial data serve as the basis for predicting road traffic accidents [7].

To correctly predict the occurrence of the number of road accidents, it is necessary to evaluate a set of infrastructural factors. Determining the degree of their impact on the road accidents occurrence makes it possible to plan measures to improve safety indicators.

Currently, various modeling methods are used for forecasting purposes. To model stationary series, econometric modeling is used, including extrapolation models, regression modeling methods and spatial economics models.

To forecast non-stationary time series, system-dynamic modeling is used for cases where cause-and-effect relationships are clearly established. Cognitive modeling is used to model weakly structured systems; the goal is to identify significant factors. Fuzzy logic relies heavily on three fundamental components: fuzzy sets, membership functions and production rules. Fuzzy logic applications are diverse and span three main areas, including consumer products, industrial/commercial systems, and decision support systems [8].

These modeling approaches can be used to assess occurrence of RTAs and manage road safety indicators.

The purpose of this study is to develop and test an algorithm for assessing road infrastructure factors based on the fuzzy linear regression method, which allows us to identify the degree of impact of a set of factors on the frequency and severity of RTAs.

2 Algorithm for Assessing Infrastructure Factors Based on the Fuzzy Linear Regression Method

Figure 1 presents an algorithm demonstrating the main steps for identifying the degree of impact of infrastructural factors on road accidents.

Step 1 Collection of initial data and determination of the parameters under study.

At the first stage, the geographic location of the research area and the coordinate system are determined for the placement of traffic management facilities and other infrastructural factors. Next, it is necessary to determine a limited list of factors, the influence of which on the characteristics of road accidents is to be studied. At the next step, it is necessary to verify information about traffic patterns on the studied sections of the road network, including through a field survey. Then it is necessary to collect information about road traffic accidents that occurred in the study area, and the period under study must be at least 1 year to take into account the seasonality factor.

Fig. 1 Flowchart of the
algorithm

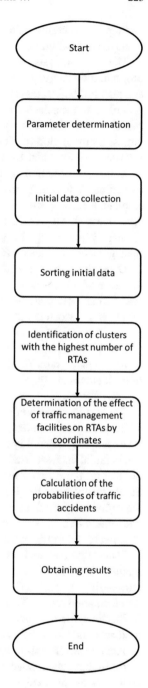

Step 2 Processing data on the locations of accidents and the location of traffic management facilities.

The data analysis code demonstrated in this article is written in the Python programming language, using the numpy, sklearn, geopy, pandas, geopandas, folium and matplotlib libraries (code listings are given below). The data under study is sorted according to the selected area and the period under study. After sorting the data, the model is trained to identify clusters with a mark at the center of coordinates of the largest accumulation of RTAs. Accidents are clustered to determine sections of the road network for which the measures to improve road safety are most relevant. At the next stage, information about traffic management facilities is specified using geoinformation system or based on the results of a field survey. For each infrastructure element, a zone of influence is specified: for a point object—in the form of a radius, for a linear object—in the form of equidistant lines.

Step 3 Assessing the degree of impact of the factors of location of traffic management facilities on RTAs.

Depending on the coordinates of the accident, we determine whether they are in the zone of influence of any element of the road infrastructure. For each group of road infrastructure elements, the total number of road accidents within the zone of influence is assessed and then the probability of an accident occurrence taking into account the impact of various road infrastructure factors is calculated. The results of the quantitative impact of traffic management facilities on RTAs are presented in the form of a histogram. Additionally, the influence of traffic density on the occurrence of accidents is analyzed.

Step 4 Results.

The results of the study are the following: a list of road accident clusters with the addresses of their centers, the level of influence of traffic management facilities on RTAs and the probability of accidents occurrence under given conditions.

As initial data, we used data obtained from the official website of the state project dtp-stat.ru, the primary source of which is the official website of the State Traffic Safety Inspectorate [6]. Road Accident Map is a non-profit project designed to solve the problem of road accidents in Russia. This is a traffic accident data collection platform, a free and open traffic accident analytics service. In this work, data on the location of the traffic management facilities were obtained through a field survey.

Using libraries such as numpy, sklearn, geopy, pandas, geopandas, folium and matplotlib, the original json file with information about RTAs throughout St. Petersburg during the period from January 2015 to February 2023 was converted. The file was formatted for the Vasileostrovsky district during the period 02/01/2021–02/01/2023. Since the transport infrastructure changes over time, it was decided to study road accidents over the past two years.

Also, from the entire data structure, only those that are relevant to the transport infrastructure were selected, namely: "light"—time of day, "road_conditions"—condition of the road surface, "nearby"—type of crossroads at which the accident occurred and/or objects that were located in close proximity, "point"—the coordinates of the accident. Table 1 shows a fragment of formatted data in xlsx format.

Table 1 Fragment of formatted data for analysis

Light	Road_conditions	Nearby	Point
Night-time, lighting is on	['Dry', 'Inappropriate location or poor visibility of roadsigns']	['Public transport stop', 'Signalized crosswalk']	{'lat': 59.95428, 'long': 30.218925}
Daytime	['Dry', 'Absence or poor visibility of horizontal road marking', 'Lack of road signs in proper locations']	['Non-signalized intersection of unequal streets (roads)', 'Marked crosswalk', 'Other educational organization']	{'lat': 59.929813, 'long': 30.270209}
Night-time, lighting is on	['Dry', 'Lack of pedestrian guardrails in proper locations']	['Public transport stop', 'Signalized intersection', 'Signalized crosswalk']	{'lat': 59.938343, 'long': 30.280702}
Daytime	['Wet']	['Public transport stop']	{'lat': 59.942454, 'long': 30.231435}
Night-time, lighting is on	['Wet', 'Lack of pedestrian guardrails in proper locations']	['Signalized intersection', 'Signalized crosswalk']	{'lat': 59.938741, 'long': 30.263407}
Daytime	['Dry']	['Apartment buildings', 'Inner yard area']	{'lat': 59.953372, 'long': 30.234311}
Night-time, lighting is on	['Dry', 'Lack of road signs in proper locations', 'Inappropriate location or poor visibility of roadsigns']	['Non-signalized intersection of unequal streets (roads)', 'Public transport stop', 'Marked crosswalk']	{'lat': 59.933984, 'long': 30.265381}
Night-time, lighting is on	['Wet', 'Lack of pedestrian guardrails in proper locations']	['Apartment buildings', 'Signalized intersection', 'Signalized crosswalk']	{'lat': 59.94685, 'long': 30.257109}

It should be noted that in this work we examine all columns except road_conditions, since the data specified in it will not determine the influence of the location of the traffic management facilities.

Next, using the folium library, a geomap of the Vasileostrovsky district was obtained with all road accidents for the studied period and the coordinates of each road accident (Fig. 2).

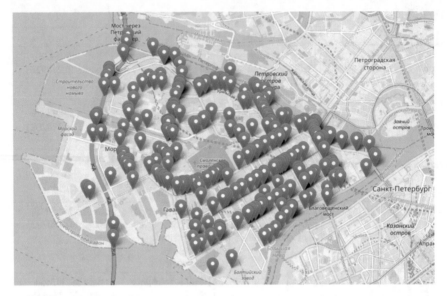

Fig. 2 Map of road accidents in the Vasileostrovsky district for the period 01.02.2021–01.02.2023

Based on the data, a program code was created using machine learning to automatically determine clusters with the highest number of accidents by coordinates with a scale of eps = 0.01. Through programming, we received a map with red marks—the centers of the clusters of the largest accumulation of accidents and a list of clusters—the number of accidents. The largest clusters are on the following nodes (cluster number, number of accidents): [(2, 6), (6, 4), (9, 4), (0, 3), (8, 3), (1, 3), (3, 3), (4, 3), (7, 3), (5, 3), (10, 3)]. It can be seen that most frequently accidents occur in the cluster of 2–6. The exact addresses of the center of coordinates of the clusters are given in Fig. 3.

Figure 4 shows a geomap with red marks—the centers of clusters with the highest number of accidents (more than three).

Cluster -1 is placed at Krutikovskaya dorozhka, Smolenskoye pravoslavnoye kladbishche, Vasileostrovskiy District, Saint Petersburg, 199057, Russia
Cluster 2 is placed at Bolshoy Vasilyevskogo Ostrova Avenue, Saint Petersburg, 199004, Russia
Cluster 0 is placed at Kamskaya Street, Saint Petersburg, Saint Petersburg, 199178, Russia
Cluster 8 is placed at Beringa Street, Saint Petersburg, 199397, Russia
Cluster 1 is placed at Srednegavanskiy Avenue, Saint Petersburg, 199106, Russia
Cluster 3 is placed at Bolshoy Vasilyevskogo Ostrova Avenue, Saint Petersburg, 199004, Russia
Cluster 4 is placed at Sredniy Vasilyevskogo Ostrova Avenue, Saint Petersburg, 199178, Russia
Cluster 7 is placed at Maliy Vasilyevskogo Ostrova Avenue, Saint Petersburg, 199048, Russia
Cluster 5 is placed at Bolshoy Vasilyevskogo Ostrova Avenue, Saint Petersburg, 199026, Russia
Cluster 10 is placed at Nalichnaya Street, Saint Petersburg, 199406, Russia
Cluster 6 is placed at Morskaya Embankment, 199397, Russia
Cluster 9 is placed at Bolshoy Vasilyevskogo Ostrova Avenue, Saint Petersburg, 199026, Russia

---

Fig. 3 List of all identified clusters with addresses

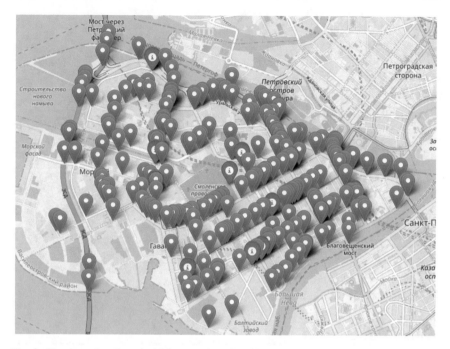

Fig. 4 Cluster marks where the highest number of accidents was found

The cluster where the highest number of accidents occurred is shown in Fig. 5. Cluster address is the crossroads of Bolshoi Avenue of V.O. (Vasilyevsky Island) and 8 and 9 lines V.O. (Vasilyevsky Island).

The next step was to upload data on the coordinates of the location of speed bumps, pedestrian crossings and lighting in the Vasileostrovsky district in xlsx format. Let's set the radius of influence to 50 m, and create a matrix of 0 and 1 on the impact of the traffic management facilities on road accidents. The impact of traffic accidents traffic management facilities will be determined as "in an accident zone with a radius of 50 m". Thus, we get all the accident marks in the color of the impact specified in the method. Table 2 shows the coding for each color.

Figure 6 shows a geomap with marks.

From the results obtained, it can be noted that the map has mainly green marks, which indicates the great influence of the location of the pedestrian crossing on road accidents.

All geomaps are obtained in html format. They open in any browser and can be scaled. An example of a scaled geomap at the crossroads of Nalychnaya Street and Nakhimov Street is presented in Fig. 7.

Calculation of the probabilities of accident occurrence, based on the location of traffic management facilities and conditions, plays an important role in predicting road traffic accidents. Let us consider the probability of an accident occurring at night when the lights are on; during daylight hours; at signaled and non-signaled

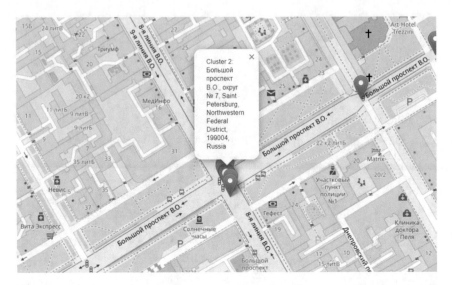

Fig. 5 Cluster 2

Table 2 Marks and color coding

Color of the mark	Coding
Red	Influence of lighting in the conditions of darkness
Green	Influence of pedestrian crossing location
Black	Influence of the location of speed bumps
Orange	Influence of location of lighting
Blue	Lack of influence

pedestrian crossings. According to formula (10), we write down all the conditions and print out the resulting values. Figure 8 shows the conditions and formulas for each of the parameters, and Fig. 9 shows the calculation results.

Based on the results obtained, it can be noted that the probability of an accident occurring during daylight hours is three times higher than the probability in the dark when the lights are on. This may indicate a reduced intensity of traffic in the dark in general.

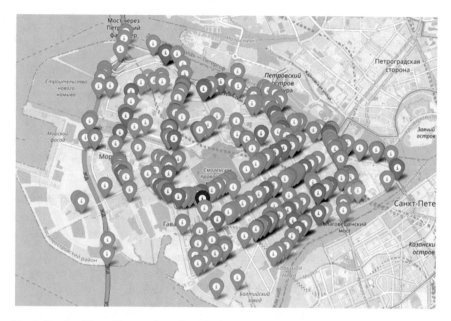

Fig. 6 Road traffic accident marks, colored by impact detection

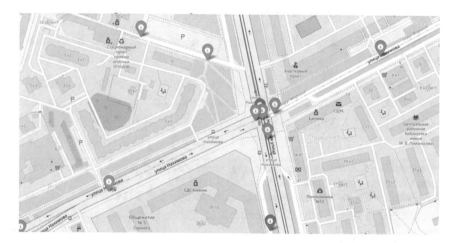

Fig. 7 Marks at the intersection of Nalychnaya street and Nakhimov street

The probability of an accident at a signaled pedestrian crossing is 6% higher than at a non-signaled one. This is due to the fact that pedestrians are more cautious when crossing a signalised crossing; people feel safer and do not pay enough attention to approaching cars. When crossing a non-signalised crossing, pedestrians are more attentive as they feel more responsible for their own safety.

```
dark_light_count = sum(1 for accident in dtp_data if 'В темное время суток,
освещение включено' in accident['light'])

light_light_count = sum(1 for accident in dtp_data if 'Светлое время суток' in acci-
dent['light'])

crossing_count = sum(1 for accident in dtp_data if 'Регулируемый пешеходный
переход' in accident['nearby'])

non_crossing_count = sum(1 for accident in dtp_data if 'Нерегулируемый пеше-
ходный переход' in accident['nearby'])

total_count = len(dtp_data) probability_dark_light = dark_light_count / total_count

probability_light_light = light_light_count / total_count

probability_crossing = crossing_count / total_count

probability_non_crossing = non_crossing_count / total_count

print("Probability of a traffic accident in the night-time with lighting switched on: ",
probability_dark_light)

print("Probability of a traffic accident in the daytime: ", probability_light_light)

print("Probability of a traffic accident at a signalized crosswalk: ", probabil-
ity_crossing)

print("Probability of a traffic incident at a marked crosswalk: ", probabil-
ity_non_crossing)
```

Fig. 8 Conditions for calculating probabilities

```
Probability of a traffic accident in the night-time with lighting switched on: 0.2448
Probability of a traffic accident in the daytime: 0.7379
Probability of a traffic accident at a signalized crosswalk: 0.2414
Probability of a traffic incident at a marked crosswalk: 0.1793
```

Fig. 9 Calculated probabilities

It should be noted that the data obtained directly depends on the clarity of the accident report.

In conclusion, we present a histogram showing the dependence of the influence of the location of the traffic management facilities on the number of accidents (Fig. 9).

In the histogram, you can see the predominant influence of lighting over other traffic management facilities when determining the impact on accidents by location.

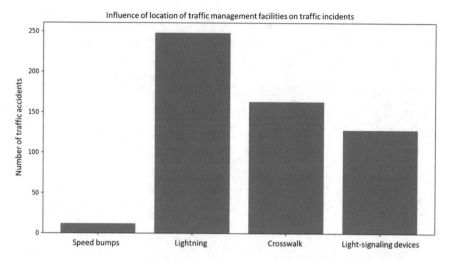

Fig. 10 Histogram of the influence of the location of traffic management facilities on the number of accidents

Table 3 Color coding for traffic density

Color	Coding
Green	Low
Yellow	Medium
Red	High
Dark-red	Very high

Figure 2 shows the colored marks of all road accidents for the period under study. After obtaining the map of the streets where the accidents occurred, it was necessary to study the traffic density on these streets. The process involved collecting data from Yandex.Maps based on the colors of segments in a certain time period. Time periods are as follows: 07:00–10:00; 10:00–16:00; 16:00–19:00; 19:00–21:00; 21:00–07:00. Figure 10 shows a map of the Vasileostrovsky district with the segments marked in color, and Table 3 presents the coding for each color (Fig. 11).

Fig. 11 Map of the Vasileostrovsky district with colored sections indicating traffic density

It should be noted that there are no segments of very high traffic density in the district, and the busiest streets are: lines 8–9 of Vasilievsky Island and Uralskaya Street, which can be seen in Fig. 13, it shows the map at 17:30 Moscow time.

Table 4 presents the example of data on traffic density on the street during a certain time period.

The collected data on traffic density will serve as an additional parameter when assessing the impact of the location of the traffic management facilities on road traffic accidents.

Based on the results of the program, the map was obtained with traffic accident color marks corresponding to Table 3. The map is presented in Fig. 12.

From the collected data in Table 4, it can be seen that streets with low traffic density in the Vasileostrovsky district are predominant in all time periods. Figure 12 shows at what traffic density the accident occurred.

For comparison, we present a histogram of the dependence of the number of accidents during the study period on the traffic density at the exact time of the accident (Fig. 13).

Table 4 Example of traffic density data on the streets of Vasileostrovsky district

Streetname	Start	End	21:00–07:00	07:00–10:00	10:00–16:00	16:00–19:00	19:00–21:00
Leytenanta Shmidta embankment	59.928733, 30.268571	59.934037, 30.277335	Low	Low	Low	Low	Low
	59.934037, 30.277335	59.936423, 30.288130	Low	Low	Low	Low	Low
Birzhevaya square	59.941538, 30.305063	59.942587, 30.306628	Medium	Medium	Medium	Medium	Medium

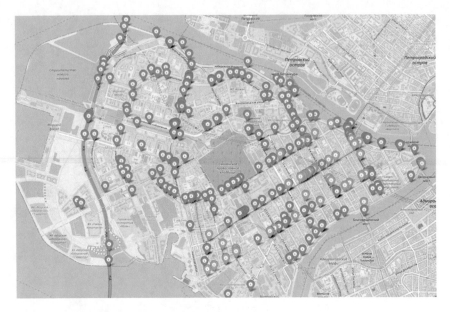

Fig. 12 Map with traffic density marks during an accident

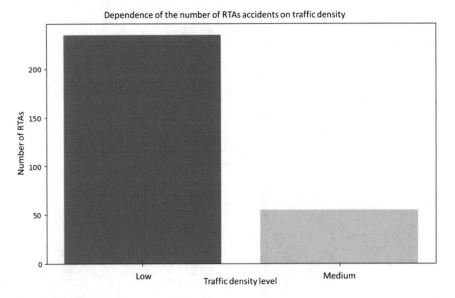

Fig. 13 Histogram of the dependence of the number of accidents on traffic density

3 Conclusion

From the analysis made, we can conclude that traffic density in the study area has an insignificant influence on the number of accidents.

The developed method is based on the program code and allows us to obtain the following results: geomaps with color coded marks, the probability of accident occurrence, the map with marks indicating the impact of the location of traffic management facilities on the location of an accident, and a histogram for the best representation of data. The example of the Vasileostrovsky district of St. Petersburg illustrated the effectiveness of the method. These results can be further developed in the studies related to the use of publicly available data from the Road Accident Map website with the analysis of specific areas, cities, city districts, as well as transport hubs.

The method enables us not only to assess the effect of road infrastructure on traffic accidents, but also to identify the areas of the largest number of accidents to predict probable bottlenecks using machine learning techniques.

Acknowledgements This research was funded by the Russian Science Foundation (project No. 23-78-10176, https://rscf.ru/en/project/23-78-10176/).

References

1. Xing, J., et al.: Traffic state estimation of urban road networks by multi-source data fusion: review and new insights. Phys. A: Stat. Mech. Appl. **595**, 127079 (2022)
2. Vonglao, P.: The solution of traffic signal timing by using traffic intensity estimation and fuzzy logic. Edith Cowan University, Research Online. https://ro.ecu.edu.au/theses/50 (2007)
3. Petrov, A.I.: Philosophy and meanings of the information entropy analysis of road safety: case study of Russian cities. Information **14**(6), 302 (2023)
4. Das, S.K., et al.: Knowledge and practice of road safety measures among motor-bikers in Bangladesh: a cross-sectional study. medRxiv, p. 2023.07. 25.23293162 (2023)
5. Najafi Moghaddam Gilani, V., et al.: Data-driven urban traffic accident analysis and prediction using logit and machine learning-based pattern recognition models. Math. Probl. Eng. 1–11 (2021)
6. Santos, D., et al.: Machine learning approaches to traffic accident analysis and hotspot prediction. Computers **10**(12), 157 (2021)
7. Yang, Y., et al.: Identification of dynamic traffic crash risk for cross-area freeways based on statistical and machine learning methods. Phys. A: Stat. Mech. Appl. **595**, 127083 (2022)
8. Pedrycz, W.: An introduction to computing with fuzzy sets. IEEE ASSP Mag. **190** (2021)

Ecosystem of Tax Administration as an Element of the Digital Economy in the Functionality of "Smart Government": An Integral-Systemonomic Approach

Irina A. Zhuravleva🔟

Abstract The Chapter is devoted to the prospects for the development of the Russian environmental friendliness of tax administration with elements of the digital economy and the effective use of "Smart Government". Taking into account the existing domestic and foreign scientific studies of systemic system analysis, an integral syste-monomic approach to building a tax administration ecosystem model is proposed. For this, three research strata have been identified: systemic, platform, integral-systemonomic. From a scientific and analytical point of view, the tax administration ecosystem model (TAEM) is presented in the legal field and symbolic form. The author's definitions of two concepts are proposed: the tax administration ecosystem, the digital platform of the ecosystem used by the "Smart Government". The concept of systemonomy and the relevance of the application of the scientific industry— taxonomy in order to improve the efficiency of the "Smart Government" are disclosed. Theoretical, legal, instrumental and technological support of TAEM has been system-atized in the supporting part of the proposed model—the main elements used by Smart Government in a synergistic effect have been identified. A variant of the TAEM based on philosophical systemonomic principles is presented and described, which makes it possible to increase the efficiency of the "Smart Government" and the flexibility of adaptability to changing internal and external circumstances. Conclusions are drawn, questions for discussion are proposed, and directions for further develop-ment of TAEM are formulated. The Chapter was prepared based on the results of research carried out at the expense of budget funds under the state order of the Federal State Budgetary Educational Institution of Higher Education "Financial University under the Government of the Russian Federation"

Keywords Digital economy · Integral-systemonomic approach · Nalogonomy · Smart government · Tax administration ecosystem · TAEM

I. A. Zhuravleva (✉)
Financial University Under the Government of the Russian Federation, Moscow, Russia
e-mail: sia.mir67@mail.ru

1 Introduction

Any financial and economic system for its functioning in the conditions of complexity and uncertainty of world development requires a modern approach to analyze it as a super-complex system. "Smart government" can be safely attributed to one of the elements of the systems. However, neoclassical theory reproduces the functioning of simple systems within the framework of system methodology, and real objects have long been working on the principles of complex and supercomplex systems. At the same time, the use of digital technologies is a hot topic in these systems or their derivatives.

Therefore, within the framework of the requirements of the general theory of systems, each object in the course of its study must be analyzed as a large and complex system, as well as an element of a more general system. "Smart government" was created, an element of the system for managing the State with its subjects. At the same time, it is necessary to integrate developments in the areas of systems theory, cybernetics, systems analysis, systems engineering, operations research and synergetics, and systemonomy [1]. Understanding such a systematic approach also includes the importance of theoretical assimilation and practical application of a number of its aspects: systemic-complex; system-historical; system-target; system-structural; system-functional; system-resource; system integration; system-communication and systemonomic [2].

The tax system as one of the economic systems and elements of Smart Government is no exception. It requires a constant and diverse analysis, taking into account social transformations caused by simultaneously increasing challenges of the external environment, global trends in technological development, including the digitalization of the economy.

The proposed study continues the author's developments in terms of the formation of a new type of such a system as an element of "Smart Government"—a tax administration ecosystem model (TAEM) in the context of sanctions based on a systemonomic approach [3].

There are few publications in this area and, as a rule, they deal with the tax administration ecosystem in the applied aspect [4, 5].

The purpose of this study is to form a model of the tax administration ecosystem in the context of the development of the digital economy and information technology in order to improve the efficiency of Smart Government. The study is theoretical and methodological in nature.

2 Methods

When building a tax administration ecosystem model, a systemic approach is used, when the object is not divided into elements/components, but is depicted as a whole, where the enlarged components emerge in relation to the TAEM itself. Such a system is able to function as a whole structure focused on achieving the highest potential of goals and objectives, the implementation of the set requirements, created in the form of structures that combine elements/components and links between them using digital technologies, taking into account external and internal factors, as well as the position "Smart Government", by introducing the principles of client-centricity into the practice of public services and services in order to increase the efficiency of their functioning.

The model of such a system is as follows (formula 1):

$$\text{Smart government} \equiv < Z + GS + METD + FAKT + YP \tag{1}$$

where:

Z—integrity/integrity, a set of structured systemically built goals and objectives, identified problems;

GS is a unity of functionality or a set of public services, structures, ministries and departments, services, business entities that implement the goals and objectives and are included in the structure of the "Smart Government";

METD is a set of information, legal and financial-economic, digital mechanisms and technologies that implement TAEM and ensure the effective functioning of Smart Government [21];

FAKT—the conditions for the existence of TAEM, that is, institutional and legal factors (internal and external), industry specifics that affect its creation and the functioning of "Smart Government";

YP—observers/legislators—subjects authorized to make and execute decisions, fulfill and structure goals and objectives, select methods and means of modeling— "Smart Government" [21].

Note that in hypothetical modeling, an important principle of research is a systemonomic approach to the tasks and problem solving and the point of view—the legal and organizational position of the YP observer/legislator. Thus, at different stages of analysis, various considerations can be invested in the concept of a system, as if the existence of a system can be implied in different forms—strata (according to M. Mesarovich) [6]. Therefore, it is proposed to analyze the TAEM from three positions: the evolution of existing systems, the platform approach, the integration of technologies based on digitalization. Accordingly, in the proposed working TAEM, three research strata are identified: *systemonomic, platform, integral-systemonomic*. Let's consider them in more detail.

2.1 Systemonomic Strata

To substantiate and develop the TAEM, a systemonomic strata is initially analyzed, built on the principles of systemonomy, where each element/component of which is characterized by its applied feature and scientific and practical tools. Depending on the level of complexity in science, systems are divided into simple, complex and supercomplex systems.

At the same time, a stratum is a multidimensional phenomenon that cannot be distinguished by any one feature. In addition to the systemonomic strata for the development of TAEM, other signs of stratification should be identified, highlighting, for example, financial, economic, political, customs and tariff, legal and personnel aspects. In the conditions of complexity and uncertainty of development, scientific research is rapidly moving apart in space and time, giving a powerful impetus to the digital, technological, system integration outlines [7].

Fundamental in modeling is also the main idea in systems theory—the recognition of the principle of *isomorphism* [8]. For example, the entire history of accounting is the history of the development of models: from simpler isomorphic to more complex homomorphic models that make management in an organization simpler and more efficient [9]. Hence the methodological significance of this principle for the legiti-macy of transferring the knowledge gained in the study of any isomorphic system (structure) to another, including the system of tax administration and the "Smart Government" as a whole.

In the 1930s, L. Bertalanffy considered the system as an independent scientific category, defining it as a complex of interacting elements or subsystems that have their own needs to be satisfied [10]. The systemic and hierarchical nature of technolo-gies, the systemonomic approach and the basis make them a generator of innovative development of the TAEM and Smart Government systems.

Systemonomy is a science that studies the principles, algorithms, structure, logic of functioning of natural systems of Laws that underlie the existence of the World and ways of working with them. [11] Systemonomy has a complex component and includes the following areas: general systems theory (GTS)—scientific and theoret-ical basis of the fundamental systems approach, developed by L. Bertalanffy; system philosophy, which is the worldview basis of systemonomy, developed by Academi-cian of the Russian Academy of Natural Sciences Yu. A. Urmantsev; nomology, as the doctrine of law; systemology—methodology of systematics; OTS-method, OTSU-technology (OTS of academician Urmantsev), methods—technological approaches; systemonomy is an integral doctrine of systems of laws based on the genetic unity of the World. Theoretically, systemological analysis implies an analysis of the presence of all the components of the system: goals, elements, unity of links between elements, compositions. The structure of systemonomic analysis is as follows: System of General Laws of Cognition Comprehension; goal (system life program); system elements; the unity of links between the elements of the system; composition of systems [2]. Note that the structure of systemonomy is based on systemonomic principles:

1. The principle of systemic (evolutionary) antecedence;
2. The principle of the need for system completeness;
3. The principle of sufficiency (economical) of the system of laws;
4. Principle of manageability;
5. The principle of self-organization;

6. The principle of evolution and variability;
7. The principle of hierarchical consistency;
8. The principle of unity and infinity (universality) [12].

"Smart government" will a priori function effectively if it is based on an integral systemonomic approach. The branch institute of nalogonomy proposed to the scientific community—as a branch of financial and economic science that studies the self-organizing structural and functional composition of the primary elements of taxation and the tax system as a whole on the basis of their common links with the aim of creating, developing and evolving the country's tax system—includes the ecosystem of the tax administration. Tax administration is one of the elements of nalogonomy, and accordingly, on the basis of legal and functional principles, it is included in the structure of the "Smart Government", having a direct impact on its development and improvement.

It is from the above positions that a different view of the scientific approach, as well as the concept and tools of the TAEM in its development, has developed.

Continuing the study, using formula (1), the following definition of TAEM is proposed: The tax administration ecosystem model is a differentiated whole (TAEM), focused on achieving the goals and objectives (Z) based on systemonomy, implemented in the form of various static data and dynamic structures (GS), combining elements and links between them using digital technologies (METD), taking into account external and internal factors (FAKT), as well as the position of the observer (UP). It should be noted that in TAEM in its structure it consists of two parts: scientific, theoretical and practical component (NPS) and providing (OB), which, by their structure, contain diverse subsystems and components of systemic scientific and practical tools. This allows us to derive a single formula (2):

$$\text{TAEM} = [\text{NPS}; \text{OB}] = [\{\text{NPS} = 12\}; \{\text{OB} = 10\}] \qquad (2)$$

where
NPS—12 representations of the content of the TAEM;
OB—10 supporting subsystems of scientific and practical tools.
Details of the NPS and OB subsystems will be given further in Tables 1 and 2.

The following starting definition for the study of the content of the TAEM: the ecosystem of tax administration is a super-complex systemonomic functional model of interaction between the state and the taxpayer, built on a legal basis and in a certain way interconnected digital platforms and programs based on updated information technologies.

Here, a digital platform (CPU) is a real-life model that uses system technologies for combining and interacting people, organizations and resources into a modern ecosystem. It is the core of TAEM and is used as a digital online system for the interaction of participants in tax relations, as a technological environment for developers and as a cloud solution for integrating applications, which is reflected in the elements of "Smart Government" and allows it to expand functionally.

2.2 Platform Strata

As a first approximation, we will use a simplified classification of digital platforms (CPUs) in the economy [13], highlighting three bases within the framework of the platform stratum:

(1) instrumental CPU—development of software and hardware-software solutions, as well as applied and information technologies;
(2) infrastructure CPU—providing IT services and information for decision-making, as well as access to data sources implemented in the infrastructure of this ecosystem of tax administration, the stability of the legal space;
(3) applied CPU—a business model for providing the opportunity to exchange certain values between a large number of independent market participants by conducting transactions in a single information environment (single tax account, single tax payment, digital ruble).

It should be noted here that the platforms that actually exist and are being created in taxation and tax administration will not fully meet these criteria. Then, using the previously obtained results of the author's research [1], we will complete the picture of meaningful representations of the TAEM from the standpoint of the platform stratum and the systemonomic approach (Table 1).

Based on the synthesis of particular definitions, the desired concept is formulated in a broader sense, depending on the purpose of the study and the author's vision of current problems. Based on the definitions of the last two ideas, we can give a broader interpretation: the Integral CPU of the tax administration ecosystem, as an element of "Smart Government", is a hypothetical systemic model that is open to innovative technologies, dynamically improving in a multidimensional financial, economic, legal field, flexible under conditions of uncertainty. It is from these positions that we will analyze the tools, implying the use of an integral technological stratum for the supporting subsystem of the TAEM.

2.3 Integral-Systemonomic Strata

The proposed structure of the supporting part of the TAEM is only a preliminary (intelligence) study, which, in the context of ongoing system-technological, legal and digital transformations, is a kind of foundation for substantiating and choosing future versions of tools and mechanisms for the effective functioning of Smart Government with an element of the tax administration ecosystem model. Here technology is a complex developing system of artifacts, production operations and processes, resource sources, subsystems of the social consequences of information, management, financing and interaction with other technologies [17].

Table 1 Particular definitions of the CPU in the context of meaningful representations of the model [14]

№	Private definition of CPU
1.Genetic	The platform is a cognitive base for the implementation of multifaceted scientific and technical cooperation in the theory and practice of tax administration
2. Predictive	Infrastructural CPU for predicting and designing the future of the tax system in the context of the development of foresight technologies and futorology, achieving its highest multifaceted potential
3. Target	Applied CPU as a business model of a dynamic goal-setting system for the phased formation of the tax service as a service company and one of the elements of "Smart Government"
4. Hierarchical	Infrastructural CPU to substantiate, develop and implement modern adaptive mechanisms for taxation and tax administration and structures of tax authorities in the context of total digitalization in order to increase the efficiency of the Smart Government
5. Functional	Applied CPU is an eco-model of the system for the execution by the tax service of the traditional and new functions assigned to it in the digital economy based on a systemonomic approach and the implementation of the goals and objectives of nalogonomy [1]
6. Organizational	Application CPU for the distribution of tasks, own and third-party resources in order to create new values and ensure the competitive advantages of the national tax system, as an element of "Smart Government"
7. Procedural	Instrumental CPU for the development of software and hardware solutions, regulation of procedures and actions of actors of tax relations, with subsequent evaluation of the results achieved
8. Interactive	Applied CPU—an eco-model for making transactions and automatically calculating tax payments at the time of transactions based on digital technologies and evaluating the effectiveness of applying tax preferences by "Smart Government"
9. Irrational	Tool CPU for creating a virtual environment, modeling taxes and taxation, making fundamentally new decisions, the effectiveness of a risk-based approach to tax control measures, the environmental friendliness of tax administration [15]
10. Homeostatic	Infrastructural CPU for maintaining the homeostasis of the TAEM in the face of uncertainty and the human factor, tax, financial, transactional, currency, digital and other risks
11. Innovative	Applied CPU is a model of complex subsystems open for innovation in the context of a systemic, dynamic transformation of society and mechanisms of both taxation and tax administration based on digital technologies
12. Probabilistic	Applied CPU is a simulation model for the implementation of the taxation-2050 project under conditions of uncertainty, a systemic and systemonomic approach and the human factor, artificial intelligence and digitalization of the economy, and optimization of the tax burden on business [16]

3   Results

Below is the decomposition of the supporting part of the proposed TAEM, that is, the allocation of 10 sections of the study as two types of support. This allows us to correlate each type of theoretical and practical developments from different fields of knowledge, to outline the contours and innovations of the modern TAEM tools, as an element of nalogonomy in the structure of "Smart Government" (Table 2).

4   Discussion

The digital tools actively used in the tax industry, as well as the new systemonomic one, proposed by the author and taking into account global development trends. It can be assumed that Data Mining, the intellectual analysis of big data, can become a promising tool for the effective use of TAEM by "Smart Government", because this technology allows you to identify hidden patterns in the tax sphere, which will lead to the emergence of new areas and topics of scientific research. At the heart of a significant startup is either a new technology, or the integration of technologies and the functionality of an industry institute of taxonomy. At the same time, the system dynamics is incredibly accelerating, and in 2045–2050 a technological singularity is coming. This is a new direction of scientific research in the field of mechanisms of taxation and tax administration and the entire tax system of the country, its elements/ components. The regulatory guillotine as a large-scale revision and repeal of normative legal acts and the regulatory sandbox, a special legal regime that allows organizations to experiment with the introduction of financial innovations without the risk of violating the current legislation, can become important instruments of tax law [18]. Without budget damages and minimization of shortfalls in state revenues or tax expenditures of the country, stagnation of aggressive tax planning, successful implementation of a risk-based approach in tax control measures.

Other innovations include the following:

- application of neural networks in mathematical modeling, which can be used in solving management problems and analyzing a large database, as well as in fiscal forecasting, preventing tax evasion schemes and aggressive tax planning;
- transpositional virtual model, which adds a fundamental property: the ability to partially, completely or exceed the replacement of physical reality by computer models with subsequent analysis and forecasting.

The very idea embodied in virtualization models opens up wide possibilities for their use together with simulation models, subject to a systemonomic analysis and foundation.

It is also worth highlighting separately communication trends in the B2B and B2C market segments, such as omni-channel, automation and personalization.

Table 2 Schematic outline of the supporting part of the TAEM [14]

№	Type of support: tools	Innovations
1	Scientific and practical support. Scientific approaches: systemic, integral-systemonomic. Theory and practice of taxation, environmental taxes, environmental friendliness of administration. Global and local databases. Websites—Federal Tax Service of Russia, Statista, ICT Moscow	System-technological innovations: global modeling, AQAL model, noospherism, system dynamics models, quantum technologies, nalogonomy;
2	Personnel and educational support. The knowledge economy: a competency model. S-education: lifelong learning, PLE, connectivism, MOOCs. Digital transformation: mobile learning, BYOD, intelligent learning systems. Gigonomics: outsourcing and crowdsourcing, self-employed. Noonomics	Machine learning, cognitive systems, VR/AR, B2C, B2B, C2C, B2G technologies
3	Information and communication support. Unified database of the tax service: data center, AIS "Nalog-3", "ASK VAT-2", information technology. Portal of the Federal Tax Service of the Russian Federation: reference and open information, software tools, statistics and analytics, electronic document management, online cash registers, a single tax account, a single tax payment	Digital transformation, web integration of systems (elements), technology export;
4	Technical and technological support. Ecosystem approach to technology. Data center architecture: servers, workstations, communication channels, cloud solutions, virtualization, FTS blockchain platform. Tools: PCs, smartphones, tablets, online cash registers, programs and web services, automated simplified taxation system	Data mining, technology integration, technological singularity, system economic analysis
5	Political and legal support. Internet resources and online services on tax law. Multifunctional center "My Documents", AIS MFC. Taxcom website: IT-services ecosystem. SMEV. Website of the Federal Tax Service of the Russian Federation: BFO resource, documents, programs, State services, "Smart Government"/Open Government	Regulatory guillotine, regulatory sandbox, technology integration
6	Management and analytical support. Trendspotting. Tax management tools. Tax administration technologies: compliance, tax monitoring, self-employed tax, risk-based approach. Internet resources and services of the Federal Tax Service, web reviews on tax disputes, pre-trial settlement of tax disputes, tax monitoring	Foresight technology, digital transformation, integration of systems and technologies

(continued)

Table 2 (continued)

№	Type of support: tools	Innovations
7	Regulatory and supervisory support. Reform of the control and supervision activities of the Federal Tax Service, control work of the Federal Tax Service. Technologies in other areas: financial (FinTech), regulatory (RegTech), supervisory (SupTech) and CovidTech	Integration of AIS technologies/elements/platforms, implementation of tax agreements and rules
8	Model-virtual software. Models: mathematical, statistical, simulation, hypothetical, as well as for environmental taxation. Tools: simulation, human–machine dialogue, expert forecasting, OLAP technologies	Neural networks, virtual models, transpositional virtual model
9	Human-robotic software. Human-AI Interaction: A Soft Singularity. Trends: Internet transformation and the digital person. Toolkit: machine learning, neural network, intelligent systems and robots	Biometrics, speech recognition, IBM Watson, neurotechnologies, Neuralink technology, technology integration
10	Risk-based security. Tax and legal aspects of risks, measures of tax security and tax homeostasis. FTS tools: risk-based approach, mobile applications and proactive services	Innovation perspective: conceptual model of noospherism, socio-economic-environmental taxation in the context of sustainable development, combined SFC + IO models

Omni-channel implies a comprehensive, deep and close integration of communication channels by combining them into a single common system. Thanks to omni-channel, it is assumed that the necessary information is found at any stage of the user's journey, both in the process of searching for information and in the process of performing planned actions.

In this type of communication, real-time interaction takes place, as a result of which users are given full access to products and services when they need it.

Due to the omni-channel of communication, it is possible to bridge the gap between online and offline, for example, online platforms and offline government departments provide service users with greater convenience due to service updates and the ability to sign up online, which also saves time for end users of services.

With comprehensive data collection and deep analysis across a wide range of channels, the omni-channel communication model takes customer-centricity to the next level. The omni-channel concept offers many advantages (Fig. 1).

Further, as the next communication trend, it is worth highlighting automation. An automated simplified taxation system has been introduced in Russia. The decision to introduce an automated taxation system began to operate as an experiment from July 2022 to December 2027. The application of this system applies to small organizations and individual entrepreneurs (hereinafter referred to as individual entrepreneurs).

Due to automation, small commercial organizations and individual entrepreneurs are exempted from filing tax returns, and also receive simplifications in a number of other aspects of tax administration. The territorial distribution of the system concerns

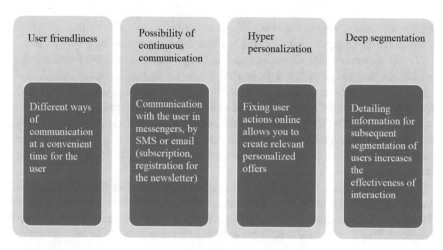

Fig. 1 Advantages of the omni-channel communication model

individuals and legal entities registered in Moscow, Moscow Region, Tatarstan and the Kaluga region [19].

The automation system in the tax administration ecosystem works as follows: the taxpayer selects the online payment mode (which requires authorization on the website of the Federal Tax Service of Russia), then the taxpayer specifies the object of taxation (for example, income or income reduced by the amount of expenses).

The new system establishes a tax period equal to a month (an ordinary calendar month). You can change the object of taxation annually, with notification of the tax authority about the change in the object of taxation no later than December 31.

Thanks to the transition to a new automated taxation system, individuals and legal entities get the opportunity to save time on filing a tax return through online services. Within the framework of the new system, employees of the tax authorities calculate taxes based on the data on accounting for the expenses and incomes of taxpayers. It is important to note that this system implies accounting only for those expenses that were made in non-cash form.

The final trend of communication, considered in the work, carries a personal approach, that is, it implies personalization in communication in various segments. Personalization allows you to take into account the unique characteristics of service consumers, due to which it is possible to maintain an individual connection.

Personalization is implemented through a variety of promotion tools:

– in text format (email-mailings);
– in the format of banners (pictures);
– in video format.

The most relevant communication format is personalized communication through video format, since text is perceived worse than a picture or video. So, for example, the human brain processes video 60,000 times faster than text, and high-quality video with the ability to go to a subdomain page increases conversion by 80% [19, 20].

Thus, we can say that all institutional and legal factors affect the functioning of the tax administration ecosystem, however, it is worth highlighting digitalization, which directly interacts within the framework of economic development. It is the development of digitalization of the tax administration ecosystem that is necessary in the context of the digitalization of the economy. So, for example, high-quality and prompt communication is impossible, provided that only one of the participants has switched to electronic document management, and the other leaves documentary exchange in a classic printed form.

Other factors in the functioning of the tax administration ecosystem, such as the level of taxation, the tax regime, the taxation system, as well as tax control, have a direct impact on the implementation of economic development tasks in the context of sanctions. Proper use of these factors can have a direct impact on market participants in its various segments, which directly determines future trends in the development of the economy.

In conclusion, it is worth pointing out that economic factors, such as the rate of economic growth and inflation, subsequently have an impact on the implementation of economic development tasks under sanctions, since the rate of economic growth and inflation affect the size and dynamics of tax revenues to the federal budget. That is, these are mutually influencing factors that act in conjunction, influencing each other. With a slowdown in economic growth, there is often a reduction in tax revenues to the federal budget, since the level of income decreases during an economic downturn [2]. And when inflation rises (depending on the pace of economic development), nominal tax revenues also increase, while real ones fall.

5 Conclusions

When substantiating and developing the TAEM for the effective functioning of the "Smart Government", the key principles were: the systemic and systemonomic approach as a specification of the methods of the general theory of systems based on the systemonomic approach and philosophical principles; study of taxation from the standpoint of the complexity of the mechanism for calculating tax administration systems; integration of the potential of modern theories, concepts and technologies, as well as practical innovations, information technologies.

The authors proposed the allocation of three lines of research—strata: systemic, platform (digital), and integral-systemonomic.

From the standpoint of "Smart Government", TAEM should be considered as a super-complex system (systemic strata), which is built on the basis of a platform approach, and the list and content of digital platforms (platform strata) is determined by the purpose and functional purpose of the tax administration ecosystem. It should be noted that the proposed version of the TAEM in the future will allow us to start developing a simulation model in the field of research and management of the tax administration ecosystem.

The study of the TAEM tooling was carried out using structural analysis and decomposition of the subsystem into 10 parts—types of support, is systemonomic in nature and uses philosophical principles. This made it possible to correlate each type of provision of developments from different areas of knowledge and practical experience, to identify technologies and their integration potential, and to outline the contours and guidelines for the technological development of TAEM in the future, keeping a focus on systemonomy and dynamically implementing the industry institute of taxonomy.

Directions for further research in updating the TAEM tools for improving the analysis and evaluation of the effectiveness of digital platforms, as well as technologies for their formation and increasing the efficiency and consolidation of the Smart Government functions. Undoubtedly, these studies will be based on an integral-system-economic approach.

In organizational terms, it is necessary to develop a simulation model of a highly complex ecosystem of tax administration as soon as possible and scientifically based: national priorities for the transition to a new progressive technological order and digital economy, taking into account the needs of "Smart Government", flexibility of thinking and operational adaptability. Since, in practical terms, the medium-term target setting of the Federal Tax Service of the Russian Federation is already being successfully implemented—the tax service as a high-tech service company. In connection with the spread of the digital economy, the elements of the tax system have gained faster momentum, they have become more mobile, that is, the transfer of certain information about taxes between the subjects of tax legal relations has accelerated. That is why the tax authorities, automatically receiving information about a particular situation in the life of taxpayers, at the same time simplify not only their activities, but also make life easier for citizens and business entities by providing them with all the necessary documents, information and notifications [21].

Positive trends in the work of the tax authorities, the use of fundamentally new forms of tax administration are of great importance in the field of accelerating the development of the economy.

The statistics of the Federal Tax Service of Russia convincingly indicate an increase in the efficiency of administration using digital technologies: the information environment of the tax service processes information from 4.5 million legal entities, 3.8 million individual entrepreneurs and 155 million individual taxpayers. In total, 4 petabytes of data are stored and processed in the information system of the Federal Tax Service of Russia.

In addition, this system plays an important role for continuous monitoring and analysis of the state of socio-economic processes, assessing the effectiveness of economic entities. An effective tax administration institution is able to quickly and accurately identify cases of tax evasion, which is a significant step in the digitalization of the economy. The effective functioning of the tax administration ecosystem makes it possible to largely control the receipt of tax revenues in the Russian budget system, especially under the influence of global external shocks.

References

1. Zhuravleva, I.A.: Philosophical foundations of the systemonomic model for the development of the tax system. Audit Financ. Anal. **6**, 15–27 (2019)
2. Zhuravleva, I.A.: The direction of reforming the tax system on the basis of the scientific systemonomic author's model: nalogonomy. In: Yonk R.M., Bobek, V. (eds.) Perspectives on Economics Development. Public Policy, Culture, and Economic Development, pp. 125–145 (2020)
3. Viktorova, N.G., Evstigneev, E.N.: Tax ecosystem in the context of innovative and technological transformations of the economy. Belarus. Econ. J. **2**(95), 36–45 (2021)
4. Ilyinsky, A.I.: Agent-based modeling of the development of a complex tax ecosystem in the event of erosion of the tax base during the implementation of Suptech and Regtech. Chronoeconomics **32**, 62–65 (2021)
5. Rukina, S.N., Denisova, I.P.: Characteristics of the structural components of the ecosystem of tax control. In: In the Collection: Innovative Development of Modern Society: Topical Issues of Theory and Practice, pp. 48–50. Penza (2021)
6. Mesarovich, M.: General theory of systems and its mathematical foundations. In: Sadovsky, V.N., Yudin, E.G. (eds.) Research on General Theory of Systems. Sat. translations, 520 pp. (1969)
7. Vylkova, E.S.: The Versatility of Modern Pandemic Reality: A Collective Monograph. Vylkova, E.S. (ed.). SPb., 300 pp. (2021)
8. Bertalanfy, L. von: General System Theory—A Critical Review. General System, vol. VII, pp. 1–20 (1962)
9. Sokolov, Ya.V., Sokolov, V.Ya.: History of Accounting: Textbook, 3rd ed., revised and additional, 287 pp. (2009)
10. Bartalanfi, L. von: General Systems Theory—Critical Review. In the book: Research on General Systems Theory, pp. 23–24 (1969)
11. Maslova, N.V.: Periodic System of the Universal Laws of the World. Institute of holodinamics, 184 pp. (2005)
12. Maslova, N.V.: Periodic System of the General Laws of Knowledge and Comprehension. Institute of Holodinamics, 180 pp. (2007)
13. Sokolov, Ya.V., Sokolov, V.Ya.: History of Accounting: Textbook, 3rd ed., revised and additional. 287 pp. (2009)
14. Viktorova, N.G., Evstigneev, E.N.: Tax ecosystem as a digital economy element: a system–integral approach to construction. Econ. Prof. Bus. **2**, 28–33 (2022)
15. Zhuravleva, I.A., Nazarova, N.A.: Systematic model of tax management: problems of functioning. Audit Financ. Anal. **1**, 7–14 (2020)
16. Zhuravleva, I.A., Nazarova, N.A., Eremin, I.V.: The model of managing the company's tax burden: problems and prospects. Audit Financ. Anal. **4**, 7–23 (2020)
17. Frolova, I.T. (ed.): Philosophical Dictionary, 560 pp. (1991)
18. What is the Regulatory Guillotine? https://knd.ac.gov.ru/about/. Accessed 01 December 2022
19. An automated taxation system is being introduced in Russia—Cnews. https://www.cnews.ru/articles/2021-08-16_kak_ustroena_aisna_kotoroj_vystroena. Accessed 05 August 2023
20. Reasons Why Video Is More Effective than Text—Idea Rocket. https://idearocketanimation.com/17385-reasons-video-effective-text/. Accessed 05 August 2023
21. Innovative development of modern society: current issues in theory and practice. In: Gulyaev, G.Yu. (ed.) Collection of Articles of the All-Russian Scientific and Practical Conference. ICNS "Science and Enlightenment", Penza, 166 pp. (2021)

Development of a Device to Reduce the Toxicity of Road Transport Emissions

Yulia Tunakova, Idgai Mingazetdinov, Alsu Ablyasova, Elnara Mukhametshina, and Ksenia Novikova

Abstract The chapter reveals topical issues related to the implementation of the "smart urban transport" approach. The aim of the research is to create ways to reduce the toxicity of motor transport emissions in highway zones, for the development of the process of digital transformation of the transport complex. The research task is to modernize the device of exhaust gas neutralization with the help of a catalytic converter to reduce the toxicity of vehicle emissions. The compositions of emissions during the operation of various types of internal combustion engines during fuel combustion have been analyzed. The factors leading to an increase in the content of emission components during the operation of internal combustion engines are considered. The components of emissions from vehicles are given depending on the operating modes and the type of internal combustion engine (ICE). The advantages of using catalytic converters, analysis of the structure, and disadvantages of work are shown. Modernization of the catalytic converter design through the use of recirculation systems is proposed, which makes it possible to increase the completeness of fuel combustion. The proposed modernization makes it possible to significantly reduce nitrogen oxide emissions by installing an exhaust gas recirculation valve after the thermal afterburning chamber. The calculated volume of exhaust gases for restarting allows us to reduce the concentration of nitrogen oxides in exhaust gases, for example, in gasoline internal combustion engines—up to 40% at the maximum degree of recirculation of 25%.

Keywords Smart transportation · Digital transformation · Vehicle emissions · Catalytic neutralization · Device upgrade

Y. Tunakova (✉) · I. Mingazetdinov · E. Mukhametshina · K. Novikova
Kazan National Research Technical University Named After A. N. Tupolev, Kazan, Russia
e-mail: juliaprof@mail.ru

A. Ablyasova
Kazan State Power Engineering University, Kazan, Russia

1 Introduction

Emissions from vehicles are carried out with exhaust gases, blow-by gases and fuel vapors. The share of exhaust gas emissions is 95–99% and is an aerosol of complex composition, depending on the quality of the fuel and the engine operating modes. The main components of emissions from internal combustion engines are carbon monoxide, hydrocarbons, particulate matter and nitrogen oxides. Based on the reversibility of chemical reactions, the process of fuel combustion theoretically cannot reach the end, the completeness of combustion is determined by the conditions of the process. The lack of an oxidizer in the local zones of the flame in the combustion chamber, a decrease in the temperature of the gases, leads to incomplete oxidation of the fuel and an increase in the concentration of hydrocarbons, carbon monoxide and soot particles in the exhaust gases. In the model combustion of a mixture of fuel with air, only nitrogen, carbon dioxide and water are present in the combustion products:

$$C_nH_m + \left(m + \frac{n}{4}\right)O_2 + 3.76N_2 = mCO_2 + \frac{n}{2}H_2O + \left(m + \frac{n}{4}\right)3.76N_2 \quad (1)$$

In real conditions, emissions contain, in addition to the products of complete combustion, products of incomplete combustion—carbon monoxide (CO), hydrocarbons C_mH_n, aldehydes, solid carbon particles, peroxide compounds, hydrogen and excess oxygen, as well as products of thermal reactions of interaction of nitrogen with oxygen—nitrogen oxides (NO_x). An excess of oxidant at high temperatures and pressures leads to the intensive formation of nitrogen oxides [1–3]. CO is formed during combustion with an excess of oxidant, as a result of cold-flame reactions (in diesel internal combustion engines), during the dissociation of CO_2, taking into account high temperatures. C_nH_m (more than 200 items) are formed in the chamber as a result of excess air, lack of oxidant, low temperature, poor atomization, leakage of the exhaust valve and crankcase ventilation system. Table 1 shows the mass content of emission components during combustion of 1 kg of fuel.

Gasoline-type ICEs emit 7 times more CO, and 3 times more aldehydes than diesel ICEs. Diesel engines emit more C_nH_m, and sulfur dioxide 10–15 times [4, 5].

A significant reduction in the toxicity of emissions from internal combustion engines is achieved with the use of exhaust gas converters (liquid, catalytic, thermal and combined), which improve the cleaning efficiency. Liquid neutralization of exhaust gases from internal combustion engines is one of the first methods to reduce the toxicity of emission components. It is based on the removal of soot, particulate

Table 1 Mass content of toxic substances

ICE	CO, g	C_nH_m, g	NO_x, g	SO_x, g	Soot, g	Aldehy-des, g
Petrol	225	20	55	1.5–2	1–1.5	0.8–10
Diesel	20	4–10	20–40	10–30	3–5	0.8–1

matter by passing exhaust gases through water or aqueous solutions of chemical reagents.

The absorption of nitrogen oxides can be up to 10–30% of the initial concentration during liquid neutralization. Up to 80% carbon monoxide can be absorbed, depending on the composition of the reagent. Aldehydes are absorbed by fresh reagents by 50%. However, the emission components after eight hours of operation cease to be removed due to the saturation of the reagent solutions, which requires their replacement [6–8]. The cleaning of gases from soot is carried out by 10–80%, depending on the operating mode of the engine. In addition, the disadvantages of the considered neutralizers are: large weight and dimensions, the impossibility of using in conditions of negative ambient temperatures [9–11].

Thermal neutralization is based on the principle of additional oxidation of the products of incomplete combustion, provided that high temperatures are maintained and an additional supply of oxidant is provided. Thermal reactors are reaction chambers built into the engine exhaust system. Nevertheless the complexity of the design and the need for additional fuel consumption, combined with a high fire hazard, did not make this method popular for use [12, 13]. The method of catalytic neutralization is one of the most common; it is used to transform hydrocarbons, carbon monoxide CO, nitrogen oxides into less toxic compounds [14, 15]. The essence of this method lies in the flameless oxidation of the combustible components of the exhaust gases on the catalyst surface to 75–90%.

The catalytic converter of exhaust gases increases the efficiency of cleaning exhaust gases, which carries out the simultaneous reduction of NO_x to neutral molecules along with the oxidation of CO, hydrocarbons, soot aerosol particles.

Reduction of nitrogen oxides into elemental nitrogen and oxygen:

$$NO_x \rightarrow N_x + O_x. \tag{2}$$

Oxidation of carbon monoxide to carbon dioxide:

$$CO + O_2 \rightarrow CO_2 \tag{3}$$

Oxidation of hydrocarbons into carbon dioxide and water:

$$C_xH_{4x} + 2\,xO_2 \rightarrow xCO_2 + 2\,xH_2O \tag{4}$$

As a result, the toxicity of components of automobile emissions is reduced [3].

The analysis of technical solutions has shown that expensive precious metals are used as catalysts in catalytic converters; with a variation in the structural design of catalytic converters. The most common are neutralizers with a honeycomb element (39.08%) or catalyst deposition on a film (20.69%) [4–6]. One of the main disadvantages of catalytic converters is the inefficiency of operation at low temperatures, which occur when the vehicle is operating in the start-warm-up mode, idling and small share loads [7–10].

Analysis in the field of designs of exhaust gas neutralizers showed that the best cleaning results can be obtained by combining cleaning agents. The combined method of thermocatalytic neutralization with exhaust gas recirculation is the creation of a circulation line for a part of the exhaust gases in a closed cycle and mixing with air and fuel supplied to the engine. The exhaust air from the engine enters the thermal neutralization chamber, from where it is fed to the catalytic neutralization chamber. A decrease in the combustion flame temperature leads to a decrease in the formation of nitrogen oxides, which is achieved with a lack of oxygen in the air. Achieving a lack of oxygen in the air can be obtained from the exhaust gas after the thermal neutralization chamber, where the main part of the oxygen has already burned out, and, taking it through the pipe, send it to the engine to dilute the air through the recirculation line.

2   Research Methods

The studies carried out were based on systemic and integrated approaches, methods of studying special literature, theoretical analysis and synthesis of the obtained experimental material, inductive and deductive methods of generalization of empirically obtained data, mathematical and statistical methods of processing the obtained experimental materials, as well as to establish quantitative relationships between the phenomena under study. The theoretical basis of the work was based on the use of the main provisions of the theory of working processes of heat engines, methods of statistical processing of test results and computer modeling, as well as scientific research in the field of engine building, thermodynamics and heat engineering.

Exhaust gas recirculation consists in bypassing a part of it into the engine intake system and then returning it to the combustion chamber. Since the exhaust gases contain very little oxygen after the combustion process, the maximum temperature and pressure during the combustion process are reduced, which reduces the emission of nitrogen oxides.

Reducing NO_x emissions is carried out by reducing the concentration of oxygen in the combustion chamber, as well as reducing the consumption of exhaust gases, lowering the temperature due to the higher heat capacity of gases that do not participate in the reaction (for example, CO_2).

3   Results and Discussions

The maximum allowable NO_x limits for Euro-5 and Euro-6 are 0.18 g/km and 0.08 g/km. There are known technologies to reduce NO_x content in exhaust gases to harmless N_2, such as SCR, SNCR (selective non-catalytic reduction), and LNT. The paper [16] presents an analysis of different after-treatment methods applied in the last decades. Technologies such as SCR, SNCR, and LNT reduce NO_x content in exhaust gases to harmless N_2. The difference between SNCR and SCR technology is that SNCR does not require a catalyst. For SNCR, the achievable level of NO_x emission reduction is no more than 50%. Therefore, we consider NO_x reduction using a catalyst to be promising, which can reduce NO_x content by 95% or more.

The LNT method has low-temperature activity due to NO_x adsorption but has limitations in use at high temperatures. LNT catalysts are based on ceramic substrates and layered porous plugs, which is expensive, therefore requires the use of precious metals such as platinum, palladium, and rhodium [17]. The SCR method has high NO_x reduction efficiency. Despite the complex parameters such as urea dosing, mixing, and proper decomposition to NH_3 and preparation for spraying, the SCR technology has been successfully introduced in the designer passenger car market [18]. Among them, SCR is a sustainable technology in which the chemical reaction takes place at low temperatures in the range of 200–400 °C. In urea SCR, NH_3 is used as a reducing agent. Cu-Z and Fe-Z vanadium-based catalysts are widely used as SCR catalysts. These catalysts are preferred because of their resistance to temperature variations [19]. A catalytic converter typically consists of a structured substrate (metal-based with a high geometric surface area to minimize the amount of expensive catalyst deposited on it. The main disadvantages of the catalytic converter are:

– significant dependence on temperature;
– high sensitivity to fuel quality.

The analysis of known designs of thermocatalytic neutralizers has shown that in the proposed devices there is no provision and maintenance of the necessary temperature. At that time, it is known that thermocatalytic reactions proceed only in a certain temperature range. If the gas temperature is lower than necessary, catalytic reactions are not realized, and at temperatures above the permissible temperature the catalyst material burns out, is covered with oxide film and catalysis does not take place. For automobile engines lower temperature is more characteristic, because the gas passing from the combustion chamber to the exhaust pipe, closer to which the catalytic device is installed, cools down considerably. With cooled gas in the afterburning chamber, the afterburning process may not occur [20–22]. Therefore, we have proposed a design and described the principle of operation of a thermocatalytic neutralizer, which does not have these limitations.

The formation of nitrogen oxides depends on the operating conditions of the engine, therefore, it is necessary to regulate the amount of gas supplied to the recirculation through the highway. Regulation is carried out by the regulator. In the line for removing gases from the thermal afterburner there is a branch pipe, which is part of the line, and has a movable connection with the exhaust gas line after the thermal neutralization chamber. Gas for recirculation is taken from the main exhaust gas line after the thermal afterburner chamber through a branch pipe that can rotate in the nodes. Rotation of the branch pipe in the starting and idling mode is the smallest, the generated exhaust gases are minimal and the bushing, due to the spring, occupies the lowest position on the branch pipe, providing the smallest passage of gases through the hole, which enter the engine combustion chamber through the cuff and the branch pipe. The number of revolutions of the engine shaft increases with increasing power

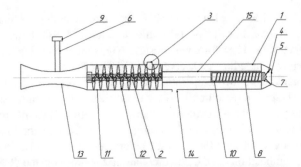

Fig. 1 Thermocatalytic converter. 1—catalytic neutralization unit; 2—IR emitter; 3—exhaust gas recirculation valve; 4—bracket; 5—exhaust branch pipe; 6—air supply branch pipe; 7—thrust bearing; 8—spring; 9—flow regulator; 10—glass; 11—thermocatalytic converter; 12—auger; 13—ejector; 14—temperature sensor; 15—porous catalyst (MnO_2 on Al_2O_3 substrate)

of the engine and through the gear system the number of revolutions of the branch pipe, which receives more exhaust gases, increases. The rocker arm with the load begins to spin when the number of revolutions of the branch pipe increases. The centrifugal force, untwisting the load, moves it along the visor, lifting the sleeve upwards, ensuring its movement along the vertical channel by the pin. The bushing, rising up, lifts the damper, the hole opens more and more exhaust gases begin to flow into the engine combustion chamber. Exhaust gas recirculation significantly reduces nitrogen oxides. Further, the exhaust gases, after thermal neutralization, enter the catalytic neutralization chamber, where the hydrocarbon residues are oxidized to CO_2 and H_2O due to the active catalyst layer. Exhaust particulate matter can become trapped in the replaceable fabric filters. Figure 1 shows a thermocatalytic converter, Fig. 2 shows an exhaust gas recirculation valve.

The volume of the gas-air mixture was determined by the formula (5):

$$G = ((V_f + V_{air}))/t \tag{5}$$

where V_f—fuel consumption, m^3; V_{air}—volume of air required for combustion of 1 kg of fuel, m^3; t—time, s.

One of the main disadvantages of catalytic converters is inefficiency at low temperatures. Low temperatures occur when the vehicle is idling and when the throttle is fully open. It is in such modes that the movement of vehicles at intersections is carried out [7–10]. A recirculation rate of about 30% can be maintained at any engine operating mode, as shown in [10]. Once configured, this will be kept constant. The EGR ratio is calculated using the formula:

$$\mathcal{K}_{EGR.} = M_{bp}/M_{total} \tag{6}$$

M_{bp}—mass of bypassed exhaust gases.
M_{total}—total mass of gases entering the combustion chamber.

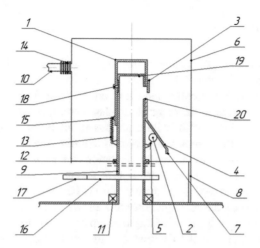

Fig. 2 Exhaust gas recirculation valve. 1—bushing, 2—weight, 3—damper, 4—visor, 5—rocker arm, 6—regulator body, 7—limiter, 8—body support, 9—exhaust gas outlet, 10—exhaust gas outlet from the regulator, 11, 12—a movable unit with a seal, 13—a return spring, 14—a bellows, 15—an attachment point, 16—a gear wheel of a driven exhaust pipe, 17—a driving gear, 18—a pin, 19—a hole for pressure equalization, 20—a hole for gas passage

In this case, the mass of bypassed exhaust gases is estimated by the formula:

$$\mathcal{K}_{bp} = M_{total} - M_{ex} \tag{7}$$

M_{ex}—mass of exhaust gases entering the engine intake manifold.

The calculation of the exhaust gas recirculation system is carried out using the indicator of the engine cycle filling—G_{ecf}, which is estimated by the formula:

$$G_{ecf} = M_{fc} \times I \times \mathcal{K}_{EGR.}/1, 2 \times n \tag{8}$$

M_{fc}—mass filling of the cylinder, I—engine stroke rate, n-crankshaft speed.

On the basis of research work [23], the exhaust gas bypass is determined to be no more than 30%, so as not to worsen the engine operating parameters. The calculation using the formulas given above made it possible to evaluate the efficiency of the exhaust gas recirculation system to reduce the concentration of nitrogen oxides for various types of engines. So for gasoline internal combustion engines with a maximum recirculation ratio of 25%, a decrease in the content of nitrogen oxides occurs by 40%.

4   Conclusion

The relevance of the chapter is due to the need to radically improve the environmental performance of automobile engines. Nitrogen oxides are the most significant gaseous toxic component of exhausted diesel engines. Mechanisms and models of nitrogen oxide formation in the combustion chamber are considered. The ways of nitrogen oxide emission reduction with emissions are analyzed. A combined method of thermocatalytic neutralization was proposed to reduce the concentration of toxic components in exhaust gases during the operation of internal combustion engines. It was proposed to increase the completeness of fuel combustion through the use of recirculation systems, as well as the neutralization of exhaust gases using a catalytic converter. Determining the recirculation ratio allows you to quantify the amount of bypass of the exhaust gases. Limiting the bypass of exhaust gases is necessary to improve the environmental friendliness of the internal combustion engine and optimize its operating modes.

Acknowledgements Scientific research was carried out with the financial support of the Ministry of Education and Science of Russia as part of the fulfillment of obligations under the Agreement № 075-03-2023-032 dated 16/01/2023 (topic number FZSU-2023-0005).

References

1. Timberlake, K.C.: Chemistry: An Introduction to General, Organic, and Biological Chemistry, 10th ed. Prentice Hall Higher Education, Upper Saddle River (2008)
2. Petrucci, R.H., Harwood, W.S., Herring, G.E.: General Chemistry: Principles and Modern Applications, 9th ed. Prentice Hall, Upper Saddle River (2006). Biological Chemistry, 10th ed. Timberlake, K.C.: Chemistry: An Introduction to General, Organic and Biological Chemistry
3. Patel, B.S., Patel, K.D.: A review paper on catalytic converter for automotive exhaust emission. **7**(11) (2012)
4. Abhinesh, A.K.: Minimization of engine emission by using non—noble metal based catalytic converter. **4**(11), 2663–2468 (2014)
5. Shah, M., Mistry, C., Shimpi, N.: Emission norms (case study). **5**(3) (2016)
6. Borrás, E., Tortajada, L.A., Vázquez, M., et al.: Polycyclic aromatic hydrocarbon exhaust emissions from different reformulated diesel fuels and engine operating conditions. Atmos. Environ. **43**, 5944–5952 (2009)
7. Pardiwala, J.M., Patel, F., Patel, S.: Review paper on catalytic converter for automotive exhaust mission (2011)
8. Kataliticheskij nejtralizator [Catalytic converter]/Lozhkin, V.N., Shul'gin, V.V., Gulin, S.D., Zolotarev, G.M.: Utility model patent № 2204027C1, Published on 10.05
9. Gilmanov, E.S., Kamaletdinov, R.R., Kozeev, A.A., Kostarev, K.V., Yamaletdinov, M.M.: Kataliticheskij nejtralizator otrabotavshih gazov [Catalytic converter of exhaust gases] Useful model patent No. 2727122C1. Published on 20 July 2020
10. Kataliticheskij nejtralizator [Catalytic converter]/Mingazetdinov I.Kh., Tunakova Yu.A., Shipilova R.R., Galeev K.I. Utility model patent No. 203184, bulletin No. 9 dated 25.03.2021
11. Official website of the manufacturer of the catalyst with a heater Emitec. http://www.emitec.com/en/. Accessed 13 August 2021

12. Presti, M., Pace, L.: An alternative way to reduce fuel consumption during cold start: the electrically heated catalyst. SAE Int. http://www.emitec.com/fileadmin/user_up-load/Bibliothek/Vortraege/11ICE_0255_Final_Manuel_Presti.pdf (2011). Accessed 13 August 2021

13. Weiss, M., et al.: Analyzing on-road emissions of light-duty vehicles with portable emission measurement systems (PEMS). JRC Sci. Tech. Rep. http://ec.europa.eu/clima/policies/transport/vehicles/docs/2011_pems_jrc_62639_en.pdf (2011). Accessed 13 August 2021

14. Mollenhauer, K.: Hand Book of Diesel Engine, p. 636. Mollenhauer, K., Tshoeke, H. (eds.). Springer, Heidelberg (2010)

15. Strategies and Issues in Correlating Diesel Fuel Properties with Emissions. US Environmental Protection Agency: EPA420-P-01-001, p. 28, July (2001)

16. Blinov, A.S., Malastovsky, N.S., Bykov, A.E.: Performance evaluation of static mixers in the urea injection pipe for SCR system. In: Proceedings of the 5th International Conference on Industrial Engineering (ICIE 2019), Platov South-Russian State Polytechnic University (NPI), pp. 1465–1473 (2020). https://doi.org/10.1007/978-3-030-22063-1_15

17. Sathish Sharma, G., Sugavaneswaran, M., Prakash, R.: NOx Emission Control Technologies in Stationary and Automotive Internal Combustion Engines Approaches Toward NOx Free Automobiles, pp. 223–253 (2022). https://doi.org/10.1016/B978-0-12-823955-1.00008-5

18. Praveena, V., Leenus Jesu Martin, M.: A review on various after treatment techniques to reduce NOx emissions in a CI engine. J. Energy Inst. **91**(5), 704–720 (2018). https://doi.org/10.1016/j.joei.2017.05.010

19. Mahyon, N.I., Li, T., Martinez-Botas, R.: A new hollow fibre catalytic converter design for sustainable automotive emissions control. Catal. Commun. **120**, 86–90 (2019). https://doi.org/10.1016/j.catcom.2018.12.001

20. Trofimenko, Y.V., Komkov, V.I., Potapchenko, T.D., Donchenko, V.V.: Period. Eng. Nat. Sci. **7**(1), 465–473 (2019)

21. Lozhkin, V., Gavkalyk, B., Lozhkina, O., Evtukov, S., Ginzburg, G.: Transp. Res. Procedia **50**, 381–388 (2020)

22. Lozhkina, O., Lozhkin, V., Vorontsov, I., Druzhinin, P.: Transp. Res. Procedia **50**, 389–396 (2020)

23. Erokhov, V.I.: Sistema retsirkulyatsii otrabotavshikh gazov sovremennykh dvigateley [Exhaust gas recirculation system of modern engines]. Altern. Fuel Transp. **4**(34), 36–43 (2013)

Artificial Intelligence and Smart Railway Transport as a Tool to Accelerate the Transformation Process in Uzbekistan

Zumrat Gaibnazarova

Abstract Worldwide the transport sector faces several issues related to the rising of traffic demand such as congestion, energy consumption, noise, pollution, safety, etc. Trying to stem the problem, the European Commission is encouraging a modal shift towards railway, considered as one of the key factors for the development of a more sustainable European transport system. This contribution proposes a synthetic methodology for the capacity and utilization analysis of complex interconnected rail networks; the procedure has a dual scope since it allows both a theoretically robust examination of suburban rail systems and a solid approach to be applied, with few additional and consistent assumptions, for feasibility or strategic analysis of wide networks (by efficiently exploiting the use of Big Data and/or available Open Databases). Artificial intelligence and cloud are leading to the digitalization of logistics and transport sectors and the development of smart railway terminals. Smart railways are made possible with next-generation technologies that bring new levels of safety, connectivity, and sustainability. The chapter scientifically founded ways to reduce costs based on the "wireless electro pneumatic brake" that is, energy–efficient innovative technology that reduces the main energy raw materials in the railway transport of Uzbekistan. Ways to reduce and increase energy efficiency on the basis of digital information and communication technologies are shown. Since currently very much attention is paid to environmentally friendly technologies, this type of technology is very environmentally friendly and safer than the current type of brakes. While this type of brakes can qualify for international standards. If their implementation takes several months, the benefit of their implementation can benefit from the year to the railway industry. On the basis of wireless electro-pneumatic brakes, an intra–train information and diagnostic system of a freight train is being implemented, which allows for continuous control and predictive monitoring of all critical and running parts of cars ("intelligent" freight train).

Z. Gaibnazarova (✉)
Tashkent State University of Economics, 49 Islam Karimov Str., Tashkent 100066, Uzbekistan
e-mail: z.gaibnazarova@tsue.uz; falsezumrat59@rambler.ru

© The Author(s), under exclusive license to Springer Nature Switzerland AG 2024
M. Sari and A. Kulachinskaya (eds.), *Digital Transformation: What are the Smart Cities Today?*, Lecture Notes in Networks and Systems 846,
https://doi.org/10.1007/978-3-031-49390-4_19

Keywords Artificial intelligence · Smart railway transport · Digital technology · Pneumatic brake · Wireless electro pneumatic brake · Smart freight train

1 Introduction

Artificial intelligence and smart railway transport accelerates transformation for Uzbekistan, and logistics companies can replicate this approach across the railway transport industry with following efficiency improvement:

- Enhancing productivity by 20%
- Reducing opex by 40%
- Reducing workplace accidents
- Improving identification accuracy
- Improving energy conservation
- Significantly improving the workplace environment.

Artificial intelligence and smart railway transport of the transportation and logistics industry is ushering in a new era of collaboration, and companies must evaluate how they can provide exceptional customer experiences by acquiring new capabilities and creating differentiated global partnerships.

Governments and industry leaders must guide industry modernization and set the standards for making freight more resilient and sustainable.

The 2011 EU Transport Whitepaper presents ambitions for modal shift: 30% of road freight over 300 km should shift to other modes such as rail or waterborne transport by 2030, and more than 50% by 2050. This should be facilitated by efficient and green freight corridors and by optimizing the performance of multimodal logistic chains, including by making greater use of more energy–efficient modes [1]. This ambitious objective is set against the backdrop of a declining rail freight market share for inland (intra-EU) transport, having decreased from 20.2%, in 1995, to 17.4%, in 2011. In contrast the share of road freight has increased from 67.4%, in 1995, to 71.8%, in 2011 [2].

Since the first Shinkansen (bullet train) was introduced in Japan in 1964, high-speed trains such as the TGV have been an undeniable technological, commercial and popular success. As seen in Table 1, many countries have invested in what has become a vast network of high-speed rail (HSR) lines, on which trains can run, in best cases, at speeds up to 350 km/h. In 2016, there was about 35,000 km of HSR line running worldwide, to which approximately 23,000 km is projected to be added [3].

At the same time, it is one of the important tasks for the production of production networks, taking into account the urgency of innovative ideas, technologies and projects, taking into account the urgency of foreign investment.

President of the Republic of Uzbekistan Sh. M. Mirziyoev noted, we need to "master and condition digital knowledge and modern information technologies for

Table 1 Economic calculation WEPB for locomotive

Average consumption in one locomotive in one day (ton, diesel)	4		4.5		5	
Average consumption in 200 locomotives in one day (ton, diesel)	800		900		1000	
This size fuel is enough to walk 800–1200 (km)						
Fuel economy as a result of the introduction of WEPB (%)	5.0	8.0	5.0	8.0	5.0	8.0
Fuel cost 1 kg (sum)	8		8		8	
Savings in tonnes	40	64	45	72	50	80
Savings in kgs	40,000	64,000	45,000	72,000	50,000	80,000
Savings in the form of cost (thousand sum)	320,000	512,000	360,000	576,000	400,000	640,000
Annual cost savings (thousand sum)	116,800,000	186,880,000	131,400,000	210,240,000	146,000,000	233,600,000
WEPB installation cost (thousand sum)	17,250,000	17,250,000	17,250,000	17,250,000	17,250,000	17,250,000
Profit from the introduction of the WEPB (thousand sum)	99,550,000	169,630,000	114,150,000	192,990,000	128,750,000	216,350,000

development. This allows us to go to the shortest path of rise. After all, today information technology in all areas is deeply entering. Although the country has risen to 8th in 2019, we are still very back. Digital Technologies Increases the quality of goods and services, reduces excess expenditures. The management of the state and society, in the social sphere can also introduce digital technologies and improve its impacts, improve people's lives" [4].

Many regulations are adopted for the introduction of energy economical digital information technologies. According to them, the Resolution of the President of the Republic of Uzbekistan dated May 26, 2017 Decree of the President—3012 "On the program of the measures to further develop the energy efficiency, energy efficiency in the social sphere, and energy efficiency in the social sphere in 2017–2021 years", April 27, 2018 Decree of the President—3682 "On measures to further improve the appliance, technologies and projects", February 1, 2019 Decree of the President—5646 "On measures to radically improve the fuel and energy sector", August 22, 2019 Decree of the President—4422 "On increasing energy efficiency of sectors of the economy and the social sphere on rapid measures to the introduction of energy-saving technologies and the development of renewable energy sources", July 10, 2020 Decree of the President—4779 "On increasing energy efficiency of the economy and attracting existing resources on additional measures to reduce the dependence of the economy in fuel and energy products" normative documents Program in the development and implementation of innovations in this direction.

To underline the importance of both the levels of the problem, it is worth to remind how the European Commission with the railway packages and the related directives, after having fully opened to competition the markets for rail freight services and for international passenger transport (long distance, i.e. large-scale), currently is focusing also on national markets for domestic passenger transport services (i.e. regional, small scale) which remain largely closed and are still considered the bastions of national monopolies [5].

Currently, the main source in the transport sector in the effective organization of innovative activities is to form a portfolio of own funds of enterprises in this area. To this end, the JSC "Uzbekistan Temir Yullari" should introduce the high–efficiency technologies that are high effective technologies. The ability to save "Wireless electro pneumatic brake", which contains a "wireless electro pneumatic technology", its economic efficiency and its implementation, scientific research creates the content of our work.

The purpose of the study is to analyze the current types of brakes in railway transport and the offer of a new type of brake for Uzbekistan.

The use of this type of brakes is more effective than the current appearance. In this regard, a new type of brakes is given and preliminary efficacy in high-quality and quantitative terms is calculated.

The objectives of the chapter are to define and describe the essential advantages of the BEPT system which performs the following functions for freight cars:

- Continuous pressure monitoring in the brake line (TM) and brake cylinder (TC)
- Automatic detection of the state of auto brakes on each car

- Brakes not activated (disabled)
- Slow-release brakes
- Slow release of brakes
- Monitoring and emergency telesignalization in case of unscheduled pressure changes in the brake line and brake cylinders.

The results of the accomplished studies are useful in identifying the appropriate method for transport project valuation in a certain framework. Therefore, this paper focuses not on a comparative examination of the project assessment. The aim is to develop a model which could supplement any current applied method [6].

2 Theory and Practical Application

B. S. Erasov in the process of studying the problem, "the creation of mechanisms of the wagon repair complex, changes in the relations between production organizations at Joint Stock Company "Russian Railways" (RZD) were made mainly to adapt to changes in external and internal media. This allowed the wagons to receive real-time solutions to increase the effectiveness of the repair industry. The current progressive forms of production processes, the introduction of resource-saving technologies, the norms of new technical-saving technologies, will remain related to new developing organizational and economic relations between students and repair services" [7] did. That is, it is necessary to re-equip wagons and adapt to the need of modern.

"Russian Railways" Joint Stock Company is a Russian fully state-owned vertically integrated railway company, both managing infrastructure and operating freight and passenger train services. The company was established on 18 September 2003, when a decree was passed to separate the upkeep and operation of the railways from the Ministry of Railways of the Russian Federation. In March 2016, RZD approved an updated version of high-speed rail development program until 2030. The 5 trillion ruble program includes the construction of Moscow-Kazan-Yekaterinburg, Moscow-Adler and Moscow-Saint Petersburg high-speed lines, as well as other high-speed lines connecting regional cities. Between 2021 and 2025 RZD plans to build Rostov-Krasnodar-Adler, Tula-Voronezh high-speed rail and the extension of Kazan-Yelabuga high-speed rail, as well as other regional high-speed rail links.

During the 2026–2030 third phase of the program, Russian Railways will build Moscow–Saint Petersburg high-speed rail section; the railway line will be extended from Yelabuga to Yekaterinburg, and from Voronezh to Rostov-on-Don. The main activities of Russian Railways involve freight and passenger traffic. In Russia, railways carry 42% of the total cargo traffic, and about 33% of passenger traffic. Some passenger categories, such as pensioners, members of parliament, and holders of Soviet and Russian state decorations, receive free or subsidized tickets.

O. V. Selina was one of the first to develop the economic foundations for energy management outside the railway industry as a use of modern brakes. According to him, "… one of the strategic directions of the railway transport is now increasing the level of modernization of trains and the level of modernization of the moving content. Work is underway to investigate the technology that ensure the safety of movement and introduce resource-saving technologies, technical re-equipment of the material base, and the introduction of scientific and technical developments, intellectual technologies in intellectual technology. It is mentioned that the need to impose high-speed movement. Currently, taking into account the safety of trains and especially those mentioned above, it is proposed to install wireless electro pneumatic brakes for wagon and locomotives" [8].

M. Luchak is about the introduction of modern and energy-efficient trains in their work "… as the stage of scientific and technological progress in railway transport, it is worth the replacement of steamers with dieseling" [9].

R. Cul and P. Jonson have scientifically based on the following views. That is, in Spoornet, the effectiveness of ECP is characterized by the following quantitative indicators:

- Reduction of the brake path by 60–27%
- Decrease in 37% of the voltage regime and 23% of the brake's regime
- Reduction of the moving content by 9%
- Reduction of electricity by 23%
- The dynamic brake can lead to a decrease in the same amount of work wagonried out by pneumatic brakes.

Technical activation of energy-efficient wagons from our country and increasing their effectiveness can be introduced to wireless electro pneumatic brakes to wagon go and passenger transport. It has the following basic advantages:

- Reduction of the stop road to 15–70%
- 2 times reduction of the longitude forces in the stopping
- Increase their service life by 25% decreased by 25% of brakes
- Decrease in damage to the brakes of wheels
- Reduction of fuel and electricity to 5–8%
- Rising by 20–25% of average movement
- Reduction of trains cycle time by 5–9%
- Decrease in "Damage" to RELEs during the stop process
- Reduction of the range of technical inspections to 3–5 times
- Removal of the limit one to the length of wagon go on management trains.

The wireless electro pneumatic brake has the technical requirements and structural scheme for wagons, which will be effective in the above-mentioned quantities and effectiveness due to proper placement and start [10] (see Fig. 1).

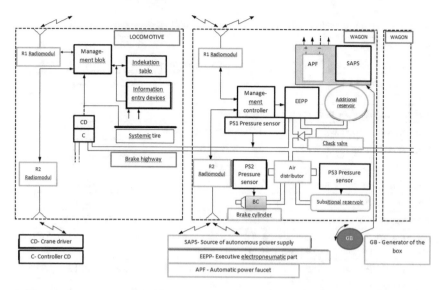

Fig. 1 The structural scheme of the WEPB

3 Research and Discussion

The following energy efficiency can be achieved due to the implementation of the study to the rail transport of our country. The conditions of 3 different scenarios were organized based on the capacity of the dieselots. The largest power mercolovis spends 5 tons of diesel fuel to cross 1000 km. We can see in Table 1 that the profit from the introduction of the WEPB is at least 99,550,000 thousand sum for 200 locomotives of workers in the country.

WEPB is an innovative development in terms of control of pneumatic brakes on freight trains in Russia and neighboring countries, and at the global level. The introduction of wired electro-pneumatic braking systems on route coal and ore trains in the USA, Canada and some other countries has significantly reduced the length of the braking distance. In addition, it was possible to reduce the harmful effect of braking on the track and rolling stock, to exclude the "pulling" and "squeezing" out of the track of empty and partially loaded cars, significantly increase the distance between brake inspections, and significantly facilitate the work of drivers. And the further most important prospect for improving such structures is the introduction of wireless electropneumatic brakes for freight cars (WEPB) on the railways.

We can see from Table 2 that as a result of the introduction of a WEPB, the economic effect, which is calculated in the braking, is 991241580,7 sum.

When developing the principles for constructing a wireless electropneumatic brake (BEPT) for freight trains, the following defining prerequisites were taken into account:

- The need for a fundamental increase in the speed and controllability of auto brakes

Table 2 Economic calculation from the introduction of WEPB to passenger wagons

Passenger turnover passenger/km	4,350,100,000
Action length is in a km for an average of 100 passengers per year	43,501,000
Action length per 1 passenger wagon km for an average of 100 passengers per year	62,144
The service life of the pads is for passenger wagon	110,000
The increase in service life of the pads is for passenger wagon	137,500
Pad's service is up to the WEPB for passenger wagon	1.77
The service life of pads is in the year passenger wagon after the WEPB	2.21
The value of service to the WEPB	354,015
The value of service after the WEPB	442,519
After the WEPB, the economy is the source of the value of the service value	88,504
After the WEPB, the economy accounted for the increase in the value of the service 1 (16 pads) for 1 passenger wagon	1,416,059.4
After the WEPB, the economy for the increase in the value of the service is 700dona passenger wagon (for 1 cycle using pads)	991,241,580.7
The cost of installing WEPB in all passenger wagons	10,500,000,000

- The need to reduce the volume and timing of periodic maintenance of brake equipment, increase the warranty areas several times
- Minimum changes (at this stage) to existing technologies for forming freight trains and controlling brakes, technical means and maintenance of auto brakes in freight trains
- Existing qualifications and mentality of freight train brake service personnel;
- Electromagnetic compatibility with all types of electronic and radio engineering products used in railway transport
- Use of a frequency range that provides stable radio communication under any weather conditions and in artificial structures (bridges, tunnels, etc. p.) to the specified distance
- Maintenance of the system's operability in case of equipment failure on individual cars and when putting into separate unequipped cars
- The possibility of rapid expansion of the system's functions in the direction of intratrain diagnostics of the technical condition of all critical and running parts of each car—the creation of an intra-train information and diagnostic system, an "intelligent" freight train.

Table 3 can extend the pad's service period as a result of the introduction of WEPB in freight wagons to 0.21 year or 75 days. Due to the life cycle of each pad, only freight wagons can be saved from the only loads from its freight wagons from the same year 7,232,876,712.33 sum. In addition, when using the innovation, the service life of the brake pads increases by a quarter, the wheels are damaged less often, and the fuel and electricity consumption for train traction is reduced by 5–8%.

Table 3 Economic calculation for the introduction of WEPB to loads

Cost pad for cars type M 659 composite	200,000.0
Service life in kilometers for freight wagons	90,000.0
Saving from WEPB Pads Increased service life, %	25.0
Service duration of the freight wagons to introduce the WEPB, km	112,500.0
1 freight wagons daily movement distance, km	300
Pads service duration to the WEPB day on the freight wagons	300.00
The service life of pads is on the freight wagons after the WEPB	375.00
The difference obtained as a result of the WEPB	75
The difference obtained by the WEPB is in km	22,500.0
Freight wagons movement distance in 1 km	109,500.0
Pads's service life to the WEPB year in freight wagons	0.82
The service life of pads is in freight wagons after the WEPB	1.03
The value of service to the WEPB	164,383.6
The value of service after the WEPB	205,479.5
After the WEPB, the economy is the source of the value of the service value	41,095.9
After the WEPB, the economy accounted for the increase in the value of the service is 1 in the sum for freight wagons (8 pads)	328,767.1
After the WEPB, the economy accounted for the increase in the value of the service is in the sum for 22,000 freight wagons	7,232,876,712.33
The cost of installing the WEPB in all freight wagons in the sum	330,000,000,000

At the same time, the speed increases, the harmful effect of the rolling stock on the track during braking decreases. When creating the equipment, scientists used parts with a diamond-like coating. Thanks to this, it was possible to reduce energy consumption and increase the service life of the devices by several times.

The analysis of Table 4 can save 94,000,000 sum a year, which will lose the crawlers on the unintiquities in the railways in our country.

Table 4 Economy analysis

Number of crawlers	4000
The cost of repair of one slider, sum	100,000
The effectiveness of the WEPB is in %	76.5
The number of sliders after the WEPB	3060
The difference in the middle after the WEPB is introduced	940
The cost of all sliders repairs, sum	400,000,000
The cost of repair of all sliders after the WEPB, the sum	306,000,000
The profit from the introduction of the WEPB is a year, sum	94,000,000

Locomotives of the modern type are practically quite "smart", only the equipment for the intra-train exchange of telemechanical information is lacking. But this requires "smart" freight cars equipped with the appropriate equipment. The main automatic properties of such cars should be:

- Control of the condition of axle boxes
- Heating, shift, bearing defects
- Determination of wheel defects
- Sliders, dents, uneven rolling
- Control of the operation of the air distributors and the inexhaustibility of the brakes
- Assessment of the resource (actual state)—control of the load, the actual mileage sensor, the recorder of mechanical influences;—control of the axle or bogie derailment
- Active identification, control of the safety of the cargo (the integrity of the containers).

The cost of installing WEPB in the economic insecurity for electricity for electricity in Table 5 is 30 million sums. But this is a one-time cost. The current economy is 60,500 sum per day, this value can reach at least 22,082,500 sum for the full 1 year.

Table 6 show that the average weight of the 5600 tons is worth 90,000,000 sum. The introduction of the WEPB is 93,500 sum per day, this value can reach at least 34,127,500 sum per whole for 1 year.

The need to increase the technical speed of freight trains, especially accelerated container trains, to 95–100 km/h will lead to the fact that their braking distances will not fit into the existing lengths of block sections, which is shown by the corresponding calculations. Brake control is also problematic, providing accurate speed control. According to our calculations, due to an excessive decrease in speed when fulfilling the restrictions and the subsequent pulling up to the established value, Russian Railways lose about 1.5 billion rubles in year. These losses are caused not

Table 5 Economic calculation for electric locomotives

For electric locomotive of Uzbekistan		
Average weight, ton	3200	
Movement distance, km	230	
Energy consumption signs, KWT	5500	
Price of 1 kW energy consumption	220	
1 day total expenses of distance, sum	1,210,000.0	
Cost of installation of WEPB to electricity	30,000,000.0	
The economy energy consumption is the economy from the WEPB, %	5	8
Economy day made of WEPB, sum	60,500.0	96,800.0
The economy of the introduction of a WEPB is a year, sum	22,082,500.0	35,332,000.0

Table 6 Installation of WEPB to electricity

Average weight, ton	5600	
Movement distance, km	230	
Energy consumption signs, KWT	8500	
Price of 1 kW energy consumption	220	
1 day total expenses of distance, sum	1,870,000.0	
Cost of installation of WEPB to electricity	90,000,000.0	
Economy day made of WEPB, sum	93,500.0	149,600.0
The economy of the introduction of a WEPB is a year, sum	34,127,500	54,604,000

by the qualifications of the drivers, but by the impossibility of accurately performing the desired stage of braking (aggravated by the accuracy class of locomotive pressure gauges) and its accurate maintenance.

The benefits of implementing WEPB should be enormous. According to our calculations, only in one year of its operation on domestic highways in closed routes due to energy savings for train traction, reduced wear of brake pads and reduced turnover time of block trains, an effect can be obtained.

True, even with such a record payback of the system (less than a year), its mass implementation still requires a government decision. This is confirmed by the rich experience of a number of other countries, which at one time introduced wire electropneumatic brakes.

Joint-stock company "Uzbekiston Temir Yullari" was formed on November 7, 1994 on the basis of the former Central Asian railroad located in the territory of the Republic of Uzbekistan. The total developed length of the main roads of the company makes about 3645 km today. More than 54.7 thousand people work in the company. Annual goods turnover of the company makes about 90% of total goods turnover of all means of transport.

Now the structure of the company is reformed radically. The special attention is paid to privatization and privatization of separate areas.

The main activities of the company are:

- Forwarding and cargo delivery by rail
- Repair and maintenance of railway cars
- Passenger, tourist traffic
- Service, updating of locomotive and carriage park.
- The priority direction for the company is implementation of the following investment projects:
- Strengthening of communication channels with application of fiber-optical lines
- Purchase of a new rolling stock (electric locomotives and cars), and also modernization of the existing
- Construction of new railway lines
- Electrification of the railroad

- Implementation of projects of capital track repair, the organization of release at plants of the company of elements of the top structure of a way, spare parts, etc.

The company has some projects demanding investment support and promising to bring big benefit to the investor at their implementation.

JSC "Uzbekiston Temir Yullari" since 1993 is the member of the Organization of the Commonwealth of the Railroads (OCR). The company has close connections with the international union of the railroads (IUR) and the Economic commission of the UN for the Pacific Rim (ESCAP). JSC "Uzbekiston Temir Yullari" carries out collaboration with the TRASEKA (a transport corridor of Europe-Caucasus-Asia) of the TASIS program of the Commission of the European Union.

As we can see from the following economic based Table 7, the consumptions of JSC "Uzbekiston Temir Yullari almost 364 billion soums or 38.3 mln. dollars. By introducing an innovative technology of WEPB, will coverages period in 3.35 year or 1223 days.

JSC "Uzbekistan temir yullari" increased cargo turnover volume by 22 billion in 2010–2018 tons per km. However, the increase in investment in the sector to raise profits can also have a positive impact on the volume of freight turnover. This means that by 2022, the volume of freight turnover by rail will increase to 25 billion tons per km [11].

Table 7 The final investments and their efficiency rates for the introduction of the WEPB transport to the railway transport

Price of installation to passenger and freight wagons	340,500,000,000
Cost of installation to locomotives	17,250,000,000
Price of installation in electric locomotives	6,000,000,000
Effects taken from the introduction to the diesel	99,550,000,000
After the WEPB, the economy accounted for 22,000 freight wagons (for 1 cycle of pads)	7,232,000,000
After the WEPB, the economy for the increase in the value of the service is 700 passenger wagon (for 1 cycle of pads)	991,000,000
Loss of sliders in wagons, the benefits taken from the introduction of WEPB in the year, sum	94,000,000
Effect from introduction to electric locomotives, sum	700,000,000
Coverage period, years	3.35
Coverage period, in days	1222.9

4 Conclusions

Technology integration is integral to the transportation and logistics industry, combining the complexities of warehouse management systems, transportation systems, fleet tracking systems, and data and analytics capabilities.

However, the transportation sector is challenged by rising labour costs, high fuel costs, maintenance expenses, and regulatory compliance. Supply chains are becoming more complicated as more vendors, suppliers, distributors, and manufacturers need to be synchronized and coordinated globally.

Also, workplace conditions in the railway sector are harsh, which limits the number of years people can work. This results in the railway industry having difficulty attracting new and younger talent.

Lastly, the logistics and transportation industry are having a drastic impact on the environment, and environmental sustainability is a top concern to reduce waste, emissions, fuel, and energy consumption.

Companies with advanced, unified digital strategies are currently at a significant advantage, led by transportation leaders that recognize they need to change to keep pace with a new, evolved logistics and transportation landscape.

Research indicates that there is a reconfiguration of supply chains occurring, and freight companies must move onshore or nearshore to "alleviate supply chain challenges and improve their competitive positioning". Supply chain companies have to reimagine their infrastructure and vendor interactions.

It's not just supply chain companies that must rethink their strategies and positioning. Delivery, distribution, warehousing, and manufacturing companies are all adapting to this new environment. At the same time, new players are entering the space, like new manufacturing and warehouse locations, that change the network of how goods and services are connected.

This will not only be short-term investment, which will have an intensive investment to increase the technical and economic efficiency of existing equipment. The economic impact will be on average more than 108 billion sums of money a year. The organization has provided 100 billion sum in the number of years since the 4th year. The amount of economic investments can be used to purchase new freight wagons, to finance the purchase of new freight wagons, to finance other investment projects.

In our opinion, the concept of a "smart" freight train should be defined as follows: a freight train containing a locomotive and modern-type wagons, possessing the properties of automatic monitoring of the technical condition of critical and running parts by high-tech devices with an assessment of their functioning and safety, having an intra-train wireless telemechanical exchange channel. information, as well as external radio channels for receiving and transmitting command and notification information.

Currently, specialists have a development that will largely correct the situation. Electropneumatic brakes are characteristic of railway transport. No devices that are temporarily installed on the tail or along the train's composition will provide the functions that EPT initially has. In the presence of EPT, the process of objective testing of brakes is really automated, with documented registration and the exclusion of the human factor during testing. To date, there are developed technical solutions that have been tested in operational conditions. The feasibility of wireless EPT and its high viability with parameters far exceeding the current indicators are proved. The use of BEPT along with an intra-train information and diagnostic system with predictive monitoring allows you to switch to a new technological platform and get rid of unproductive labor.

We will have the opportunity to increase energy consumption, due to the use of this innovative technology, due to reduction of fuel consumption in the railway network of the country. As a result, it is possible to mobilize the quality of production and service of new types of products through the use of the energy necessary for consumption and other industries. At the same time, it allows the prevention of technological breaks in the wireless electro pneumatic brakes, as well as the possibility of future electric brakes, as well as to increase the goods and passengers to their destinations.

References

1. European Commission: Transport white paper—roadmap to a single European transport area— towards a competitive and resource efficient transport system, vol. 144, Brussels (2011)
2. Zunder, T.H., ZahurulIslam, D.M.: Assessment of existing and future rail freight services and technologies for low density high value goods in Europe. Eur. Transp. Res. Rev. (2018). https://doi.org/10.1007/s12544-017-0277-1.P2
3. Blanquart, C., Koning, M.: The local economic impacts of high-speed railways: theories and facts. Eur. Transp. Res. Rev. (2017). https://doi.org/10.1007/s12544-017-0233-0
4. Refail on the President of the Republic of Uzbekistan Shavkat Mirziyoyev on 28 January of 2020
5. Francesco, R., Gabriele, M., Stefano, R.: Complex railway systems: capacity and utilization of interconnected networks. Eur. Transp. Res. Rev. (2016). https://doi.org/10.1007/s12544-016-0216-6
6. Raicu, S., Costescu, D., Popa, M., Rosca, A.: Including negative externalities during transport infrastructure construction in assessment of investment projects/Raicu et al. European Transport Research Review (2019). https://doi.org/10.1186/s12544-019-0361-9
7. Erasov, B.S.: Comparative Study of Civilizations: Tutorial for University Students, p. 556. Aspect Press, Moskow (2018)
8. Selina, O.V.: Economic assessment of the modernization of rolling stock on railway transport. Dissertation thesis on the degree of candidate of economic sciences, p. 106. Yekaterinburg (2017)
9. Luczak, M.: Railway Age **1**, 69–73 (2016)
10. Cull, R., Johnson, P.: Int. Railw. J. **8**, 19–25 (2020)
11. Saitkamolov, M.S.: Econometric model of operating joint-stock company "Uzbekiston Temir Yullari" and possible directions of development. Econ. Innov. Technol. **2019**(6), November–December, 112–122 (2019)

A Demand-Response Approach for HVAC Systems Using Internet of Energy Concept

Nikita Tomin⬤, Irina Kolosok⬤, Victor Kurbatsky⬤, and Elena Korlina⬤

Abstract Rapid urbanization, increasing electrification, the integration of renewable energy resources, and the potential shift towards electric vehicles create new challenges for the planning and control of energy systems in smart cities. Energy storage resources can help better align peaks of renewable energy generation with peaks of electricity consumption and flatten the curve of electricity demand. The chapter proposes an approach to decentralized demand response (DR) of a distributed multi-energy network on the Internet of Energy (IoE) structure, receiving electricity from a centralized power grid. IoE is implemented as a multi-agent hardware and software environment where HVAC building agents are able to alter the overall electricity demand curve by controlling the heat storage of each agent. In addition, it is shown that the net-centric DR management structure is well suited for prosumers and active consumers who are IoE users and are connected by smart contracts to their aggregator cluster. The proposed DR approach are tested in power district of nine simulated HVAC-associated buildings located in climate zone D. The buildings have diverse load and domestic hot water profiles, PV panels, thermal storage devices, heat pumps, and electric heaters. The multi-agent reinforcement learning method achieves superior results for the DR problem by reducing the average daily peak load, ramping, and increasing the load factor.

Keywords Demand response · HVAC · Internet of energy · Reinforcement learning · Multi-agent system · Distributed multi-energy systems

1 Introduction

Demand response (DR) of retail electricity consumers in the long term is an important tool for maintaining and regulating the balance of supply and demand in the electricity market. DR, as a new flexibility source in the power system, competes with

N. Tomin (✉) · I. Kolosok · V. Kurbatsky · E. Korlina
Melentiev Energy Systems Institute SB RAS, Irkutsk 664033, Russia
e-mail: tomin@isem.irk.ru

© The Author(s), under exclusive license to Springer Nature Switzerland AG 2024
M. Sari and A. Kulachinskaya (eds.), *Digital Transformation: What are the Smart Cities Today?*, Lecture Notes in Networks and Systems 846,
https://doi.org/10.1007/978-3-031-49390-4_20

both the least efficient and most expensive part of the existing generation, and with investments in new generation and network facilities, reducing the need for them. At the same time, DR allows mass consumers to have income in this market, making their useful contribution to improving the operation of the power system. HVAC (Heating, Ventilation and Air Conditioning) systems that operate on the demand side of power systems have tremendous potential as a demand-side mechanism. The significant energy consumption of the HVAC systems, along with their direct influence on the user's well-being, highlights the necessity for effective HVAC management algorithms that reduce the power consumption in the home buildings, taking into account the end-user's comfort.

However, the involvement of mass consumers in the DR practice is associated with significant integration and transaction costs. The smaller and more mass the consumer is, the higher the specific and total size of these costs. The solution to this problem in many countries, and more recently in Russia, is being implemented through the architecture of the Internet of Energy (IoE) [1], which allows the deployment of various services with their own business models and control algorithms. In addition, this architecture provides a suitable hardware platform for implementing B2G (building-to-network) peak regulation.

1.1 Papers Review and Problem Statement

The smart grid is envisioned as the evolution of the current power grid, which faces important challenges, such as blackouts caused by peaks of energy demand that exceed the energy grid capacity [2]. A proposed approach to alleviate this problem is to incentivize the consumers to defer or reschedule their energy consumption to different time intervals with lower expected power demand. These incentives of DR are based on smart (or dynamic) pricing tariffs that consider a variable energy price [3]. For instance, in real-time pricing (RTP) tariffs, the price of the energy will be higher at certain time periods, where the energy consumption is expected to be higher, for example, during the afternoon or in cold days. Other types of smart pricing tariffs are critical-peak pricing (CPP) or time-of-use pricing (ToUP) [3–5]. Energy scheduling algorithms are the state-of-the-art methods to manage the energy consumption of loads within a smart pricing framework [3, 4, 6]. These techniques assume a specific smart pricing tariff and various time periods. For each of these time intervals, the DR aggregator determines the operational power of each appliance to minimize the energy consumption cost.

It's useful to note that the appliances that can be controlled by the energy scheduler can be categorized into three classes: (i) nonshiftable, (ii) time-shiftable, and (iii) power-shiftable, whose operational power can be changed. The latter include HVAC modules, which are the most energy-intensive devices in residential buildings. HVAC systems represent the largest end-users of electricity in buildings for which more than 50% of the building energy consumption [7]. According to studies [8], they represent the 43% of residential energy consumption in the USA and the

61% in UK and Canada. In Russia, the volume of the HVAC market is still small, and traditionally accounts for 11–12% of the European [9]. However, the project of the System Operator of the UES of Russia, launched in 2019, involves the increasing involvement of retail consumers in DR practice, primarily industrial enterprises and commercial real estate, equipped with HVAC systems [10]. Apparently, the significant energy consumption of the HVAC systems, along with their direct influence on the user's well-being, highlights the necessity for effective HVAC management algorithms that reduce the power consumption in the home buildings, taking into account the end-user's comfort.

HVAC systems can participate in DR events by pre-heating or pre-cooling the indoor spaces to achieve some degree of load shifting (passive energy storage in the thermal mass of the building), or by regulating the amount of thermal energy stored in specialized thermal energy storage systems or in domestic hot water (DHW) storage devices [11]. At the same time, energy flexibility of buildings on both individual and aggregated level can be defined as a dynamic function suitable for control [12]. Through proper controls [13] or machine learning techniques [14], HVAC systems buildings have great potentials for implementing DR. Particularly in [15], integrating multiple buildings as a group will improve the demand response performance of multi-building.

In [16] it is shown, that under demand dependent prices and presence of multiple buildings, either a multi-agent or centralized needs to be implemented. It is assumed that buildings can exchange information with the power grid, such as their demand and electricity prices, while they can potentially establish some degree of dynamic coordination among each other (competitive or cooperative), and/or even share information (Fig. 1).

For instance, a game theory-based decentralized control strategy for power demand management of building cluster can provide enhanced performance and robustness in real applications [17]. In [18] proposed a hierarchical coordinated

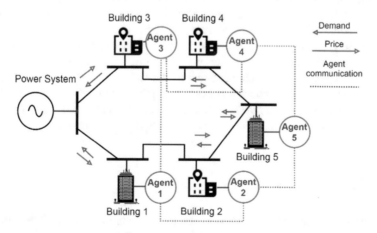

Fig. 1 Multi-agent coordination for DR, adapted from [15]

demand response control for building cluster with improved performances in computation, peak demand reduction and cost saving.

Methods of multi-agent reinforcement learning (MARL) can be promising in this regard, given its adaptability and capacity to learn preferences of the user through interaction and without an explicit mathematical model. In [19] was implemented this decentralized multi-agent approach not only to reduce or shift the peak demand, but also to directly maximize renewable energy use in a non-stationary environment with up to 90 households and simulated in GridLAB-D. In [20] was proposed a multi-agent deep RL controller for the control of HVAC systems in multiple buildings in a competitive environment. Buildings share the same electricity prices, which increased with demand. Therefore, buildings learned to coordinate with each other to avoid simultaneous energy consumption.

The energy sector is undergoing its largest transformation since the days of Thomas Edison. Smart grid is considered as one of the main Internet of Things (IoT) applications and it has attracted a great interest during the last few years [21, 22]. With this approach, "communicating smart energy things" from individual storage devices and electric vehicles to buildings and energy communities can themselves determine the optimal required amount of their own consumption, turn off or reduce the power of devices that are unnecessary at a particular time, choose the most economical modes of their operation, distribute the load and reduce consumption during peak hours, when the cost of energy is maximum [23]. A striking example of energy projects in this area is RESPOND H2020 aims to bring DR programs to Europe for smart buildings produced by IoT systems [24].

An extension of IoT technology for organizing the energy of the future is IoE—a cyber-physical infrastructure for information systems based on decentralised intelligent control of energy systems, energy hubs, power supply systems and integration of distributed active energy consumers, distributed energy sources and energy flexibility. This paradigm is being developed and tested in different countries of the world. It is the framework for the formation of a technological vision for the National Technology Initiative "EnergyNet", which are developing from 2014 in Russia and serves as the basis for the currently ongoing development of IDEA (Internet of Distributed Energy Architecture), i.e. a type of decentralised power system in which smart distributed control is implemented through energy transactions between its users [1]. In [9], a variant of the technical implementation of the DR aggregator based on IDEA is proposed. This approach makes it possible to implement new qualities of DR practices by installing a unified means of connecting to the IoE at end users—the "IDEA agent", which is a control set-top box with the appropriate software.

1.2 Paper Contribution

In the presented paper an approach in the IoE structure to decentralized DR of a distributed multi-energy systems (DMES) with HVAC systems, receiving electricity from a centralized power supply system have been developed. The proposed IoE model for controlling the end consumer's equipment of HVAC building is implemented based on the MARL and a network-centric energy management structure. The proposed approach allows HVAC building agents to modify the overall electricity demand curve in a DMES by controlling the heat storage of each agent. Agents, through built-in IoE ports and protocols, gain access to heat pumps and electric heaters, which allow them to control the storage of domestic hot water and chilled water.

2 Methods

2.1 DR-Aggregator Based on Net-Centric Management

In the context of intellectualization of the electric power industry, active consumers acquire new properties, one of which is net-centricity. The net-centric management system (NCM) is more preferable in comparison with the traditional hierarchical one. The property of "net-centricity" means the ability to control energy supply and energy consumption based on an extensive energy network, where each element of the system gets the opportunity to interact with any other element through a telecommunications network, which becomes the basis of control [25].

Figure 2 shows the structure of a DR-aggregator with NCM, where the power supply company (PSC), oil and gas company (OGC), metallurgical enterprise (Metallurgical plant) are represented as aggregators-clusters, solar and wind power plants, which include energy storage systems (ESS, batteries) and diesel generator sets (DGS). In such a structure, a group of commercial consumers with HVAC systems (hotels, social facilities), whose goal is to reduce electricity consumption, can also be consider.

The net-centric structure is suitable for prosumers and active consumers who connected by smart contracts to their aggregator-cluster. They may be IoE users. In the event of a default, the resource-deficient consumer as an IoE user turns to other over-resource users through machine-to-machine interaction for support. This structure increases the DR mechanism flexibility, which means here as the ability of the system (object) to continue to function with a partial or temporary loss of its link. It is possible to increase the flexibility of the power system operation not only by reserving weak links (nodes, equipment, additional operating personnel), but also by a MAS built correctly. This chapter discusses the DR mechanism as applied to power systems.

In [26], the net-centric architecture of the UES is presented, which includes an information level that coordinates all management actions (formation of the basic infrastructure, inter-system interaction, risk assessment, arbitrage), and the territorial level where the indicated tasks. Net-centrism according to [26] is the organization of the interaction of MAS agents, and the net-centric architecture of an urban agglomeration is a smart city.

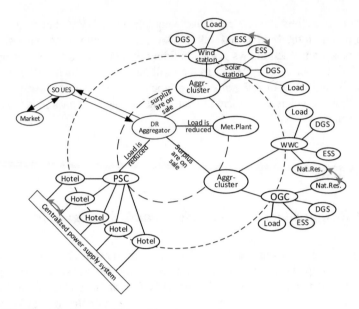

Fig. 2 DR-Aggregator based on Net-centric management

One of the important aspects (qualities) of the NCM concept is self-synchronization. According to [27] self-synchronization is a requirement for the system to continue performing its tasks in an extreme or critical situation, using "not only its own, but also not fully loaded forces and resources of neighbors". The self-synchronization process allows reconfiguration within well-organized links due to dynamic connections that can change over time. To manage the self-synchronization process, an agent is originated in each link. The community of such agents (multi-agent system) must act in a coordinated manner both in time and according to the goal, since they belong to a single hierarchical control system (are embedded in it), but each agent himself determines the goals and strategies of self-synchronization. Thus, the MAS fully meets the requests of the DR mechanism. The diagram (Fig. 3) schematically shows the relationship of the DR mechanism, NCM and MAS. Section 3 will present their interactions in the IoE.

Fig. 3 Options for
interconnecting DR, NCC
and MAS in the field of IoE

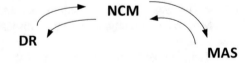

2.2 IoE Architecture for DMES Demand Response

In this section, we present the decentralized DR approach for DMES based on the IoE-NCM framework described in the previous section. The considered DMES model receives electricity from a centralized power supply system. By DMES, in this case, we mean a distribution network of energy cells (Grid of cells), i.e. a local energy system composed of energy cells connected to distribution networks of a centralized (external) energy system and exchanging energy with each other and with a centralized (external) energy system. By stand-alone energy cell we mean a stand-alone MES, which is a building equipped with an HVAC climate control system with rooftop solar panels and electrical and thermal energy storage. At the same time, the IoE-NCM is implemented as a multi-agent software and hardware environment, where agents of HVAC buildings are able to change the aggregated curve of electrical demand by controlling the storage of energy by every agent (Fig. 4).

A certain group of DMES energy objects, i.e. a certain type of smart building with HVAC modules (e.g. multi-storey residential building) equipped with the appropriate interface with IoE platforms is energy cells. Communication with the external main grid is carried out through an energy router (ER), which is a two-way power flow regulator that is used to control the power flow in order to carry out energy transactions and is installed in the section between the external power system and the DMES, in which it is necessary to ensure the regulation of the magnitude and direction of the active and/or reactive power flow in order to perform MARL functions. At the same time, MARL in the IoE structure is both part of the NG platform and IoT due to the fact that MARL is responsible for both energy and information interactions. In the proposed MARL model, each IoT user acts as an agent and learns a policy by interacting with the urban DMES environment that guides them to select optimal

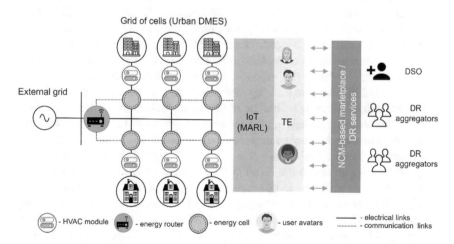

Fig. 4 Implementation of the proposed IoE-NCM DR approach in DMES based on MARL controllers

actions. Apps that organize the work of DMES and implement various energy services operate on the basis of two platform systems—IoT, TE. The TE system provides informational interconnection between user avatars entering into peer-to-peer (p2p) smart contracts, user apps, and digital wallets.

p2p and peer-to-operator (p2o) transactions are used to sell from generation facilities to energy cells and buy energy from the power system, respectively. At the same time, the flexibility service allows you to compensate for power fluctuations in the power system through multi-agent control of HVAC modules within the framework of the DR program. This allows you to smooth out consumption peaks, increase the efficiency of using the power of the DMES own generation, as well as save on the purchase of electricity and power from the power system.

p2p and peer-to-operator (p2o) transactions are used to sell from generation facilities to energy cells and buy energy from the power system, respectively. At the same time, the flexibility service allows you to compensate for power fluctuations in the power system through multi-agent control of HVAC modules within the framework of the DR program. This allows you to smooth out consumption peaks, increase the efficiency of using the power of the DMES own generation, as well as save on the purchase of electricity and power from the power system.

2.3 MARL-Based DR Management

Buildings are complex systems that are affected by weather changes, various loads, human behavior and, in the context of DR, external power signals. Accounting for this dynamics in a simulation model based on the physics of such processes, which can facilitate the DR procedure, is a very difficult task. This is due to the highly non-linear behavior of thermodynamics and the fact that no building is identical to another in a heterogeneous building stock [16]. Reinforcement learning (RL) usually offers a model-free approach, when an agent (or agents) learn to imitate this dynamics through interaction with a specially created environment or a real object, and therefore find optimal behavior strategies (for example, reducing power consumption).

RL is one of the machine learning methods, during which the system under test (agent) learns by interacting with some environment, for example, a dynamic DMES model. Reinforcement signals are the response of the environment to the decisions made. Based on such interaction with the environment, the agent learning with reinforcement must find a policy, $\pi : S \times A \rightarrow [0, 1])$, where $\pi(s, a)$ is the probability of choosing an action $a \in A(s_t)$ in state s. This policy maximizes the reward $R = r_0 + r_1 + \cdots + r_n$ in the Markov decision-making process [28]. At the same time, MARL is an extension of the one-agent model and refers to systems with several agents/players. The goal of cooperative MARL is to maximize the total global reward $R_{g,t} = \sum_{i \in \mathcal{V}} R_{i,t}$, where $R_{i,t} = \sum_{k=o}^{T} \gamma^k r_{i,t+k}$ denotes the cumulative reward for the i-agent.

In the chapter, we use the MARL algorithm, which is a multi-agent extension of actor-critic RL methods and allows you to coordinate the actions of agents through the distribution of rewards, as well as the mutual exchange of information. This algorithm is the result of a generalization of a number of works [28]. In the proposed MARL model, the IoT user acts as a learning agent and interacts with the environment. For coordination, each IoT agent needs to share two variables with one other IoT agent, which makes such an algorithm scalable, since the number of variables needed by each agent does not increase with the number of agents. Given the novelty associated with MARL in this area, the behavior of multiple energy consuming IoT agents (e.g. buildings) subjected to demand-driven power system signals is an area that requires more research and standardization. One of the tools for such standardization can be the concept of the Internet of Energy.

In the proposed multi-agent DMES demand management approach, the agent, through the built-in ports and protocols from among those used to form the IoE, gets access to the end equipment of the HVAC consumer of the building. In this case, these are heat pumps (HP) and electric heaters (EH), which allow controlling the storage of DHW, and chilled water (for sensible cooling and dehumidification) (Fig. 5). In this setting, the agent actually integrates the functions of building load control.

In this setting, the agent actually integrates the functions of building load control. In addition to HP and EH, the DMES environment includes models of solar PV panels, batteries and pre-computed building energy loads, driven by air conditioning, appliance operation, hot water and solar generation. The MARL controller sends a control signal for how much energy it wants to charge or discharge from thermal energy storage devices (chilled water and DHW). The state of charge (SOC) is calculated at each time step for both cooling energy storage and DHW energy storage as follows:

$$SOC_{t+1} = SOC_t + max\{min\{a_t \cdot C, Q_{t_{max}} - Q_b\} - Q_{dem}\} \tag{1}$$

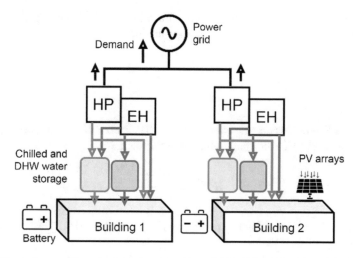

Fig. 5 Thermal energy flows and electric demand for urban DMES environment

where $C \geq 0$ is maximum energy storage capacity of storage device; Q_b—energy storage capacity of battery; $-\frac{1}{C} \leq a_t \leq \frac{1}{C}$—the actions of MARL-controller; Q_{dem}—building energy demand for cooling or DHW; $Q_{t_{max}} \geq Q_{dem}$—maximum heat energy of heating device. In this case, the sign of $Q_{t_{max}}$ may change over time (for example, in a HP), due to the outside temperature affecting the coefficient of performance (COP_t) and thus changing the maximum value. Важно отметить, что C determines the capacity of the storage device and is defined as a multiple of the maximum thermal energy consumption by the building.

The heat pump model is an air-to-water HP. Its efficiency for cooling and heating is calculated as:

$$COP_c = \eta_{tech} \cdot \frac{T^c_{target}}{T_{outdoorair} - T^c_{target}} \tag{2}$$

$$COP_h = \eta_{tech} \cdot \frac{T^h_{target}}{T^h_{target} - T_{outdoorair}} \tag{3}$$

where T^c_{target} and T^h_{target} are the target cooling and heating temperatures, respectively, and η_{tech} is the technical efficiency (typically 0.2–0.3). T^c_{target} is equal to the average logarithmic value of the temperature of the water supplied to the storage tank and the temperature of the water returning to the heat pump. T^c_{target} is typically between 7 and 10 °C, while T^h_{target} is often around 50 °C.

The certain amount of thermal energy based on the SOC, as well as the demand for thermal energy by the building Q_{dem}.

$$Q^{hp}_{t+1} = C \cdot (SOC_{t+1} - SOC_t) + Q_{dem} \tag{4}$$

This HP model takes electrical energy from the grid E^{hp} and supplies thermal heating of cooling energy Q^{hp} to the building and/or the storage device following the equation:

$$E^{hp}_t = \frac{Q^{hp}_t}{COP_t} \tag{5}$$

In turn, EH provides heating energy, Q_{heater}, (i.e. for DHW) consuming electricity from the grid, E^{heater}, according to the following equation:

$$E^{heater} = \frac{Q^{heater}_t}{\eta_{eh}} \tag{6}$$

where η_{eh} is the heater efficiency and is usually greater than 0.9. Q^{heater}_t is obtained analogous to Q^{hp}_t in equation.

Unlike thermal storage, battery capacity is not defined as a multiple of a building's maximum hourly heat demand. Instead, the battery capacity C is defined in kWh, its rated power in kW. The battery model includes the capacity loss factor c_{loss}, which is defined as the ratio of capacity lost during charging and discharging, expressed in units of $\frac{1}{cycle}$. For example, $c_{loss} = 1e^{-5}$ would mean that 0.001% of the battery's initial capacity is lost in each charge and discharge cycle. The capacity of the new battery C_{new} is given by the equation:

$$C_{new} = c_{loss} \cdot C_0 \cdot \frac{\left|E_{in|out}\right|}{2C} \tag{7}$$

where C_0 is the initial battery capacity as defined by the user in kWh, C is the current capacity, and $E_{in|out}$ is the amount of energy in kWh that has been charged or discharged.

If the RL agent takes an action that discharges the battery by more than the building's electricity demand requirement, the excess power goes into the DMES. Such surplus electricity can be used by other buildings (and reduce the amount of electricity consumed from the main feeder), or can flow into the main grid (DMES overgeneration). This behavior is described by the following equation:

$$E_{net}^{DMES} = \sum_{i=0}^{n}\left(E_{b_i} - E_{bat}\right) \tag{8}$$

where n—the number of buildings in the DMES, E_{b_i}—the total net electricity consumption of the building, including the energy supply systems and E_{net}^{DMES}—the net electricity consumption of the DMES.

The proposed IoE model is implemented through the MARL method, where agents learn and communicate with each other to maximize the total reward function, i.e. minimizing the total net electrical demand.

$$r_i = sign(E_i)\left(|E_i|max\left(0, \sum E_i\right)\right)^2 \tag{9}$$

where E_i is the net electric demand of building and i, $\sum E_i$ shows the net electric demand of DMES, $sign(E_i)$—the sign of net demand (for example, negative when IoE users are selling electricity. Function (10) is a combination of the individual net electric demand E_i and the collective component, $\sum E_i$, i.e. total net electricity consumption of the entire DMES energy region, and is used to exchange information between agents, which rewards them for reducing the coordinated energy demand. Thus, Eq. (9) is the agent's multi-objective function, which to minimize an average of the entire load shaping metrics: ramping, load factor, maximum peak electricity demand, average daily peak net demand, and the total amount of electricity consumed.

Table 1 Buildings and descriptions in the power district

Building number	Type	Type details	Cooling storage[1]	DHW storage[1]	PV (kW)
1	Commercial	Company	3	3	120
2	Commercial	Company	3	3	N/A
3	Commercial	Company	3	N/A	N/A
4	Commercial	Company	3	N/A	40
5	Residential	Multi-family	3	3	25
6	Residential	Multi-family	3	3	20
7	Residential	Multi-family	3	3	N/A
8	Residential	Multi-family	3	3	N/A
9	Residential	Multi-family	3	3	N/A

3 Experimental Results

3.1 Description of the Research Object

In this case study, we have considered the DMES model, which includes different types of buildings with HVAC systems, combined in a multi-agent demand control system based on MARL. The buildings have diverse load and DHW profiles, PV panels, thermal storage devices, HPs, and EHs. In addition to its own distributed sources of electricity, DMES is connected to an external power system through a distribution network. The DMES model is based on a fragment of the 110–6/ 0.4 kV power distribution network, which includes 4 commercial and 5 residential multi-storey buildings as potential MESs (Table 1).

Thus, the buildings have HVAC systems, as well as their own energy sources, which will transform the traditional distribution network into DMES. This potentially forms the basis for the deployment of the IoE architecture described in Sect. 3 in this power district (Fig. 3).

3.2 Emergency Mode in the Structure of the IoE-NCM

Let us project the considered example of DMES, i.e. 5 residential multi-storey buildings and 4 companies for 2 energy cells, which are included in ER (aka Aggregator-cluster) (Fig. 7). At the request of the main DR Aggregator to reduce the load, the Aggregator-cluster generates a signal in the ER, after which the proposed MARL implements the corresponding regulation.

Let us consider the case of self-synchronization in the ER operation. Suppose that at the agreed DR time of the load reduction, two out of 4 companies left without power supply due to cable damage during the repair of the highway separating the

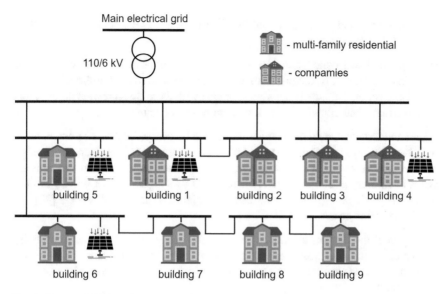

Fig. 6 The test DMES model, which includes different types of buildings with HVAC systems

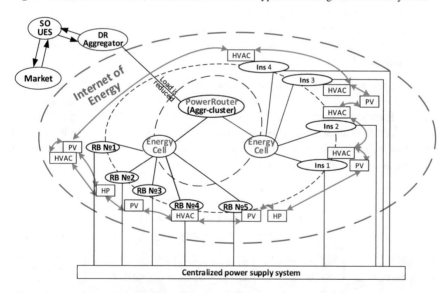

Fig. 7 4 companies, 5 residential buildings—in the structure of the NCM (the interaction of agents is shown by red double-headed arrows). The presence of autonomous HVAC systems, PV solar panels, HP heaters is simulated

residential area and the commercial building territory. The characteristics of these institutions are as follows:

- Company No. 1 conducts round-the-clock calculations on a super-computer; the main email server of the power district is also located in the building.
- Company No. 2 conducts theoretical studies also there are a greenhouse complex, and an artificial climate station for growing exotic plants.

In an emergency, both companies should not be completely de-energized, therefore, the buildings are equipped with a backup power supply. However, both companies are unable to fulfill DR's contractual obligations. In this regard, the problem arises of finding measures for the redistribution of obligations that will be fulfilled by the ER participants in the NCM, using the hierarchy analytical process [29].

The selection criteria are follows:

1. which of the ER participants can maximally reduce their energy consumption;
2. which of the ER participants at the moment has force majeure circumstances that are not presented above (in one of the institutes there is a conference at the moment);
3. what is the financial consequences of the ER participants from the provision of excess capacity to DR Aggregator;
4. what is the temperature outside (summer/winter);
5. whether emissions are exceeded outside the window, requiring increased ventilation of the premises.

The result of analytic hierarchy process for RB1; RB2; RB3; RB4; RB5; Ins1; Ins2; Ins3; Ins4 is 0.21; 0.08; 0.08; 0.1; 0.17; 0.04; 0.04; 0.08; 0.17 accordingly (Fig. 8). It shows that the first of the possible candidate to cover the deficit have to be Residential Building No. 1, which has autonomous HVAC and PV; the second candidate is Residential Building No. 5 with autonomous HP and PV or Commercial Building No. 4 that allows employees to work remotely.

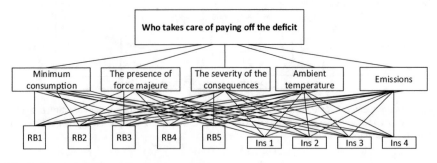

Fig. 8 The analytic hierarchy process

3.3   MARL Optimal Management

For modeling DMES, the CityLearn open-source library was used, which is an OpenAI Gym environment [30], which allows you to simulate various types of buildings in the context of the DR problem. Adaptive MARL-based controller trained and tested on various datasets. Each dataset contains year-long hourly information about the cooling and DHW demand of the building, electricity consumed by appliances, solar power generation, as well as weather data and other variables. The energy demand for each building has been pre-simulated using EnergyPlus for climate zone D (cold continental climate), which also corresponds to some cities in the Irkutsk region. Realistic profiles of electrical loads, hot water supply, and temperature settings for heating and cooling buildings were obtained from open sources and grouped in [31].

In this structure, each building, which consists of four energy systems and four control solutions, has its own RL agent (aka IoT agent), which uses 18 state-space values, such as: outdoor air temperature, current state of charge, net electricity consumption, solar radiation level, time of day, etc., to predict an optimized set of actions corresponding to the accumulation of thermal and electric energy, namely: (1) cooling storage, (2) heating storage and (3) electric storage.

At each time step, actions corresponding to energy accumulation are sent to the appropriate device object (HP, EH, DHW storage, battery) to charge or discharge their storage reserves. In this case, the energy storage device serves two loads: the first is the required thermal loads calculated by EnergyPlus for occupant comfort, P_{demand}, and the second is an external energy storage device for building energy management, $P_{requist}$. Regardless of the charge/discharge signal sent by the RL agent, the thermal load is satisfied first, and only then the storage load is taken into account with the remaining available power of the device.

At the start of the simulation, during the random two-epoch phase, we collect all rewards and calculate their mean and standard deviation. We then normalize all collected rewards and any future rewards by subtracting the computed mean from them and dividing by the standard deviation. Rewards normalization using constant mean and standard deviation values is necessary because using other values recalculated during simulation may invalidate Q-values that agents have already learned. This normalization of rewards ensures that RL agents can be tuned with similar hyperparameters. For example, the choice of the optimal learning rate and temperature depends on the magnitude of the reward.

Table 2 shows individual scores, i.e. individual components of the reward function (10), analyzed to evaluate the effectiveness of MARL. Here RBC is the reference hand-optimized rule-based controller to which all results are normalized. A significant improvement over baseline RBC occurred for ramping (about 35% better than baseline RBC), which demonstrates the ability to flatten the demand curve. The algorithm performed well when reducing the metric 1—load factor and the average daily peaks (by about 10–20%). The net electricity demand remains at the same level as RBC, which is a good result since the RL reward is aimed at leveling the net load,

Table 2 Individual scores of MARL after training and normalized by the scores of the baseline rule-based controller (RBC)

Number	Individual scores	Value
1	Net electricity demand	0.99
2	Annual peak demand	0.89
3	Average daily peak	0.86
4	1—load factor	0.90
5	Ramping	0.67

rather than reducing the total amount of electricity consumed. Finally, MARL has reduced its annual peak demand by about 3–11%, which is likely to be significantly improved.

On Fig. 9 shows how MARL-based controllers make different storage decisions over the course of a full summer week, adapting to the different energy profiles of their buildings, some of which have rooftop solar panels.

The total net load for the system of 9 buildings is reduced. This is because as a result of training, the RL agents (HVAC systems) of the building find the optimal cooperative policy for managing the thermal storage units, which leads to a decrease and smoothing of power consumption compared to the scenario when the thermal storage control is not implemented (Fig. 10). It is also clearly seen that the use of own PV generation in buildings, as expected, leads to a decrease in power consumption in the considered DMES. Figure 10 is shown the profiles of EH (as a heating storage device) and HP (as a cooling storage device), as well as battery for Building 9 during 4 days of MARL control.

It is also important to note that MARL-based controllers learn a unique control strategy specific to each building and its associated electricity consumption patterns.

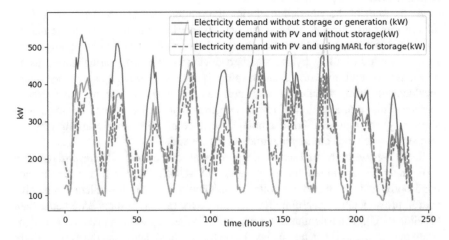

Fig. 9 A comparison of district DMES electricity consumption for **i** no PV and load shifting, **ii** no load shifting and **iii** MARL control

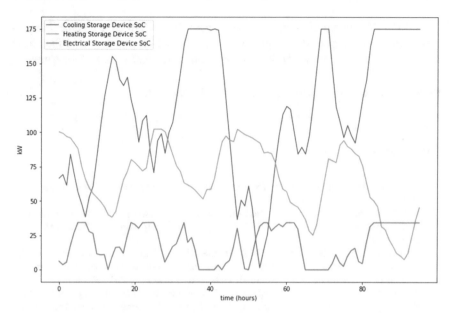

Fig. 10 The results of MARL control for heating, cooling, and electrical devices for Building 9 during 4 days

In addition, MARL also provides some additional coordination between buildings as IoT agents in the Internet of Energy platform.

4 Conclusions

Districts and cities have periods of high demand for electricity, which raise electricity prices and the overall cost of the power distribution networks. Flattening, smoothening, and reducing the overall curve of electrical demand helps reduce operational and capital costs of electricity generation, transmission, and distribution networks. The development of DR systems from smart home management and IoE to the creation of new players in the wholesale energy and capacity market (distributed energy aggregators and storage systems) can optimize energy demand and effectively increase the flexibility of the power system in the decarbonization era.

This paper presents a DR approach for HVAC systems using IoE concept and NCM structure using MARL. In such statement, DR is the coordination of electricity consuming agents (i.e. buildings) in order to reshape the overall curve of electrical demand. MARL allows the implementation of reinforcement learning agents in a multi-agent setting to reshape their aggregated curve of electrical demand by controlling the storage of energy by every agent. For an effective technical implementation of this approach, we propose to use the IoE technology and the NCM structure, which determines the scientific and practical significance of the study.

Funding This work was supported by the RSF (No. 19-19-00673).

References

1. EnergyNet Infrastructure Center: The architecture of the internet of energy. White paper. Moscow, p. 77 (2021) (in Russian)
2. Joseph, A., Balachandra, P.: Smart grid to energy internet: a systematic review of transitioning electricity systems. IEEE Access **8**, 215787–215805 (2020). https://doi.org/10.1109/ACCESS. 2020.3041031
3. Mohsenian-Rad, A.-H., Wong, V.W.S., Jatskevich, J., Schober, R., Leon-Garcia, A.: Autonomous demand-side management based on game-theoretic energy consumption scheduling for the future smart grid. IEEE Trans. Smart Grid **1**(3), 320–331 (2010)
4. Zhu, Z., Lambotharan, S., Chin, W.H., Fan, Z.: Overview of demand management in smart grid and enabling wireless communication technologies. IEEE Wirel. Commun. **19**(3), 48–56 (2012)
5. Nguyen, H.T., Nguyen, D., Le, L.B.: Home energy management with generic thermal dynamics and user temperature preference. In: Proceedings of the IEEE International Conference on Smart Grid Communications (SmartGridComm'13), pp. 552–557, Vancouver, Canada, October 2013
6. Tsui, K.M., Chan, S.C.: Demand response optimization for smart home scheduling under real-time pricing. IEEE Trans. Smart Grid **3**(4), 1812–1821 (2012)
7. Avci, M., Erkoc, M., Asfour, S.S.: Residential HVAC load control strategy in real-time electricity pricing environment. In: Proceedings of the IEEE Energytech, pp. 1–6, Cleveland, Ohio, USA, May 2012
8. Perez-Lombard, L., Ortiz, J., Pout, C.: A review on buildings energy consumption information. Energy Build **40**(3), 394–398 (2008)
9. Market of climate systems in 2020: volume, dynamics and trends, manufacturers and segments, types of equipment, forecasts. https://hiconix.ru/publications/articles/rynok-klimaticheskikh-sistem-v-2020-obem-dinamika-i-trendy-proizvoditeli-i-segmenty-vidy-oborudovani/
10. EnergyNet infrastructure center: demand response in the Russian electric power industry: opening opportunities. Expert and analytical report. Moscow (2019)
11. Ran, F., Gao, D., Zhang, X., Chen, S.: A virtual sensor based self-adjusting control for HVAC fast demand response in commercial buildings towards smart grid applications. Appl. Energy **269**, 115103 (2020)
12. Junker, R.G., Azar, A.G., Lopes, R.A., Lindberg, K.B., Reynders, G., Relan, R., et al.: Characterizing the energy flexibility of buildings and districts. Appl. Energy **225**, 175–182 (2018)
13. Shan, K., Wang, S., Yan, C., Xiao, F.: Building demand response and control methods for smart grids: a review. Sci. Technol. Built Environ. **22**(6), 692–704 (2016)
14. Pallonetto, F., Rosa, M.D., Milano, F., Finn, D.P.: Demand response algorithms for smart-grid ready residential buildings using machine learning models. Appl. Energy **239**, 1265–1282 (2019)
15. El Geneidy, R., Howard, B.: Contracted energy flexibility characteristics of communities: analysis of a control strategy for demand response. Appl. Energy **263** (2020) https://doi.org/10.1016/j.apenergy.2020.114600

16. Vázquez-Canteli, J.R., Nagy, Z.: Reinforcement learning for demand response: a review of algorithms and modeling techniques. Appl. Energy **235**, 1072–1089 (2019)
17. Tang, R., Li, H., Wang, S.: A game theory-based decentralized control strategy for power demand management of building cluster using thermal mass and energy storage. Appl. Energy **242**, 809–820 (2019)
18. Huang, P., Fan, C., Zhang, X., Wang, J.: A hierarchical coordinated demand response control for buildings with improved performances at building group. Appl. Energy **242**, 684–694 (2019)
19. Marinescu, A., Dusparic, I., Taylor, A., Canili, V., Clarke, S.: P-MARL: prediction-based multi-agent reinforcement learning for non-stationary environments. In: Proceedings of International Conference on Autonomous Agents and Multiagent Systems, AAMAS, vol. 3, pp. 1897–1898 (2015)
20. Vázquez-Canteli, J.R., Ulyanin, S., Kämpf, J., Nagy, Z.: Fusing tensorflow with building energy simulation for intelligent energy management in smart cities. Sustain. Cities Soc. (2019)
21. Saleem, Y., Crespi, N., Rehmani, M.H., Copeland, R.: Internet of things-aided smart grid: technologies, architectures, applications, prototypes, and future research directions. IEEE Access **7**, 62962–63003 (2019)
22. Abir, S.M.A.A., Anwar, A., Choi, J., Kayes, A.S.M.: IoT-enabled smart energy grid: applications and challenges. IEEE Access **9**, 50961–50981 (2021)
23. Musleh, A., Yao, G., Muyeen, S.: Blockchain applications in smart grid-review and frameworks. IEEE Access **7**, 86746–86757 (2019)
24. Esnaola-Gonzalez, I., Diez, F.J.: Integrating building and IoT data in demand response solutions. In: Proceeding of the 7th Linked Data in Architecture and Construction Workshop—LDAC2019, pp. 92–105 (2019)
25. Kolosok, I.N., Korkina, E.S.: Application of the principle of network-centric control to the structure of the electricity demand aggregator. Ind. Energy **12**, 2–8 (2020) (in Russian)
26. Bushuev, V.V.: Network-centric approach to the organization of management of the functioning and development of the energy sector. Report at the II MEF, Moscow (2013). (in Russian)
27. Trahtengerts, E.A.: Use of network-centred principle of self-synchronization in management. In: Открытое образование, vol. 2, pp. 15–23 (2015)
28. Sutton, R.S., Barto, A.G.: Introduction to Reinforcement Learning. MIT Press, Cambridge, MA, USA (2018)
29. Saaty, R.W.: The analytic hierarchy process: what it is and how it is used? Math. Model. **9**(3–5) (1987)
30. Brockman, G., Cheung, V., Pettersson, L., Schneider, J., Schulman, J., Tang, J., Zaremba, W.: OpenAI Gym. arXiv:1606.01540 (2016)
31. Vázquez-Canteli, J.R., Kämpf, J., Henze, G., Nagy, Z.: CityLearn v1.0: An OpenAI gym environment for demand response with deep reinforcement learning. In: BuildSys 2019—Proceedings of the 6th ACM International Conference on Systems for Energy-Efficient Buildings, Cities, and Transportation, p. 356 (2019)

Improving the Efficiency and Environmental Friendliness of Diesel Engines of Mining Dump Trucks by Installing Hybrid Gas-Diesel Fuel Systems

Sergey A. Glazyrin, Gennadiy B. Varlamov⬤, Michael G. Zhumagulov⬤, Timur T. Sultanov⬤, and Olzhas M. Talipov⬤

Abstract The volume of environmental pollution has reached critical values. Industrial transport plays a huge negative role in pollution, especially mining dump trucks that operate whole day on diesel fuel. The cost of diesel fuel in Kazakhstan has more than doubled over the past few years, which, in turn, affected the cost of raw materials for the mining and processing and metallurgical industries. The objective of the research was a development of hybrid gas-and-diesel fuel system to transfer the diesel engines operation into the gas-and-diesel mode with replacement of some part of the diesel fuel by natural gas. The advantages of the developed hybrid gas-and-diesel fuel systems are as follows: the replaced amount of diesel fuel by the natural gas is up to 70%; converting simplicity and fastness; engineering or technological modifications of a basic diesel model are not required; the engine becomes universal for operation with liquid or gaseous fuel; both cycles, gas-and diesel and diesel operation possibility is maintained; the power parameters of a diesel engine operation remain at the documented level; exhaust smoke capacity is reduced by 2–3 times; the unit cost of cargo transportation is significantly reduced.

Keywords Environmental-friendliness · Economic efficiency · Diesel engine · Hybrid fuel system · Gas-diesel cycle · Atmospheric emissions · Dumper

S. A. Glazyrin · M. G. Zhumagulov (✉) · T. T. Sultanov
L.N. Gumilyov, Eurasian National University, Astana, Kazakhstan
e-mail: Zhmg_9@mail.ru

G. B. Varlamov
National Technical University of Ukraine Igor Sikorsky Kyiv Polytechnic Institute, Kiev, Ukraine

O. M. Talipov
Toraighyrov University, Pavlodar, Kazakhstan

M. Sari and A. Kulachinskaya (eds.), *Digital Transformation: What are the Smart Cities Today?*, Lecture Notes in Networks and Systems 846,
https://doi.org/10.1007/978-3-031-49390-4_21

1 Introduction

The world uses heavy-duty dump trucks with diesel internal combustion engines in the extraction of ore in the quarries of the mining industry [1]. Such dump trucks are 24 h operated. They consume a large amount of diesel fuel, while emitting into the atmosphere a huge amount of harmful substances with exhaust gases, including greenhouse gases [2, 3].

Mining enterprises in Kazakhstan use mining dump trucks of the brands BelAZ (by Belarusian Automobile Plant), Komatsu, Hitachi, Caterpillar (CAT), with a carrying capacity from 90 to 180 tons, with fuel consumption from 890 to 3200 L of diesel fuel per day with round-the-clock operation. As a result, a huge amount of expensive fuel is spent and large volumes of harmful emissions enter the atmosphere.

In recent years, dual-fuel engines have attracted increasing attention. So the authors in [4] note the reduction of harmful emissions in dual-fuel engines, and also they analytically try to estimate the compression ratio limitation according to the engine knock and engine thermal efficiency. It is noted that engine knock of natural gas and diesel fuel together in the engine is not well studied under high compression ratio. Experimental studies of dual fuel engines are also available in [5]. The studies show that different fuels have different ignition rates. The combustion of some types of fuels is accompanied by an ignition delay. It must be taken into account during the designing such engines. An attempt to increase and stabilize the combustion efficiency by creating a device based on a vortex flow was made in [6] to unify different fuels, that is to make the ignition rate similar. Today, research [7, 8] is being actively carried out even on the combustion of hydrogen in dual-fuel engines. However, hydrogen- diesel dual-fuel engines suffer from knocking combustion and higher nitrogen oxide pollution. The work [9] shows a comprehensive numerical study on the combustion and exhaust gases emission characteristics of a natural gas and diesel fuel engine operating at the high load condition. For instance it was demonstrated that at least an recycling rate of 40% should be employed to meet the NOx. Researchers from the Republic of Korea [10] determined the effect of the natural gas and diesel fuel mixture quality on the combustion process and exhaust gas emissions using high-speed flame visualization to study the combustion phenomena in an optically accessible single-cylinder engine with volume about 1.0 L. Since natural gas has a lower carbon content and higher self-ignition resistance compared to diesel fuel, the results showed a significant reduction in the diffusion flame regime, accompanied by a slow development of the flame. Publication [11] describes an attempt to find a suitable geometry for a miserable dual-fuel device. A dual injection strategy consisting of a pilot injection immediately before the main injection during the combustion of the pilot dual-fuel was evaluated in the engine under consideration.

The purpose of the research [12] is to increase the efficiency and environmental friendliness of diesel engines of mining dump trucks by developing hybrid gas-diesel fuel systems (HGDFS) for transfer the operation of diesel engines to gas-diesel operation with replacement of the main part of diesel fuel with natural gas.

2   Materials and Methods

The following methods were used in the research:

- At the initial stage-analytical methods, which made it possible to determine the initial variants of the scheme and parameters of the HGDFS;
- At the second stage-mathematical modeling using numerical methods and computer-aided design systems, which made it possible to carry out dynamic studies of the units of the fuel supply system and find their optimal parameters;
- At the final stage-conducting an experiment in order to verify the results obtained in numerical modeling.

The objects of experimental research included actuators and an all-mode crankshaft speed controller for the engine of a gas-diesel power system with microprocessor control MPSP.

Experimental studies included:

1. Determination of the rational parameters of the PID speed controller of the crankshaft gas-diesel engine for speed modes from 600 to 2500 min^{-1}.
2. Determination of the characteristics of the MUSD with a stepping motor 57 PMZ-B02 and the selection of its optimal operating parameters (working stroke, speed, initial effort).
3. Determination of the ranges of variation of the control parameters (phase and time of gas fuel injection) of the MPSP gas supply system for speed modes from 600 to 2500 min^{-1}.
4. Development of working algorithms for MPSP working using the developed interface.

3   Results

Various work and economic efficiency indicators for dual-fuel engines are shown below. These results are analyzed. Figures 1, 2 and 3 show the result of a comparative analysis of the characteristics of mining dump trucks, which are used in the quarries of Kazakhstan, in terms of engine power, carrying capacity, daily fuel consumption during 24 h operation. Figure 1 shows the power level [kW] in engines from different manufacturers.

The carrying capacity is shown in Fig. 2 for similar mechanisms. It together with power level (Fig. 1) further compared with fuel consumption (Fig. 3).

Fuel consumption is the main economic indicator. Figure 3 contains data on daily and annual fuel consumption.

Theoretical studies of the effectiveness of existing foreign HGDFS were carried out and the following shortcomings were identified [13, 14]:

- There is no separate diesel fuel control channel. The decrease in the diesel fuel feed is achieved due to the engine overload after gas supply (the engine electronics

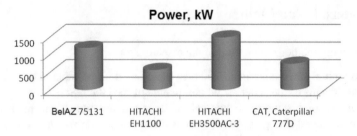

Fig. 1 Power analysis

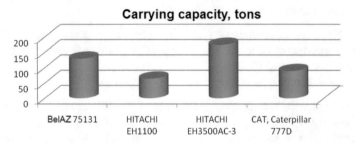

Fig. 2 Analysis by carrying capacity

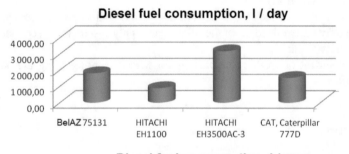

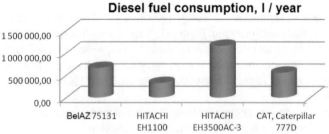

Fig. 3 Analysis of daily and annual fuel consumption

"perceives" this mode as a decrease in load and begins to reduce the diesel fuel feed);
– Real replacement of diesel fuel with natural gas up to 25–30%;
– It cannot be used in naturally aspirated engines;
– The probability of engine breakdown increases due to continuous overload conditions;
– Not suitable for use in diesel engines with diesel fuel mechanical injection.

Bench laboratory studies were carried out on the basis of the YaMZ-238M2 engine (by Yaroslavl Motor Plant) and as part of the D245 engine on the GAZ-3309 car (by Gorky Automobile Plant), and the fuel system control program was developed and tested for the controller of gas-diesel cycle.

The choice of the power control method has a significant impact on the energy and environmental performance of the gas diesel engine [15]. It includes: a qualitative method of regulation, where control of the gas damper changes the amount of gas supplied to the gas-air mixer, changing the quality of the gas-air mixture; quantitative method of regulation, where control of the gas-air damper behind the gas-air mixer changes the amount of the gas-air mixture supplied to the gas diesel cylinders; mixed control method, where the control of the gas and air dampers regulates the amount of gas and air that are supplied to the gas-air mixer.

Thus, the entire fuel supply control complex is reduced to the formation of a command impulse of a certain duration. Formation of a command pulse processes information from the system sensors. The generated command impulse of duration t is applied to the injector winding. The authors have established in previous experimental [12] and analytical studies that the optimal characteristics of gas-diesel engines are achieved with distribution and phase injection of gas (FIG) to the cylinders of the internal combustion engine (ICE). The regulation of the cyclic gas supply is carried out using electromagnetic gas injectors (EGI), which are installed in the intake valves of each cylinder of the engine. Figure 4 indicates diagrams of the operation of the gas supply system elements developed by the MPSP (system for supplying and dosing gas fuel in a universal gas-diesel power system with microprocessor control) at the intake stroke of the first cylinder of the D-245.7 diesel engine for the number of engine turns n equal to 800 min^{-1}.

The phase of fuel injection into the first cylinder is determined by the open state of the intake valve, while the control electrical impulse is applied to the EGI coil at the moment of the cyclic gas supply.

Phase sensor (FS) is installed to provide phase gas supply on the drive shaft of the high pressure fuel pump (HPFP). The sensor position is determined by the angle Ψ_{FS} relative to the top dead center (TDC) of the first cylinder of the internal combustion engine. Ψ_{FS} value is measured after the FS. ΨFS is installed on the engine and is entered by parameter k of the microprocessor control unit for the gas-diesel power system.

It is obvious that the optimal processes of phase gas injection (PGI) provide for a change in the current value of the angle Ψ_{inj} depending on the engine crankshaft speed n. For this, the dependence $\Psi_{inj} = f(n)$ is introduced into the system of parameters

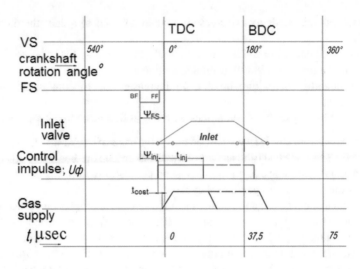

Fig. 4 The operation of the elements of the gas supply system

MPSZh (Microprocessor-based liquid fuel supply system), where Ψ_{inj} is the opening angle of the gas injector relative to the top dead center (TDC).

In principle, PGI can take place throughout the entire intake phase, but optimal values of Ψ_{0inj} of the Ψ_{inj} angle exist, at which the working processes of gas diesel engine are most effective, which is confirmed by experimental studies.

Another condition for the full implementation of PGI is the reconciliation of conflicting requirements between the limited throughput of the gas injector, the injection time and the value of the gas cycle at different engine modes. To do this, it is desirable to introduce a gas reducer-regulator (GRR) into the MPSZH gas supply system, which provides power to the gas injectors (nozzles) with variable pressure P_ϕ, depending on the ICE operating mode. GRR sets the preset pressure P_ϕ at the gas injectors inlet according to the command signal from the microprocessor unit, depending on the frequency n and the load.

The PID controller settings were carried out at the first stage of bench studies. They must ensure high-quality operation of the gas supply system. Figure 5 demonstrates the obtained performance characteristics of the PID controller of the gas supply system MPSZH for the nominal frequency n = 1000 min^{-1} when changing for the values Kp = 0.05; 0.25; 1.25. The intensity of the change in the crankshaft speed from 800 min^{-1} to 1100 min^{-1} was set at 1.1 s and 2.9 s.

A change in the proportional component (Kp) changes the slope of the regulator characteristic, and a change in the intensity of change in the rotational speed of the crankshaft practically does not affect the characteristic of the regulator.

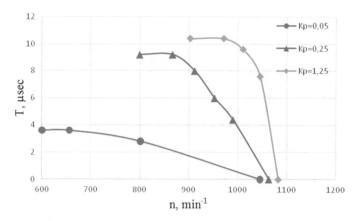

Fig. 5 PI-regulator at the position of the fuel delivery lever corresponding to the rotational speed of 1000 min⁻¹

The developed hybrid gas-diesel fuel systems were installed on the following vehicles after successful bench tests: BelAZ-75405 (YaMZ-240M2 engine); GAZ-3309 (D245 engine); KrAZ-256 (by Kremenchug Automobile Plant) (YaMZ-238M2 engine); Ikarus-256 (YamZ-238M2 engine), which are successfully operated: BelAZ-75405-more than 11 years; GAZ-3309-over 8 years; KrAZ-256-over 12 years; Ikarus-256-over 8 years.

4 Discussion

Hybrid gas-diesel fuel delivery systems are developed in two types:

- With mechanical control system-HGDFS-M;
- With microprocessor control system HGDFS-MP.

The HGDFS-M is a simple and inexpensive system used to retrofit diesel engines in legacy vehicles. Disadvantage: deterioration of the combustion process of gaseous fuel at medium and especially low loads, when the gas-air mixture is too lean, which leads to deterioration in fuel efficiency, a sharp increase in emissions of unburned methane and carbon monoxide.

HGDFS-MP is a universal system due to the use of microprocessor control of gas fuel supply, which automatically optimizes the combustion process of the working mixture at partial loads. It provides regulation of the injection phase of the cyclic portion of gaseous fuel, depending on the load and engine speed. It allows the formation of a gas-air mixture in the combustion chamber of a diesel engine, not homogeneous, but heterogeneous, with different air excess ratios. It is more economical in terms of fuel consumption. The disadvantage of the system is the higher cost compared to the cost of HGDFS-M.

5 Conclusions

The research results are confirmed by certificates of long-term industrial operation of BelAZ-75405, GAZ-3309, Ikarus-256 and KrAZ-256 vehicles in the gas-diesel cycle with replacement of diesel fuel with natural gas in the range from 50 to 70%, while maintaining the factory engine characteristics.

The patent of the Republic of Kazakhstan was obtained for utility model No. 2863 "Gas-diesel engine power supply system", registered in the State Register of the Republic of Kazakhstan on May 14, 2018.

The advantages of the hybrid gas-diesel fuel systems developed by the authors:

- The share of replacing diesel fuel with natural gas up to 70%;
- Simplicity and speed of conversion;
- Constructive or technological changes to the base diesel model are not re-quired;
- The engine becomes universal for the use of liquid and gaseous fuels;
- The possibility of full-fledged work remains in two cycles - gas-diesel and single-diesel;
- The energy parameters of the diesel engine in the gas-diesel cycle remain at the passport (factory) level;
- The smokiness of the engine exhaust gases is reduced by 2 - 3 times;
- Engine noise level in the gas-diesel cycle is reduced by 1.5 times;
- All-mode is retained to control the engine crankshaft speed in gas-diesel mode;
- If necessary, the gas-diesel engine is tuned to obtain more power (by 5–10%) than the base diesel;
- The unit cost of cargo transportation is significantly reduced.
- The annual economic effect is USD 99,797.2 with the cost of diesel fuel in Kaza-khstan USD 0.47 per liter, as well as the cost of natural gas USD 0.16 per liter, and an average operating life of 355 days per year, for one mining dump truck with a carrying capacity 130 tons per hour, with a minimum share of replacing diesel fuel with natural gas of 50%. The economic effect is achieved by reducing fuel costs while maintaining the factory characteristics of the fuel and extending the life of the engine, while reducing the harmful effect on the environment.

Acknowledgements The authors are grateful for their participation in the research to the staff of the National Technical University of Ukraine "Igor Sikorsky Kyiv Polytechnic University" Barabash Petr Alekseevich, Petrenko Valery Georgievich, Solomakh Andrey Sergeevich [16].

References

1. Siami-Irdemoosa, E., Dindarloo, S.R.: Prediction of fuel consumption of mining dump trucks: a neural networks approach. Appl. Energy **150**, 77–84 (2015)
2. Wang, Q., Zhang, R., Lv, S., Wang, Y.: Open-pit mine truck fuel con-sumption pattern and application based on multi-dimensional features and XGBoost. Sustain. Energy Technol Assess **43**, 100977 (2021)
3. Sahoo, L.K.: Santanu Bandyopadhyay, Rangan Banerjee: Benchmarking energy consumption for dump trucks in mines. Appl. Energy **113**, 1382−1396 (2014)
4. Xiong, Q., Wan, Z., Liu, L., Zhao, B.: Numerical analysis of combustion process and pressure oscillation phenomena in low-pressure injection natural gas/diesel dual fuel low speed marine engine. Therm. Sci. Eng. Prog. **42**, 101913 (2023)
5. Zheng, Z., Xia, M., Liu, H., Wang, X., Yao, M.: Experimental study on combustion and emissions of dual fuel RCCI mode fueled with biodiesel/n-butanol, biodiesel/2,5-dimethylfuran and biodiesel/ethanol. Energy **148**, 824–838 (2018)
6. Wang, H., Wang, T., Feng, Y., Zhen, L., Sun, K.: Synergistic effect of swirl flow and prechamber jet on the combustion of a natural gas-diesel dual-fuel marine engineю. Fuel **325**, 124935 (2022)
7. Hosseini, S.H., Tsolakis, A., Alagumalai, A., Mahian, O., Lam, S.S., Pan, J., Peng, W., Tabatabaei, M., Aghbashlo, M.: Use of hydrogen in dual-fuel diesel engines. Prog. Energy Combust. Sci. **98**, 101100 (2023)
8. Bhagat, R.N., Sahu, K.B., Ghadai, S.K., Kuma, C.B.: A review of performance and emissions of diesel engine operating on dual fuel mode with hydrogen as gaseous fuel. Int. J. Hydrogen Energy (2023). https://doi.org/10.1016/j.ijhydene.2023.03.251.
9. Xinlei Liu, H., Wang, Z.Z., Yao, M.: Numerical investigation on the combustion and emission characteristics of a heavy-duty natural gas-diesel dual-fuel engine. Fuel **300**, 120998 (2021)
10. Kim, W., Park, C., Bae, C.: Characterization of combustion process and emissions in a natural gas/diesel dual-fuel compression-ignition engine. Fuel **291**, 120043 (2021)
11. Park, H., Shim, E., Lee, J., Seungmook, O., Kim, C., Lee, Y., Kang, K.: Large–squish piston geometry and early pilot injection for high efficiency and low methane emission in natural gas–diesel dual fuel engine at high–load operations. Fuel **308**, 122015 (2022)
12. Glazyrin, S.A., Varlamov, G.B., Barabash, P.A., Petrenko, V.G., Solomakha, A.S., Ermolaev M.O.: Hybrid gas-diesel fuel system. Research report, State registration No 0118РКИ0395, inv. No 0218РКИ0196, Astana (2018).
13. Ter-Mcrtichan, G.G.: Converting diesel to gas engine. Homepage. https://www.science-educat ion.ru/ru/article/view?id=14894. Accessed 31 January 2021
14. Bosch dual-fuel–future of diesel engines. Homepage, http://gazeo.com/automotive/technology/Bosch-Dual-Fuel-future-of-diesel-engines,article,7831.html. Accessed 31 January 2021
15. Dolganov, K.E. etc.: Cars with Gas-gasoline and Gas-diesel Engines: Design Features and Maintenance. Tehnika, Kiev, Ukraine (1991)
16. Glazyrin, S.A., Varlamov, G.B., Barabash, P.A., Petrenko, V.G., Solomakha, A.S., Ermolaev, M.O.: Gas Diesel Engine Power System. Patent of Republic of Kazakhstan for utility model No. 2863 (2018)

Status Monitoring Automation for the Engineering Systems of the Smart Facilities

Ravil Safiullin⬡, Igor Prutchikov, Oleg Pyrkin, Ruslan Safiullin, and Vera Demchenko⬡

Abstract In the course of solving the problem of increasing the efficiency of safe life support for autonomous buildings and structures, conceptual methodological approaches were identified for automating the processes of monitoring the state of smart infrastructure technical systems, which will significantly increase their efficiency through the combined use of elements of individual control systems, opto-electronic event detection and recognition (ECO). To this end, the methodological foundations for the calculation and design of technical systems were formed on the basis of the developed methods of analysis and technical and economic modeling of objects, evaluation of the technical and economic efficiency of technical informational systems, calculation of dynamic characteristics and evaluation of indicators of the quality of energy supply of objects, as well as a comprehensive methodology for evaluating the effectiveness of multi-stage visualization systems, taking into account the priority study of issues of feasibility study and their rational construction smart infrastructure.

Keywords Autonomous buildings · Smart infrastructure · Technical systems · Security systems · Monitoring · Control · Adaptive management · Event recognition

R. Safiullin (✉) · R. Safiullin
Saint Petersburg Department of Transport and Technological Processes and Machines, Saint-Petersburg Mining University, 190005 St. Petersburg, Russia
e-mail: safravi@mail.ru

I. Prutchikov
Department of Electric Power Industry and Electrical Engineering, Saint-Petersburg State University of Architecture and Civil Engineering, Saint Petersburg, Russia

O. Pyrkin · V. Demchenko
Scientific Research Institute (Military-System Research of Logistics Support for the Armed Forces of the Russian Federation), 191123 St. Petersburg, Russia

© The Author(s), under exclusive license to Springer Nature Switzerland AG 2024
M. Sari and A. Kulachinskaya (eds.), *Digital Transformation: What are the Smart Cities Today?*, Lecture Notes in Networks and Systems 846,
https://doi.org/10.1007/978-3-031-49390-4_22

1 Introduction

Automation of technical systems installed in buildings and transport and technological machines in the world has reached a high level. In Russia, this approach also becomes more popular in various companies. The relevance of this discussion is due to the fact that the use of automation systems that effectively manage engineering systems of buildings and traffic control systems is in demand both during the retrofit of existing electrical installations and at the stage of designing new ones. Effective management of building systems such as lighting, heating and air conditioning can help reduce energy costs, as well as improve comfort during long-term stay of a person in the room.

The basis of the current theoretical research on reducing the power consumption costs during operation of engineering systems and equipment of smart buildings and structures, and, as a result, reducing the negative impact on the environment in the conditions of rapidly changing quantitative and structural parameters of smart technical means with limited changes in historically developed areas of focus became both Federal and regional target program of upgrading the power system under the "Smart City program".

Autonomous buildings and structures, despite the great variety and difference in a number of characteristics, for their normal functioning are equipped with technical systems (TS), an obligatory component of which are life support systems (LSS) and safety systems (SS).

The requirements, features and characteristics of LSS and technical safety systems (TSS) are currently being standardized [2]. Special systems for engineering systems monitoring (SESM) are also standardized for such large, sensitive and critical autonomous sites as airfields, power plants, etc. [3]. They have their life-support systems as software and hardware suites designed for uninterrupted (within the limits) performance of equipment (heat supply, water supply and discharge, power supply, gas supply, etc.) of potentially dangerous sites, buildings and structures. Thus, in terms of their purpose, life-support systems are close to technical safety systems defined according to the same source as software and hardware suites that are designed for prevention of emergencies, explosion and fire safety, protection and warning of people in emergency situations.

Currently, the conventional approach to improving the performance of technical systems in autonomous buildings and structures is the joint (comprehensive, combined or integrated) use of them. Therefore, the joint use of components of both life-support and safety systems is quite promising when building a life-support safety monitoring system (LSSM). For autonomous buildings and structures that are not permanently linked with external engineering systems, these systems are not always practical, since the resources of individual monitoring and control systems to monitor the characteristics of buildings and structures, when used comprehensively, are not enough to give a greater effect in achieving the overall purpose of their safe life support.

The methodological approach for automation of the processes of monitoring the state of technical systems in autonomous buildings and structures will significantly increase the performance of monitoring the safe life support for autonomous sites through the joint use of components of individual control systems, in particular optronic event detection and recognition means. The methodological basis of the calculation and design of the technical systems uses the following methods: analysis and performance modeling of facilities with technical systems; methods of performance assessment of technical systems; the calculation method of the dynamic characteristics and quality evaluation for power supply of sites; a comprehensive method of assessing the performance of the technical systems based on multi-pattern visualization systems for the purpose of priority study of the issues related to the feasibility study and reasonable layout of comprehensive automated systems of active optoelectronic life-support safety monitoring (LSSM) in autonomous buildings and structures.

Thus, fitting the autonomous buildings and structures with life-support safety monitoring systems based on the joint use of components of LSS and TSS should be found promising in terms of enhanced performance and safety of their operation.

2 Materials and Methods

Practical studies of the issues of enhancing the performance of buildings and structures by introducing SESM to them through the joint use of LSS and TSS components are currently promising due to the following circumstances:

- the joint use of LSS and TSS does not require the introduction of new components and, consequently, large material costs;
- options for the joint use of LSS and TSS allow for getting a total effect, both in terms of ensuring the safety and improving the life support;
- the joint use of LSS and TSS makes it possible to comprehensively improve the key performance characteristics of independent operation, safety and manageability of buildings and structures of various purposes;
- the implementation of LSS and TSS on the same basis of modern IT solutions, for example in terms of event detection and recognition, allows using the resources of both systems to achieve common targets of beneficial and safe performance of autonomous buildings and structures.

The methodology of substantiation, structural construction, calculations and optimization of the LSSM system in autonomous buildings and structures can be based on a general concept that includes the following key features:

1. The uninterrupted subsystems of the LSS (safe life-support system) are the targets of monitoring, control and management using the LSSM system;
2. The LSSM systems are built as an autonomous control subsystem using algorithms and means of remote smart control;

3. The primary objective of the LSSM system is to ensure the safe independent operation of buildings and structures;
4. Structurally, the LSSM system is made up by jointly using the components being part of safe life-support (SLSS) and technical safety systems (TSS), primarily in terms of optical electronics for event detection and recognition;
5. The SLSS and TSS components to be installed in the LSSM system have automation tools that provide the necessary level of independent operation of buildings and structures.

The methodological background and areas of focus of comprehensive research, calculation and design of the LSSM systems for autonomous buildings and structures are shown in Fig. 1.

As shown in Fig. 1, methods of analysis and performance modeling of facilities equipped with LSSM systems, methods of performance assessment of LSSM systems, the calculation method of the dynamic characteristics and quality evaluation

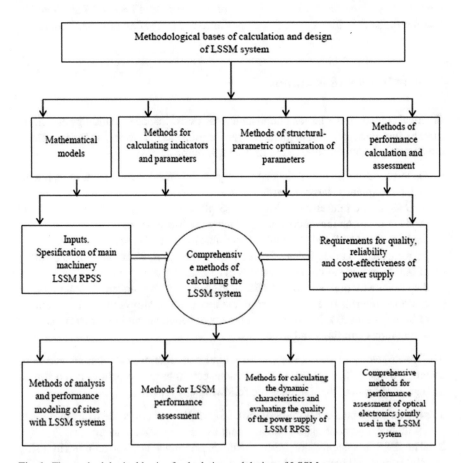

Fig. 1 The methodological basis of calculation and design of LSSM system

for power supply of sites, and a comprehensive method of assessing the effectiveness of joint use of optical electronics based on multi-pattern imaging technology in the LSSM system can serve as a methodological basis of calculation and design of LSSM systems.

3   Results and Discussions

Currently, an integrated approach is widely applied to the development of TSS providing for ensuring the security of sites in several areas, such as physical, economic, sci-tech, technological, environmental, information, engineering-and-technical safety, etc. Comprehensive and integrated safety systems can be built by combining various safety systems, such as emergency lighting, fire-fighting water supply, water-based fire extinguishing, fire alarm, perimeter security, access control and management, CCTV, etc.

According to [2], comprehensive (CSS) and integrated (ISS) safety systems, in addition to the "security" subsystems, such as security alarm systems, fire alarm systems, access control and management, CCTV and control, inspection and search, related to the feature of protection, information, etc., also includes engineering support systems for the site: electric lighting and power supply, gas supply and water supply, sewerage, heat supply, ventilation, air conditioning, etc. The analysis of the above data shows the actual availability and possibility of developing a joint application of LSS and TSS by designing CSS and engineering safety systems of various levels of integration. However, this approach can be accepted as a reasonable one only if safety matters are prioritized over the engineering support issues.

Along with the options for the joint use of LSS and TSS based on either comprehensive or integrated security system, options for creating such combined LSS-based systems can also be regarded. For example, according to [3], structured systems for monitoring and managing engineering systems of buildings and structures (SMIS) are built using software and hardware tools that monitor operational processes and machinery performance control immediately at sites, buildings and structures and communication of data on their condition via communication links to operations control duty desks (OCDD) for subsequent processing in order to assess, anticipate and eliminate the consequences of destabilizing factors. The focuses of monitoring, control, and in some cases management, are the following subsystems of life-support and safety: heat supply, ventilation and air conditioning, water supply and sewerage, power supply, gas supply, fire safety, security alarm and CCTV systems, detection and notification systems about the abnormal state of the environment, etc. However, it should be noted that SMIS systems are usually designed for potentially hazardous facilities, buildings and structures, which can include radiation hazardous sites, hydraulic engineering structures, chemically hazardous sites, large oil product storages, power plants with a capacity of over 600 MW, companies for mining and processing of solid minerals, technically complex and unique sites, the range of which is limited. So, the joint use of LSS and SS in terms of the SMIS system, as shown in [3], can be demonstrated only for potentially hazardous sites, and for most autonomous buildings and structures it seems functionally redundant.

Since the comprehensive and integrated security systems as well as the SMIS system considered above are made up on software and hardware suites, which in general can also be attributed to automated control systems (ACS), the matters related to the possibility of joint use of the LSS and SS at the level of the ACS make sense. The issues of the theory, design and creation of Automated Process Control Systems (APCS), Business Management Systems (BMS), Computer Aided Facility Management system (CAFM) based on modern information technologies are discussed in detail in [4, 5]. Currently, these systems are constantly being improved, developed and introduced at various levels in processes and on sites. For example, automated control systems based on SCADA systems, BSM systems, automated operations control systems (AOCS) of various levels from AOCS of engineering systems to AOCS of the city have become widespread. Many of these systems provide for the control and management of both safety and life-support systems. However, the implementation of combined LSS and SS systems at the level of automated control systems (AOCS) for autonomous buildings and structures also seems functionally redundant and not reasonable. More reasonable and promising should be considered the structural construction of the class of combined systems, such as LSSM, in the form of subsystems of common automated control systems (AOCS) of sites, for example, similar to SMIS, but with a limited nomenclature of controlled components to the level of uninterrupted subsystems of life-support, primarily power supply. The studies have shown that the structural construction of the LSSM system currently seems to be the most reasonable if the SLSS and event detection and recognition tools of security alarm and protection systems as part of the TSS are used jointly. Moreover, designing the LSSM system should be recognized as the most practical primarily based on uninterruptible power supply subsystems, which are made up in autonomous buildings and structures consisting of uninterruptible power supply (UPS) and reliable power supply (RPS) systems. In [6, 7], the practicability and efficiency of using systems based on the joint use of diesel generators and static power converters in autonomous buildings and structures as RPS systems is justified. These circumstances allow for considering the reasonable implementation of the LSSM systems in the form of the following structure, as shown in Fig. 2.

The practical application of the LSSM system in autonomous buildings and structures can be performed at various levels, as shown in Fig. 3. According to Fig. 3, the LSSM systems for autonomous buildings and structures can be developed at the levels of independent power sources, uninterruptible power supply installations, reliable power supply systems, as well as uninterrupted life-support subsystems. As the studies have shown, currently the most practical and promising is the implementation of the LSSM system at the level of the reliable power supply systems, since it allows for full control and management of uninterrupted power supply of the site, and monitoring of the main power-efficient processes and modes of use of power sources. The implementation of the LSSM system at the level of independent power sources and uninterruptible power supply installations is appropriate for practical development of algorithmic, software and hardware implementation of LSSM components. The

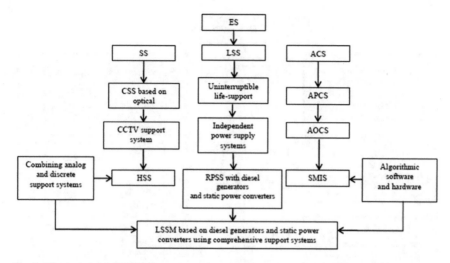

Fig. 2 The structure of LSSM system based on the joint use of SLSS and TSS components for autonomous buildings and structures

practical introduction of the LSSM system at the level of SLSS in buildings and structures can be attributed to the category of promising, requiring additional research for its introduction focusing on the development of tools and methods for automated control and emergency management of all uninterrupted life-support subsystem.

The feasibility study of the prospects for the addition of the LSSM systems to autonomous buildings and structures requires special computational and experimental studies that can be performed by known methods. The generally accepted and most commonly used to date among simple methods of evaluating the efficiency is an estimate based on the reduced costs. The efficiency assessment is carried out here by comparing the reduced costs Π_i determined for each i-th option by the formula:

$$\Pi_i = C_i + E_H \cdot K_i, \tag{1}$$

where K_i is the capital investment for *the* i-th option;

C_i—current costs (cost price) for the same option;

E_H—the standard coefficient of efficiency of capital investments.

In this case, the optimal option is taking to have the minimum value of the Π_i. These costs include capital investments and current expenses. At the same time, it is allowed in the performance calculations to quantify only that part of the reduced costs that undergo changes in the compared options. That is to say, the calculation of capital investments and operating costs for the consumer can be limited only by those components that change in the compared options.

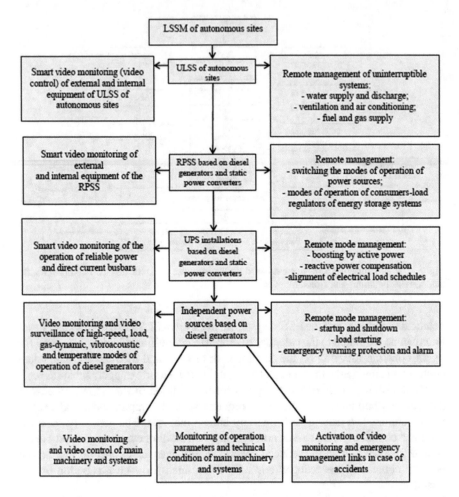

Fig. 3 The main focus of implementation of the LSSM system in autonomous buildings and structures

The selection from a variety of design or planning solutions of the optimal by the criterion minimum of the above costs is possible only if they are fully comparable, that is, the identity of the results achieved through them. Different variants of technical solutions do not always ensure the identity of the results. Therefore, there is a need to bring the options to the identical effect. The identity of the performance effect when comparing different versions of the LSSM system must necessarily take the reliability and safety into account.

In a comprehensive assessment and development of the practical application of the topologies of LSSM systems in autonomous sites according to Fig. 3, the economic performance and criteria described above can be regarded together with the indicators and results of the assessment of scientific, engineering, production, environmental, social and other effects determined by known methods. The performance assessment of the available technical, technological and infrastructural solutions for the implementation of the LSSM systems in autonomous sites of the Ministry of Defense of the Russian Federation can be performed at three levels of specification of characteristics and, accordingly, in three steps.

In step 1, the new and existing systems for remote monitoring, control and management of life support in autonomous sites can be considered in terms of ensuring the economic performance.

In step 2, it is advisable to assess the economic performance of the introduced LSSM systems in autonomous buildings and structures.

Finally, in step 3, a comprehensive performance assessment of promising infrastructure projects for the addition of life-support systems to autonomous buildings and structures having the LSSM systems already installed can be carried out, including figures and criteria for various types of effect from their practical implementation.

The analysis has shown that most of the performance assessment methods for the use of new technical solutions are based to some extent on determining and comparing the reduced costs, and similar indicators, such as payback period, efficiency coefficient, etc. This methodological approach can be maintained in steps 1 and 2 of performance assessment of the addition of the considered and future LSSM systems of autonomous buildings and structures.

Adding the RPS systems to the LSSM systems obviously requires additional costs causing an increase in the cost of power supply to the sites. However, on the other hand, addition of RPSS to the LSSM system with can significantly increase the reliability, safety and stability of power supply, significantly reduce operating costs, which ultimately can result in an improvement in the performance of power supply. A comprehensive assessment of the addition of the LSSM systems to the autonomous sites of the Ministry of Defence of the Russian Federation requires a joint assessment of performance, reliability and safety, and in general is a separate scope of research.

As an example, Table 1 shows calculation of the variable part of the reduced costs for various versions of the LSSM systems in the autonomous sites with reliable power supply systems based on 100-kW diesel generators and static power converters for the conditions of the Far North and the Arctic zone.

According to Table 1, the structure of the LSSM system using optical electronics as part of TSS and SLSS has the lowest value of the reduced costs, which is associated with a reduction in the cost to be spent for maintenance personnel and the possibility of switching the site to an unmanned mode of operation. Generally, according to the results of the computations, it should be concluded that adding the RPS system to the LSSM system causes a slight increase in the reduced costs at the level of 3–7% for remote areas and 5–12% for normal operation conditions. It means that it is quite acceptable and practicable to equip the life-support systems of autonomous sites with optical electronics, especially taking into account the effect of their addition in terms of monitoring, control and management of power supply, and increasing its reliability and stability. Details and results of computational and experimental studies of the LSSM system for autonomous sites of various purposes are given in [8, 9].

Table 1 The results of the performance assessment for installation of LSSM RPSS in autonomous sites in the Far North and the Arctic zone of the Russian Federation

References	Name and description	RPS with 100-kW diesel generator			
		RPS without LSSM	RPS with LSSM based on automated operations control system	RPS with TSS-based LSSM	RPS system with TSS- and SLSS-based LSSM system
[1]	Capital investments (variable part), $K \cdot 1$, RUB '000	4500	4950	4650	4850
[2]	Depreciation charges, S_{am}, RUB '000	225	248	233	242
[3]	Annual maintenance and repair costs, RUB '000	115	125	115	1150
[4]	Annual costs for fuel and oil, RUB '000	6930	6930	6930	5580
[5]	Annual costs for service personnel, RUB '000	1000	1000	1000	0
[6]	Other costs, RUB '000	13	13	14	4
[7]	Current (operational) expenses, RUB '000	8270	8303	8278	7173
[8]	Reduced costs, Π_i^*, RUB '000	6568	7025	6720	7043
[9]	Relative to Π_1^*, the value of the reduced costs	1	1.07	1.04	0.92

4 Discussion

The study of the existing literature and official documents revealed a gap in research in the field of engineering systems design based on the developed methods of analysis and technical and economic modeling of facilities, evaluation of technical and economic efficiency of technical systems, calculation of dynamic characteristics and evaluation of quality indicators of power supply facilities. The use of effective information technologies for supervisory control, optimal operations control, computer vision, artificial intelligence, etc., implemented in the course of computational and experimental studies provide the highest performance as compared with existing analogues up to the possibility of implementing closed video control algorithms and the operation of the site completely in unmanned mode. Therefore, the new approaches proposed and implemented in the LSSM systems to monitoring, control and emergency management of autonomous buildings and structures make it possible to bring the safety of their life support to a new level corresponding to the level currently standardized for SESM of critical civil and industrial facilities.

5   Conclusions

The addition of the LSSM system with the SLSS and TSS components to autonomous buildings and structures is currently one of the promising area for improving the efficiency and safety of their functioning in various operating conditions. Moreover, according to the LSSM topology, which includes of the RPSS with diesel generators and and static power converters as components of uninterrupted life-support subsystems and the optical electronics as components of technical safety systems, analytical, computational and experimental studies have been performed that confirm the high efficiency of their addition to autonomous buildings and structures for various purposes.

References

1. Apanaviciene, R., Vanagas, A., & Fokaides, P. A.: Smart building integration into a smart city (Sbisc). Development of a new evaluation framework. Energies **13**(9), 2190 (2020)
2. Radziejowska, A., Sobotka, B.: Analysis of the social aspect of smart cities development for the example of smart sustainable buildings. Energies **14**(14), 4330 (2021)
3. Thoben, K.-D., Wiesner, S., Wuest, T.: "Industrie 4.0" and smart manufacturing—a review of research issues and application examples. Int. J. Autom. Technol. **11**(1), 4–16 (2017)
4. Baduge, S.K., Thilakarathna, S., Perera, J.S., Arashpour, M., Sharafi, P., Teodosio, B., Shringi, A., Mendis, P.: Artificial intelligence and smart vision for building and construction 4.0: machine and deep learning methods and applications. Autom. Constr. **141**, 104440 (2022)
5. Degtyarev, I.: Classification of categorical structures of structural complexities based on prototypes in simple complexes. In: 10th International Conference on the Application of Fuzzy Systems and Soft Computing (ICAFS-2012), pp. 27–37. Lisbon (2012)
6. Kisulenko, B.V.: Automation of car control: trends of development and conditions of implementation of the international. Acad. Transp. **17**, 80–88. Moscow (2014)
7. Trinity, N.: A unified transport system: textbook for students of institutions. Professor of education, p. 240. Academy Publishing Center, Moscow (2003)
8. Safiullin, R., Epishkin, A., Safiullin, R., Haotian, T.: Method of forming an integrated automated control system for intelligent objects. CEUR Workshop Proc. **2922**, 17–26 (2021)
9. Safiullin R.N., Afanasyev A.S.: Integrated assessment of methods for calculating harm caused by vehicles in transport of heavy cargoes. In: Conference Series: Earth and Environmental Science. Issue 7 Transportation of Mineral Resources T, vol 194, p. 7 (2018)

The Urban Territory Energy Balance Spatial Model Application for the Buildings Power Supply Structure Selection Decisions Analysis

Sergey Kosiakov, Svetlana Osipova, Artur Sadykov, and Maxim Malafeev

Abstract Most of Russian cities are in the regions of a severe climate, thus the cost of energy supply of buildings is significant. The buildings are connected to the heat supply system, natural gas, and electricity supply systems of the city. When constructing new buildings, it is possible to use different types of energy resources to supply the buildings with heat and hot water. The chapter presents the results of the study of the method that allows comparing the ways to use various sources and types of energy supply of buildings according to the criteria of urban area energy balance, for example, the variation value of the total costs of power transmission, transformation, and consumption. The proposed method suggests that calculation of the cost of various types of energy carriers is performed in accordance with the normative techniques. These techniques are approved by the authorities and used in the economic calculation and tariff validation for various types of power supply of the end consumers. The authors propose the solution to assess the ways of power supply of buildings based on an energy balance spatial model. The urban energy balance spatial model is implemented in the geoinformation system environment. The chapter presents the example of the analysis of alternative way of power supply during the construction of the buildings in Ivanovo city and its impact on the urban energy balance. The developed GIS is used in the example.

Keywords Energy balance · Energy networks · Spatial analysis · Decision support

S. Kosiakov (✉) · S. Osipova · A. Sadykov · M. Malafeev
Ivanovo State Power Engineering University, 34 Rabfakovskaya St, 153003 Ivanovo, Russia
e-mail: ksv@ispu.ru

© The Author(s), under exclusive license to Springer Nature Switzerland AG 2024
M. Sari and A. Kulachinskaya (eds.), *Digital Transformation: What are the Smart Cities Today?*, Lecture Notes in Networks and Systems 846,
https://doi.org/10.1007/978-3-031-49390-4_23

1 Introduction

The population of the world is growing, so energy costs are also increasing. In this context, the issues of efficient energy use are urgent.

Currently, the approaches and methods that are related to the issues of application of information technologies, smart city [1], smart homes [2], smart networks technologies [3, 4] are used to solve this problem.

Application of these approaches provides a significant effect and reduce total energy costs in the cities partly due to the use of renewable energy sources. However, in the countries with a cold climate, cities need large amount of energy from external sources in the form of fossil fuels (coal, natural gas, oil products). Under these circumstances, when planning the construction of new buildings and urban neighborhoods, we face the problem of choosing the model of energy infrastructure.

Thus, the urban energy system should be described as a system where, a locality strives toward establishing the delivery of energy service with maximum utility of its inhabitants, using an optimum combination of resources and technologies, subject to satisfying technical, institutional, environmental and economic constraints, Farzaneh et al. [5].

The purpose of this study is to develop a method to design an urban area energy balance model considering the issues related only to the buildings (the so-called building level); and to apply this model to solve the problem of choosing a method of energy power supply of a new building according to the criterion of minimum growth of energy power supply of the city. This method to design and use the spatial models of the energy balance of urban area is new. The existing methods of modeling the energy balances of urban area use aggregated data about consumers, the proposed method considers the characteristics of the specific consumers i.e., detached buildings or neighborhoods.

2 Methods

2.1 Literature Review

The issues of optimizing energy infrastructure of the cities have been extensively studied [6]. Different reviews have examined various aspect of this issue: Koirala et al. [7], Keirstead et al. [8], Bracco et al. [9] and Scheller et al. [10] discuss different issues and approaches of energy infrastructure modelling on the municipal level. Huang et al. [11] and Liu et al. [12], as well as Mirakyan and De Guio [13], introduce and assess the respective methods and tools for energy system planning. The same applies to the articles of Mendes et al. [14], Markovic et al. [15], Allegrini et al. [16] and Tozzi and Jo [17]. A review of 37 tools for analyzing the issues of integration of renewable energy into various energy systems has been conducted by Connolly et al. [18].

Application of modern technologies of energy conversion and consumption makes it possible to meet the same demands. This problem can be solved if we consider the goals and interests of various stakeholders and use various assessment criteria. However, the goals and interests of the city authorities may not coincide with the interests of the individual buildings owners, as well as with the interests of energy companies. The city, as a single system, is interested in optimizing energy consumption in general and reducing the entire engineering infrastructure maintaining costs. The information systems that include energy system models are developed to find solutions in this area. DER-CAM-Distributed Energy Resources Customer Adoption Model [19], EnergyHub Model [20–22] are the examples of such models implementation and application. These models allow to solve optimization problems with high accuracy, but they are quite difficult to apply.

The spatial nature of these systems is one of the factors that are considered when we solve the problem of optimization of the city energy infrastructure. Geographical information systems (GIS) are used to analyze this factor [23–25].

The methods to develop and analyze energy balances models are used to analyze the production, distribution, and consumption systems of various energy types in different areas [26]. These methods deal with statistical and annual data. It means that we don't consider operating modes and technological aspects of the energy systems analysis. These methods are broadly applied to solve problems of economic planning of Russian regions. As examples, the solutions based on the energy consumption decomposition analysis in the framework of regional energy budget management systems are presented in [27] and [28].

The existing methods of modeling the urban energy balances use a matrix model, which shows the decomposition and relationship of generation, conversion, consumption of all types of energy and energy carriers. At the same time, the location of specific objects in the model is not considered in any way. In the model the block of consumption indicators shows the consolidated figures of consumption of the entire housing fond of the city, obtained based on statistical data.

At the same time, the widespread introduction of GIS into urban management system, allows obtaining detailed data on the real energy consumption of all types of energy in each building. The use of such data makes it possible to develop the energy balance model of urban area in detail, and present not the branches of the economy, but specific buildings in the block of consumption indicators. In turn, we can eliminate the opportunities to solve new analysis tasks related to the issues of assessment of the choice of locations and characteristics of buildings according to the city-wide criteria.

2.2   Setting of the Problem

Power shortage in cities are covered by supplying primary fuel and power resources from outside (natural gas, electricity, coal, oil products and other types of fuel), which

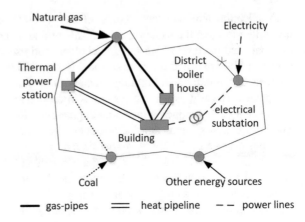

Fig. 1 The power transmission direction to the building

are transported on the city territory through distribution circuit and can be converted into other types of energy.

A simplified scheme, that illustrates possible energy transfer paths from the primary sources located outside the city to a building (see Fig. 1).

For example, based on the figure, the demand for a building heating can be satisfied by burning natural gas in individual gas boilers located in each apartment, or by transferring heat conductor through pipelines from a remote Central Heating and Power Plant, or from a nearby district boiler house, or it can be obtained by using electric heaters that are connected to the central electrical network. Other chains of energy conversion and delivery from external sources using, for example, coal, oil products and etc. to the building are also possible. Moreover, often the different needs of the building are met by different types of energy. For example, the central heating system can be used for heating of buildings, and natural gas or electricity can be used for cooking. Thus, when choosing a method of power supply for designed and reconstructed buildings, the problem of comparing alternative options for building energy balance implementing arises. A similar problem arises when new city micro districts are designed. The point of the complex requirements determination for the city various power supply systems development is appeared.

The considered problem is a multiobjective problem. The energy balance implementation decisions choice is reflected on the distribution of fuel combustion products emissions into the atmosphere, on the danger of technological disasters occurrence and other aspects of the city life. From the point of economic factors, there is a contradiction between the interests of the building owners, who tries to minimize costs for the building construction and operation, and municipal interests. In this study, the main aim is to develop a method for comparing alternative options for urban development solutions according to one criterion that assesses the decision impact on the total energy costs of the city for all types of energy consumed resources.

2.3   The Theoretical Part

The study is based on a synthesis of energy balances analyzing methods and spatial modeling in GIS. The compares the revenue and expenditure parts of all areas of production, processing, transportation, transformation, distribution and consumption of energy resources and energy in the territory from the source of their receipt to the use of energy by the consumer. The receipts and expenditures parts of production, processing, transportation, transformation, distribution and consumption areas of power resources and energy in the territory from the source of their receipt to the consumer are confronted in territory energy balance model. When choosing the optimal energy supply scheme for enterprises, regions, and economies of countries such balances are compiled. Statistical data from enterprises and organizations, as well as aggregated consumption data from various sectors of the economy are usually used.

Since in the study the range of tasks to be solved differs significantly from the task of the territory economy development planning, the only parts that are related to energy non-industrial buildings supply are highlighted in the proposed model of the territory energy balance. The focus is on the spatial aspect of the energy transmission and conversion chains and not on its industry structure. The connection between energy and production, the dynamics of economic processes, and other tasks of analyzing energy balances are not considered, since they also go beyond the scope of the tasks that are solved. These factors made it possible to reduce the volume of the studied indicators and to focus on issues related to the technical side of the problem at the level of the building's energy equipment choice.

To analyze the urban area energy balance model on a building level, its expenditure part (energy consumption) can be represented by the indicators of energy consumption for the following needs:

1. heating and ventilation;
2. hot water supply;
3. cooking;
4. other.

The above-mentioned needs can be satisfied thanks to different types of energy supply. Such obvious issue of energy consumption as lighting is not highlighted. It is considered as the "other" needs because currently there is no alternative way to electric lighting, and it makes no sense to analyze other ways of energy supply of the buildings according to this indicator. However, in terms of the method application, the additional indicators are possible to consider. In the framework of the article, "other" section is not discussed to cut the article and simplify the presentation of the material.

The following primary fuel and energy resources, that are coming to the city from the outside, are highlighted in the model:

1. natural gas;
2. electricity;

3. coal;
4. oil products;
5. other types of fuel.

It is assumed that heating energy is produced in the city. Coal, oil products and other types of fuel are not currently considered as alternative options to be used in the buildings under construction. When analyzing energy conversion paths, these types of fuel are considered as fuel for Central Heating and Power Plant and boiler houses. These types of fuel as well as natural gas are considered considering the characteristics of power-generating equipment. In the used data about Ivanovo city, these types of fuel are not discussed. Thus, this part of the energy balance model is omitted to simplify the presentation of the method.

The means of conversion and transportation of fuel and energy resources are presented in the energy balance model as the elements of systems:

1. district heating;
2. gas delivery;
3. electric power supply.

Considering the accepted terms and assumptions, the spatial model of energy balance of the urban area on the building level is presented in the form of the equation:

$$Q^{im} - Q^{ex} = \sum_{n=1}^{N} Q_n + Q^{oth} \tag{1}$$

where

- Q^{im}, Q^{ex}—the total amount of imported and exported energy of all types of energy resources;
- Q_n—the total amount of energy consumed by the nth building;
- Q^{oth}—the total amount of consumed energy for other needs and losses (not for internal consumption in the buildings).

The amount of energy consumed by each building can be divided into two parts:

$$Q_n = Q_n^C + Q_n^S \tag{2}$$

- Q_n^C—the part that does not have alternatives for delivery using various types of fuel and energy resources;
- Q_n^S—the part that depends on the chosen structure of the energy balance.

To indicate the structure of the energy supply systems of the building, the parameter S is introduced. The parameter S takes the following values:

- S = 1—the use of the central heating system for heating, for hot water supply, the gas stoves installation for food preparation;

- $S = 2$—the use of the central heating system for heating, for hot water supply, the electric stoves installation for food preparation;
- $S = 3$—the use of a central heating system for heating, installation of a gas boiler for hot water supply, the gas stoves installation for food preparation;
- $S = 4$—the installation of a gas boiler for heating and hot water supply, the gas stoves installation for food preparation;
- $S = 5$—the installation of a gas boiler for heating and hot water supply, the electric stoves installation for food preparation;
- $S = 6$—the use of the central heating system for heating, installation of a gas boiler for hot water supply, the electric stoves installation for food preparation;
- $S = 7$—the use of the central heating system for heating, the use of an electric boiler for hot water supply, the electric stoves installation for food preparation;
- $S = 8$—the installation of a gas boiler for heating, the use of an electric boiler for hot water supply, the electric stoves installation for food preparation;
- $S = 9$—the installation of a gas boiler for heating, the use of an electric boiler for hot water supply, the gas stoves installation for food preparation.

The structure of the power supply system of each existing nth building is denoted as S_n. Depending on S_n value, the total energy consumption Q_n^S can be determined by the formula:

$$
Q_n^S = \begin{cases}
\frac{P_1+P_2+R_i}{\mu_i} + \frac{H_1}{\varphi_n}, & S_n = 1, \\
\frac{P_1+P_2+R_i}{\mu_i} + W_1 + G, & S_n = 2, \\
\frac{P_1+P_2+R_i}{\mu_i} + \frac{H_1+H_2}{\varphi_n}, & S_n = 3, \\
\frac{H_1+H_2+H_3}{\varphi_n}, & S_n = 4, \\
\frac{H_2+H_3}{\varphi_n} + W_1 + G, & S_n = 5, \\
\frac{P_1+R_i}{\mu_i} + \frac{H_2}{\varphi_n} + W_1 + G, & S_n = 6, \\
\frac{P_1+R_i}{\mu_i} + \frac{H_2}{\varphi_n} + W_1 + W_2 + G, & S_n = 7, \\
\frac{H_2}{\varphi_n} + W_1 + W_2 + G, & S_n = 8, \\
\frac{H_1+H_3}{\varphi_n} + W_2 + G, & S_n = 9.
\end{cases}
\tag{3}
$$

where

- P_1—the amount of heat required for heating an apartment building using a central heating system;
- P_2—the amount of heat required for hot water supply of an apartment building using a central heating system;
- i—the source number to which the considered residential building is connected (Central Heating and Power Plant, boiler room, etc.);
- R_i—the losses in the central heating system due to heat transfer from the ith source to the building;
- G_i—the losses of the transmission of electrical energy due to connection of the building;
- μ_i—the efficiency of the equipment during heat generation;

- H_1—the natural gas consumption for food preparation;
- H_2—the natural gas consumption for hot water supply;
- H_3—the natural gas consumption for heating;
- φ_n—the efficiency of a consumer gas boiler;
- W_1—the electric energy consumption for food preparation;
- W_2—the electric energy consumption for hot water supply;

When solving the problems of influence of the energy balance structure on the consumption of all types of fuel and energy resources in the city Q^{ex}, Q^{oth}, Q_n^C can be considered constant since they do not depend on S. Then the variable (depending on the structure) part of the urban area energy balance model can be represented as:

$$\Delta Q^{im} = \sum_{n=1}^{N} Q_n^S \tag{4}$$

And the task of choosing the optimal energy balance structure for new and reconstructed buildings is presented as the task of choosing S_n, values that are ensured by a minimum growth in the number of losses and fuel and energy resources imported by the city:

$$\sum_{n=1}^{N} Q_n^S \rightarrow_{S_i} \min \tag{5}$$

All the expenses are converted into reference fuel per year according to the approved methodology of Russian Federal State Statistics Service [2]. The values of variables P_1 and P_2 can be obtained, for example, from information monitoring systems or calculated based on normative techniques [3]. According to the existing methods of calculation, the electricity and gas losses do not depend on the location of the building. At the same time, heat losses of the heating services R_i essentially depend on the length of the pipelines connecting the buildings with the sources (Central Heating Power Plant or boiler houses) [4]. Therefore, such losses are determined for each building depending on the length of the pipelines.

The developed model considers losses in the heating network that occur in the pipeline section connecting the building with the existing network. If necessary, the additional losses that occur when the load is increasing in the existing network can be considered, but these losses are low, and the calculation requires a large amount of additional data on the technical characteristics and the operating mode of the network.

Considering other assumptions made in the study, this assumption does not significantly affect the obtained results, and can be eliminated.

The total amount of energy costs and losses, determined by the formula (4), can be calculated for the whole city or for any part of it. If we define this variable for each neighborhood, we can get the spatial distribution of the debit and expenditure parts of energy balance model of the urban area.

Using GIS technology, we can get the necessary data about buildings and networks to make calculations according to the formula (3), as well as automatically aggregate the data obtained about the neighborhood, and develop thematic maps of the spatial distribution criterion (5) and its components. In addition, available resources to connect consumers to energy networks are usually presented in the form of zones in the GIS environment. This data can be used as constraints in the process of decision making.

The decision support method based on the proposed model is implemented using a specialized GIS, which ensures the following operations:

- Creating a city map and allocation of objects levels used in the study (buildings, network elements, sources, city blocks, city territory).
- Entering and storing data about the objects which are necessary for the energy balance calculation according to the formula (3) for the whole city or some part of it.
- Automatic aggregation of the data about the energy balance of the buildings in the neighborhood and in the whole city. Calculation of the criterion value (5). Saving the source data and the results of calculation in the table to compare them.
- Creating new objects on the map (buildings, network elements, sources) or changing the characteristics of existing objects.
- Implementation of the step 3.
- Comparison study according to the criterion of total consumption and other indicators of energy balance on the level of the whole city and on the level of neighborhood and presentation of the results in the form of aggregated data and thematic maps.
- If there are still alternatives for analysis, go to step 4.

Steps 1 and 2 are performed only once at the very beginning. These operations are quite laborious. Further use of the system to analyze new alternatives does not require significant costs and can be performed by any user without special training.

3   Results

The proposed method makes it possible to analyze how the decision on the choice of the energy supply system of the building may affect changes in the energy balance of the urban area. To use this method, the specialized GIS has been developed. Based on public data of the housing and public utilities website, websites of energy companies and Ivanovo engineering networks a unified map of the energy balance spatial model of one of the neighborhoods has been created. The program is implemented as a web application, and it is available for users on the Internet. The application view is shown in Fig. 2.

The developed application allows to change the energy balance structure of any building or a group of buildings, view the impact of these changes on energy balance indicators. To view and change the characteristics of a building, we need to highlight

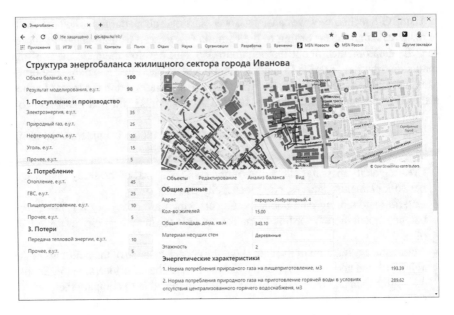

Fig. 2 Application page

them on the map. The indicators values of the selected building are displayed in the fields available for editing. During the experiments, we can change the values of these fields and, if necessary, return to the actual data. In addition, it is possible to add new buildings and delete existing ones, as well as change the scheme and characteristics of the city heat network elements.

As an example of the analysis based on the developed model, we can present the results of the study of the influence of the recently constructed multi-story building on the city energy balance model when various power supply systems are used. With the help of the developed program, all possible structures of the energy balance of the building were generated. For each option, the costs of full power supply of the building in units of conventional fuel and additional losses arising in energy networks during the transmission of energy to the building from sources are calculated. All types of energy balance structure of building S are considered. Table 1 shows the results of the total energy costs calculation for all variants.

The calculation shows that the minimum increase of the cost of primary energy resources will provide a variation of $S = 7$.

In fact, the option $S = 1$ is used during the construction of the building, which, as the table shows, is not an the optimal one according to the criterion under consideration. Probably the reasons of this decision are the use of other criteria and the absence of the calculations.

Table 1 Cost calculation results ΔQ^{im} various S_n values

Type of fuel	Type of expenditure	Natural fuels costs m³ or kWh	Losses, equivalent fuel	Equivalent fuel costs, equivalent fuel	Energy building balance structures S								
					1	2	3	4	5	6	7	8	9
Natural gas	Cooking	3713.0127		4.2848	X		X	X					X
	Hot water supply	5560.709		6.4171		X	X	X	X	X			
	Heating	53,822.77		40.827				X	X			X	X
Electric energy	Cooking	14,400	0.000113	1.7712					X	X	X	X	
	Hot water supply	3627.610	0.0000285	0.446							X	X	X
Central heating and hot water supply	Hot water supply	9906.6	3.419	11.432	X	X				X			
	Heating	14,594.305	3.419	16.842	X	X	X				X		
Amount of energy imports for structures S, equivalent fuel					35.977	33.464	30.962	51.528	49.015	28.449	22.470	43.044	45.558

4   Discussion: Practical Significance

It is obvious that the problem of development of the fuel and energy sector of the city is a multicriteria one and has an interdisciplinary nature.

The proposed method and its implementation in the form of GIS deals with only one aspect of the energy conservation problem and allows assessment using the urban energy performance criterion. The developed GIS application can be modified by adding other criteria of decision-making evaluations or transferring the results of the analysis to multi-criteria decision assessment systems.

A distinctive feature of the proposed method is the presentation of all city energy networks in the form of a single spatial system, where the impact of balance structures alternatives choice decisions is estimated on the consumer level and on the fuel and energy sector level. Thus, it offers opportunities to analyze decisions of urban development that previously were inaccessible. The article focuses on the principle of the criterion formation. To calculate the energy balance indicators the normative calculation is used. They are used to form the consumers tariffs. More accurate calculations can be obtained using data of monitoring the networks operation in real mode and the technological parameters of the energy networks. This will increase the accuracy of the calculations but will also require more data. In this case, the essence of the method will not change.

The method relies on the diverse information systems use and becomes effective when it is possible to automatically collect data on energy facilities and various types of energy consumption. These issues are not considered in the study, but currently data collection, integration and GIS application for the city and energy companies are actively developing. This fact stimulates the development of such analytical systems and methods and tends to reduce the implementation costs.

In addition to individual changes in the energy supply structure of new and existing buildings, the implementation of the method in the form of GIS application allows to graphically present on the map the various aspects of the energy balance distribution across the territory of the city, the structure of consumption and available reserves for all types of energy resources on the level of neighborhood.

Such maps help to understand the possibilities of construction of buildings and to plan the development of energy infrastructure.

5   Conclusion

A spatial model of the energy balance of urban areas is proposed in the current chapter to evaluate different options for connecting building to various types of energy networks according to a single criterion for minimizing citywide energy costs. The current model was implemented in web application in combination with GIS. The proposed model has been tested on public dataset of the Ivanovo city housing and public utilities. The results illustrate that not all buildings from dataset were connected to the energy networks in an optimal way. The reason of this problem is the using of other criteria and the lack of calculations. Combination of the study results, other methods and software can be used as a decision support system for the city and municipal government.

References

1. Dustdar, S., Nastic, S., Šcekic, O.: Smart cities: The internet of things, people and systems. Smart Cities: Internet Things People Syst. 1–268 (2017). https://doi.org/10.1007/978-3-319-60030-7
2. Chan, M., Estève, D., Escriba, C., Campo, E.: A review of smart homes-Present state and future challenges. Comput. Methods Programs Biomed. 91(1), 55–81 (2008). https://doi.org/10.1016/j.cmpb.2008.02.001
3. Gungor, V.C., Sahin, D., Kocak, T., Ergüt, S., Buccella, C., Cecati, C., Hancke, G.: Smart grid and smart homes: Key players and pilot projects. IEEE Indus. Electron. Mag. 6(4), 18–34 (2012). https://doi.org/10.1109/MIE.2012.2207489
4. Lund, H., Ostergaard, P.A., Connolly, D., Mathiesen, B.V.: Smart energy and smart energy systems. Energy, 137, 556–565 (2017)
5. Farzaneh, H., Suwa, A., Dolla, C.N., Oliveira, J.A.P.: Developing a tool to analyze climate co-benefits of the urban energy system. Procedia Environ. Sci. 20, 97–105 (2014). https://doi.org/10.1016/j.proenv.2014.03.014
6. DeCarolis, J., Daly, H., Dodds, P., Keppo, I., Li, F., McDowall, W., Pye, S., Strachan, N., Trutnevyte, E., Usher, W., et al.: Formalizing best practice for energy system optimization modelling. Appl. Energy 194, 184–198 (2017). https://doi.org/10.1016/j.apenergy.2017.03.001
7. Koirala, B.P., Koliou, E., Friege, J., Hakvoort, R.A., Herder, P.M.: Energetic communities for community energy: a review of key issues and trends shaping integrated community energy systems. Renew. Sustain. Energy Rev. 56, 722–744 (2016). https://doi.org/10.1016/j.rser.2015.11.080
8. Keirstead, J., Jennings, M., Sivakumar, A.: A review of urban energy system models: approaches, challenges and opportunities. Renew. Sustain. Energy. Rev. 16(6), 3847–3866 (2012). https://doi.org/10.1016/j.rser.2012.02.047
9. Bracco, S., Delfino, F., Ferro, G., Pagnini, L., Robba, M., Rossi, M.: Energy planning of sustainable districts: Towards the exploitation of small size intermittent renewables in urban areas. Appl. Energy 228, 2288–2297 (2018). https://doi.org/10.1016/j.apenergy.2018.07.074
10. Scheller, F., Bruckner, T.: Energy system optimization at the municipal level: an analysis of modeling approaches and challenges. Renew. Sustain. Energy Rev. 105, 444–461 (2019). https://doi.org/10.1016/j.rser.2019.02.005
11. Huang, Z., Yu, H., Peng, Z., Zhao, M.: Methods and tools for community energy planning: a review. Renew. Sustain. Energy Rev. 42, 1335–1348 (2015). https://doi.org/10.1016/j.rser.2014.11.042
12. Liu, T., Zhang, D., Dai, H., Wu, T.: Intelligent modeling and optimization for smart energy hub. IEEE Trans. Indus. Electron. 66(12), 9898–9908 (2019)
13. de Guio Mirakyan, R.: Integrated energy planning in cities and territories: a review of methods and tools. Renew. Sustain. Energy Rev. 22, 289–297 (2013). https://doi.org/10.1016/j.rser.2013.01.033
14. Mendes, G., Ioakimidis, C., Ferrão, P.: On the planning and analysis of integrated community energy systems: a review and survey of available tools. Renew. Sustain. Energy Rev. 15(9), 4836–4854 (2011). https://doi.org/10.1016/j.rser.2011.07.067
15. Markovic, D., Cvetkovic, D., Masic, B.: Survey of software tools for energy efficiency in a community. Renew. Sustain. Energy Rev. 15(9), 4897–4903 (2011). https://doi.org/10.1016/j.rser.2011.06.014

16. Allegrini, J., Orehounig, K., Mavromatidis, G., Ruesch, F., Dorer, V., Evins, R.: A review of modelling approaches and tools for the simulation of district-scale energy systems. Renew. Sustain. Energy Rev. **52**, 1391–1404 (2015). https://doi.org/10.1016/j.rser.2015.07.123

17. Tozzi Jr, P., Jo, J.H.: A comparative analysis of renewable energy simulation tools: performance simulation model vs. system optimization. Renew. Sustain. Energy Rev. **80**, 390–398 (2017). https://doi.org/10.1016/j.rser.2017.05.153

18. Connolly, D., Lund, H., Mathiesen, B.V., Leahy, M.: A review of computer tools for analysing the integration of renewable energy into various energy systems. Appl. Energy **87**(4), 1059–1082 (2010). https://doi.org/10.1016/j.apenergy.2009.09.026

19. Stadler, M., Groissböck, M., Cardoso, G., Marnay, C.: Optimizing distributed energy resources and building retrofits with the strategic DER-CA. Model. Appl. Energy **132**, 557–567 (2014). https://doi.org/10.1016/j.apenergy.2014.07.041

20. Geidl, M.: Integrated modeling and optimization of multi-carrier energy systems (Dissertation). Eidgenössische Technische Hochschule Zürichю (ETH) (2007). https://doi.org/10.3929/ethz-a-005377890

21. Krause, T., Andersson, G., Fröhlich, K., Vaccaro, A.: Multiple-energy carriers: modeling of production, delivery, and consumption. Proc. IEEE **99**(1), 15–27 (2011). https://doi.org/10.1109/JPROC.2010.2083610

22. Mohammadi, M., Noorollahi, Y., Behnam, M.-I., Hossein, Y.: Energy hub: from a model to a concept-a review. Renew. Sustain. Energy Rev. **80**, 1512–1527 (2017). https://doi.org/10.1016/j.rser.2017.07.030

23. Miles, S.B., Ho, C.L.: Applications and issues of GIS as tool for civil engineering modeling. J. Comput. Civil Eng. **13**(3), 144–152 (1999). https://doi.org/10.1061/(ASCE)0887-3801(1999)13:3(144)

24. Alhamwi, A., Medjroubi, W., Vogt, T., Agert, C.: Development of a GIS-based platform for the allocation and optimization of distributed storage in urban energy systems. Appl. Energy **251**, 113360 (2019). https://doi.org/10.1016/j.apenergy.2019.113360

25. Grover-Silva, E., Girard, R., Kariniotakis, G.: Optimal sizing and placement of distribution grid connected battery systems through an SOCP optimal power flow algorithm. Appl. Energy **219**, 385–393 (2018). https://doi.org/10.1016/j.apenergy.2017.09.008

26. Paoli, L., Lupton, R.C., Cullen, J.M.: Useful energy balance for the UK: an uncertainty analysis. Appl. Energy **228**(15), 176–188 (2018). https://doi.org/10.1016/j.apenergy.2018.06.071

27. Ratmanova, I.D., Gurfova, O.M.: Informatsionno-analiticheskoye soprovozhdeniye energeticheskogo menedzhmenta na regionalnom urovnem [Information and analytic support of regional power management]. Vestnik IGEU **5**, 59–68 (2017)

28. Bashmakov, I.A., Myshak, A.D.: Izmereniye i uchet energoeffektivnosti [Power-efficiency fiscal metering]. Akademiya energetiki **4**, 66–75 (2012)

Printed in the United States
by Baker & Taylor Publisher Services